T0137314

Contemporary Systems Thinking

Series Editor:
Robert L. Flood
Maastricht School of Management
The Netherlands

More information about this series at http://www.springer.com/series/5807

Gianfranco Minati • Mario R. Abram • Eliano Pessa
Editors

Systemics of Incompleteness and Quasi-Systems

 Springer

Editors
Gianfranco Minati
AIRS / Italian Systems Society
Milano, Italy

Mario R. Abram
AIRS / Italian Systems Society
Milano, Italy

Eliano Pessa
Department of Brain and Behavioral
Science
University of Pavia
Pavia, Italy

ISSN 1568-2846
Contemporary Systems Thinking
ISBN 978-3-030-15279-6   ISBN 978-3-030-15277-2   (eBook)
https://doi.org/10.1007/978-3-030-15277-2

This Springer imprint is published by the registered company Springer Nature Switzerland AG.
The registered company address is: Gewerbestrasse 11, 6330 Cham, Switzerland

In memory of

Professor George Jiří Klir

April 22, 1932–May 27, 2016

Preface

The Seventh National Conference of the Italian Systems Society is held in conjunction with its 21 years of activity after its foundation.

This Congress is dedicated to the memory of Prof. George J. Klir (April 22, 1932–May 27, 2016), one of the greatest scientists that Systemics can boast of.

The list of his contributions to Systemics, of the positions he held, and of the academic acknowledgments he received is so long that it would be impossible to condense it in the few words of this dedication. Some names of the domains in which Klir had worked, as a contributor rather than a true creator, are enough to give an account of the size of his interests and his volcanic activity: systems modeling and simulation, systems logic design, computers architecture, discrete mathematics, intelligent systems, fuzzy sets theory, fuzzy logic, generalized measurement theory, and soft computing.

We will try to be worthy of him and of his contributions.

The title of the conference, *"Systemics of Incompleteness and Quasi-Systems,"* aims to underline the need for Systemics and Systems Science to deal with the concept of theoretical incompleteness as incubator, probably necessary, for the establishment of processes of emergence. The subject is interdisciplinary, elaborated by various contributions, and is looking toward further research.

The topic of this Seventh conference is an evolution of the subjects of the previous conferences, namely:

2001   Emergence in Complex Cognitive, Social and Biological Systems (Minati and Pessa 2002).

2004   Systemics of Emergence: Research and Applications (Minati et al. 2006).

2007   Processes of Emergence of Systems and Systemic Properties. Towards a General Theory of Emergence (Minati et al. 2009).

2010   Methods, Models, Simulations and Approaches. Towards a General
        Theory of Change (Minati et al. 2012).
2014   Towards a Post-Bertalanffy Systemics (Minati et al. 2016).

Classical models of the first Systemics (Minati et al. 2016) are intended
to completely represent aspects of phenomena and processes, such as the
motion of a pendulum or the operation of an amplifier. They concern the
phenomena in their temporal and spatial completeness. We consider here
incompleteness as a theoretical topic of research for Systemics, related to
quasi-systems. The purpose is to consider for Systemics the concept of quasi,
used in diverse disciplines as for quasi-crystals, quasi-particles, quasi-electric
fields, and quasi-periodicity. The usual approaches in Systemics, even when
dealing with complexity, assume possible incompleteness in the modeling as
having a provisional or practical nature being still under study and that there
is no theoretical reason why the modeling cannot be complete.

There are phenomena that must be modeled by resorting to systems using multiple models according to the characteristics considered, such as their
electrical and mechanical, economic and sociological, biological and psychological aspects, their coherence being a crucial systemic theme (Minati and
Pessa 2002) also with regard to their low completeness or comprehensiveness
as considered by the Dynamic Usage of Models (DYSAM) (Minati and Pessa
2006, pp. 64–75) and Logical Openness (Licata 2012; Minati et al. 1998).

Furthermore, concepts and approaches regarding contexts and processes
for which systems modeling cannot be conceptually exhaustive have already
been introduced in the literature (see, for instance, Bailly and Longo (2011)).
We recall, first of all, fuzzy sets and fuzzy logic (Klir and Yuan 1995, 1996).
However, the difference between fuzzy and quasi we have in mind "relates
to the dynamical and structural incompleteness of the second, real identity
of the quasi, while the fuzzy relates to well-defined levels of belonging along
time" (Minati and Pessa 2018, p. 154).

Other cases, for which modeling through the use of systems is incomplete
since related to only some properties, include processes of emergence.

The last mentioned cases represent contexts and approaches where the incompleteness is intrinsic, theoretical (Minati 2016), and regards the intrinsic
impossibility of exhaustively modeled because the incompleteness itself is an
unavoidable characteristic of the process under study.

Furthermore, we consider how theoretical incompleteness, incomplete
modeling, i.e., not exhausted by using individual models, of processes and
phenomena should be explored as a conceptual coexistence of different approaches not so much with the purpose of exhausting but to conceptually
represent the structural dynamics of becoming, already considered, for instance, through the use of uncertainty and complementarity principles in
physics, without referring here to quantum physics. The focus is on the
transient, on multiplicity, and coherence that guarantee consistency.

A related concept is that of theoretical sloppiness, referring to models in
physics, biology, and other disciplines (Transtrum et al. 2015).

Examples include biological organisms, living systems, ecosystems, collective behaviors, social systems, and poly-pathologies. It is also a matter of developing knowledge for the current knowledge, information or postindustrial society still largely based on concepts, approaches, principles, and language based on completeness.

The contributions explore cases and present conceptual approaches within the novel context described above.

We conclude by observing how this setting is conducive to the use of Post-Bertalanffy Systemics as considered in the previous Conference (Minati et al. 2016).

Milano, Italy
Milano, Italy
Pavia, Italy
September 2018

Gianfranco Minati
Mario R. Abram
Eliano Pessa

References

Bailly, F., & Longo, G. (2011). *Mathematics and the natural sciences. The physical singularity of life.* London: Imperial College Press.

Klir, G. J., & Yuan, B. (1995). *Fuzzy sets and fuzzy logic: Theory and applications.* Upper Saddle River, NJ: Prentice Hall.

Klir, G. J., & Yuan, B. (Eds.). (1996). *Fuzzy sets, fuzzy logic, and fuzzy systems: Selected papers by Lotfi A. Zadeh.* Singapore: World Scientific.

Licata, I. (2012). Seeing by models: Vision as adaptive epistemology. In G. Minati, M. Abram, & E. Pessa, (Eds.), *Methods, models, simulations and approaches towards a general theory of change* (pp. 385–400). Singapore: World Scientific.

Minati, G. (2016). Knowledge to manage the knowledge society: The concept of theoretical incompleteness. *Systems, 4*(3), 1–19.

Minati, G., Abram, M. R., & Pessa, E. (Eds.). (2009). *Processes of emergence of systems and systemic properties. Towards a general theory of emergence.* Singapore: World Scientific.

Minati, G., Abram, M. R., & Pessa, E. (Eds.). (2012). *Methods, models, simulations and approaches. Towards a general theory of change.* Singapore: World Scientific.

Minati, G., Abram, M. R., & Pessa, E. (Eds.). (2016). *Towards a post-Bertalanffy systemics.* Cham: Springer.

Minati, G., Penna, M. P., & Pessa, E. (1998). Thermodynamic and logical openness in general systems. *Systems Research and Behavioral Science, 15*, 131–145.

Minati, G., & Pessa, E. (Eds.). (2002). *Emergence in complex cognitive, social and biological systems.* New York: Kluwer Academic/Plenum Publishers.

Minati, G., & Pessa, E. (2006). *Collective beings.* New York: Springer.

Minati, G., & Pessa, E. (2018). *From collective beings to quasi-systems.* New York: Springer.

Minati, G., Pessa, E., & Abram, M. (Eds.). (2006). *Systemics of emergence: Research and applications.* New York: Springer.

Transtrum, M., K., Machta, B. B., Brown, K. S., Daniels, B. C., Myers, C. R., & Sethna, J. P. (2015). Perspective: Sloppiness and emergent theories in physics, biology, and beyond. *The Journal of Chemical Physics, 143*(1), 010901.

Acknowledgments

The Seventh Italian Conference has been possible thanks to the contributions of many people who have accompanied and supported the growth and development of AIRS (Associazione Italiana per la Ricerca sui Sistemi) during all the years since its establishment and to the contribution of "new" energies from students and researchers realizing the systemic aspects of their activity.

We thank the Catholic University of Milan, the director of the Department of Philosophy, Prof. Massimo Marassi, for welcoming and hosting this event that took place within the framework of a well-established research and seminar activity in the field of Systemics, inspired and organized by Prof. Lucia Urbani Ulivi.

We have been honored by the presence of Professor Giuseppe Longo (CNRS, Département d'Informatique, École Normale Supérieure and CREA, Polytechnique), who delivered the opening plenary lecture.

Thanks are also due to all the authors who submitted papers for this conference and in particular the members of the program committee as well as the referees who have guaranteed the quality of the event.

We thank explicitly all the people who have contributed during the conference, bringing ideas and stimuli to this new phase of the scientific and cultural project of Systemics.

Contents

Part II Models of Incompleteness and Quasiness

List of Contributors

Mario R. Abram
AIRS / Italian Systems Society, Milano, Italy

Eusebia Armenia
Department of Medicine and Surgery, University of Parma, Parma, Italy

Pier Luca Bandinelli
Dipartimento di Salute Mentale ASL Roma 1, Servizio Psichiatrico Diagnosi e Cura (SPDC), Presidio Ospedaliero San Filippo Neri, Roma, Italy

Elena Bartolini
University of Milano – Bicocca, Milano, Italy

Leonardo Bich
IAS-Research Centre for Life, Mind and Society, Department of Logic and Philosophy of Science, University of the Basque Country (UPV/EHU), Donostia-San Sebastian, Spain

Lucio Biggiero
Department of Industrial Engineering, Information and Economics, L'Aquila University, L'Aquila, Italy

Natale Salvatore Bonfiglio
Dipartimento di Scienza del Sistema Nervoso e del Comportamento, Università degli Studi di Pavia, Pavia, Italy

Giordano Bruno
ISIA Roma Design, Istituto Superiore per le Industrie Artistiche, Roma, Italy

Alvaro Busetti
Freelance Consultant in "Social Enterprise" Sector, Rome, Italy

Umberto Di Caprio
ASDE (Associazione Dirigenti Enel), Roma, Italy

Francescogiuseppe Romano Maria Dossi
Catholic University of the Sacred Heart, Milano, Italy

Rodolfo Fiorini
Department of Electronics, Information and Bioengineering (DEIB),
Politecnico di Milano, Milano, Italy

Carlotta Fontana
Dipartimento di Architettura e Studi Urbani (DAStU), Politecnico di
Milano, Milano, Italy

Aldo Frigerio
Department of Philosophy, Catholic University of the Sacred Heart, Milano,
Italy

Alessandro Giuliani
Department of Environment and Health, Istituto Superiore di Sanità, Roma,
Italy

Marco Giunti
ALOPHIS - Applied LOgic Philosophy and HIstory of Science, Dipartimento
di Pedagogia Psicologia Filosofia, Università di Cagliari, Cagliari, Italy

Antonio Lizzadri
Catholic University of the Sacred Heart, Milano, Italy

Giuseppe Longo
Centre Cavaillès, République des Savoirs, CNRS, Collège de France et École
Normale Supérieure, Paris, France
School of Medicine, Tufts University, Boston, MA, USA

Maurizio Lopa
Schola Palatina, Early Music Department, La Vertuosa Compagnia de'
Musici di Roma, ISIA of Pescara, Multimedia Design, Pescara, Italy

Pier Luigi Marconi
Dipartimento di Psicologia Dinamica e Clinica, Università di Roma
"La Sapienza", Roma, Italy
AIRS / Italian Systems Society, Milano, Italy

Irune Medina
University of Rome Tor Vergata, Roma, Italy

Gianfranco Minati
AIRS / Italian Systems Society, Milano, Italy

Eraldo Francesco Nicotra
Department of Pedagogy, Psychology, Philosophy, University of Cagliari,
Cagliari, Italy

Maria Petronilla Penna
Dipartimento di Pedagogia, Psicologia, Filosofia, Università degli Studi di Cagliari, Cagliari, Italy

Italian Systems Society (AIRS), Milano, Italy

Roberto Peroncini
University of Genoa, Genoa, Italy

Eliano Pessa
Department of Brain and Behavioral Science, University of Pavia, Pavia, Italy

Emanuela Pietrocini
Schola Palatina, Early Music Department, La Vertuosa Compagnia de' Musici di Roma, ISIA of Pescara, Multimedia Design, Pescara, Italy

Rita Pizzi
Department of Computer Science, University of Milan, Milan, Italy

Roberta Renati
Dipartimento di Scienza del Sistema Nervoso e del Comportamento, Università degli Studi di Pavia, Pavia, Italy

Andrea Roli
Department of Engineering and Computer Science (DISI), Alma Mater Studiorum University of Bologna, Cesena, Italy

Dolores Rollo
Department of Medicine and Surgery, University of Parma, Parma, Italy

Roberto Serra
Department of Physics, Informatics and Mathematics, University of Modena e Reggio Emilia, Modena, Italy

European Centre for Living Technology, Ca Minich, Venezia, Italy

Andrea Spoto
Department of General Psychology, University of Padova, Padova, Italy

Francesco Sulla
Department of Medicine and Surgery, University of Parma, Parma, Italy

Guido Tascini
Centro Studi e Ricerca "G. B. Carducci", Fermo, Italy

Lucia Urbani Ulivi
Department of Philosophy, Catholic University of the Sacred Heart, Milano, Italy

Marco Villani
Department of Physics, Informatics and Mathematics, University of Modena e Reggio Emilia, Modena, Italy

European Centre for Living Technology, Ca Minich, Venezia, Italy

Editors' Biography

Gianfranco Minati, Systems Scientist, Mathematician, Founder, and President of the Italian Systems Society (AIRS); President of the European Union for Systemics; Doctoral lecturer at the Polytechnic of Milan; Member of the scientific committee of Conferences and Systems Societies. He is author of 35 chapters in books; editor of 9 books and journals; author or coauthor of 15 books; and author of 41 articles and of academic publications. His current research interest focuses on (1) Modeling processes of emergence by using Meta-Structures; (2) the emerging of a Post-Bertalanffy Systemics; (3) the Dynamic Usage of Models (DYSAM), Logical Openness; (4) Architecture and Design as the design of social meta-structures to influence processes of emergence in social systems.

Mario R. Abram, Physicist, is active in Italian Systems Society (AIRS). He worked mainly in ENEL (Italian Power Agency), at Research Department (Automatica Research Center), then in Cesi S.p.A. and Cesi Ricerca S.p.A. He is experienced in hybrid and digital simulation systems, working on models of power systems, thermoelectric and nuclear power plants and processes control, development of simulators for testing and tuning supervision and control systems. His research interests include dynamical systems, neural networks applications, modeling and simulation of processes and interactions between infrastructural networks.

Eliano Pessa, Theoretical Physicist, is Full Professor of General Psychology and Cognitive Modeling at the University of Pavia, Italy. He has already been Dean of the Department of Psychology and of the Inter-departmental Research Center on Cognitive Science in the same university. In the past he has been Associate Professor of Artificial Intelligence at the University of Rome "La Sapienza," Faculty of Psychology. He is author or coauthor of 10 books and of a large number of papers on scientific journals, books, and proceedings of international conferences. His scientific research interests include quantum theories of brain operation, computational neuroscience, artificial neural networks, models of emergence processes, quantum field theory, models of phase transitions in condensed matter, models of human memory and visual perception, models of decision making, and models of statistical reasoning.

Scientific Committee

Part I
Opening Lectures

Interfaces of Incompleteness

Giuseppe Longo

1 Introduction

In addition to being one of the main results in Mathematical Logic, Gödel's 1931 Incompleteness Theorem can also serve as a starting point for a reflection extending beyond Mathematics and the issue of its foundations in order to relate them to problems and methods pertaining to other disciplines. It is in the light of Gödel's theorem that we will present a "critical history of ideas", that is, an explicitly a posteriori reading of some marking points of modern scientific thought, times when the audacity of the propositions of knowledge would be tempered by problems demonstrated to be unsolvable and by negative or limiting results. Negative results, however, opened up new horizons for knowledge. We will examine some of the main scientific paradigms in order to find, within their respective domains, their common thread, that is, incompleteness, understood in its various meanings.

The detailed analysis, although informal, of Gödel's theorem and a reflection on Turing's work will constitute but one element of this text. Thus, we will first see the way in which incompleteness and limits of well-established approaches have been demonstrated and sometimes surpassed, starting

G. Longo (✉)
Centre Cavaillès, République des Savoirs, CNRS, Collège de France et École Normale Supérieure, Paris, France
School of Medicine, Tufts University, Boston, MA, USA
e-mail: Giuseppe.Longo@ens.fr
http://www.di.ens.fr/users/longo

A preliminary version of this paper appeared as "Incompletezza", in C. Bartocci and P. Odifreddi (Eds.), *La Matematica*, vol. 4, (pp. 219–262), Einaudi, Torino, 2010.

with Poincaré's fundamental work. This will allow to broaden our reading grid—while avoiding, we hope, inappropriate abuses and contaminations—to Laplace's scientific and epistemological approach and to the limits set by Poincaré's major "negative result", as he called it himself.

After discussing Gödel's incompleteness, we will continue with Einstein's theses on the non-"completeness" of Quantum Mechanics, to use the term employed in the very famous article written in collaboration with Podolski and Rosen which analyses this notion, (Einstein et al. 1935).

Biology has been dramatically affected by the myth of completeness of molecular descriptions, both in ontogenesis and phylogenesis: DNA as a program and "blue print" of the organism. The richness of organismal individuation and the intrinsic unpredictability of phylogenetic dynamics is then lost as well as the open ended changes of the pertinent phase space of biological evolution. Jointly to an appreciation of the role of rare events in evolution, some understanding of the limits of knowledge set the grounds for new approaches and may help to focus on more suitable a priori for the science of life.

2 From Laplace to Poincaré

As regards Laplace (1749–1827), what one must look for is the unity of the method (and of the Universe), and hence the identity between physical laws at our scale of perception and the laws which govern microscopic particles. All observable phenomena are reducible to the elementary ontology underlying matter, movement and force. And at this level, any analysis must base itself on the possibility of isolating, mathematically, a single elementary particle and of describing its movements. It must then, by means of mathematical integration operations, reconstruct the expression of the law of interaction at a distance in particle systems. The analysis of planetary systems must thus advance by progressive composition of individual movements and lead to an understanding of the "system" as the sum of the individual behaviors and their interactions, two by two, three by three ...

This mechanistic reduction is intimately linked, for Laplace, to the structure of the determination of all physical events. For all the great physicists of the eighteenth and nineteenth centuries, the systems of differential equations needed to be able to describe all important physical phenomena, starting with the description and integration of individual movements. In particular, the laws of physics, first in the form of the Lagrange equations, later in the form of Hamilton's equations, must be capable of expressing the determination of any movement, any trajectory, hence of any physical event, in the same way as the laws of Newton-Laplace determine the evolution of celestial bodies in a gravitational field.

And it is this equational determination which enables the *predictions* which measure the validity of theoretical propositions, at the center of the relation between experience and theory: observations are made, theories are produced (e.g. writing of the equations linking the observed actions and forces), predictions are made regarding the evolution of a system using these equations and finally, the predictions are compared against new observations. Effective predictions are the very objective of mathematical formalization.

The mathematical creativity of the new formalisms of the eighteenth and nineteenth centuries made scientist believe in a possible understanding of the whole Universe by dependable and progressive increases in knowledge. The ambition of the equations was to cover the whole world, to make it intelligible and predictable.

Of course, Laplace is also a great figure as regards the theory of probabilities, and this is no coincidence. He knew that many evolutions are random, thus unpredictable, for instance in the case of throws of dice which are submitted to forces and frictions that are too numerous to be known.

These systems must then be analyzed in probabilities, in a completely different way than the methods specific to the equational determinations of movement. Laplace also knew that a deterministic trajectory may depend on "almost imperceptible nuances", for example as would a marble at the peak of a mountain (a maximum of potential) which, being submitted to unobservable ("imperceptible") perturbations, can take either one particular direction or a completely different one. Laplace nevertheless considered that such situations, such "critical" initial points, are isolated, that they are rare events in what concerns the measurement of space, and that it must certainly be possible to process them using adequate mathematics in the system which is a paradigm of stability and of certitude in terms of its predictability: the solar system.

It must be possible to deduce all astronomical facts,

according to Laplace. Besides, Alexis Clairault had even computed the time of Halley's Comet's return, an extraordinary achievement for the mathematics of the second half of the eighteenth century. Determination and predictability govern the Universe, from particles to stars, with inevitable fragments of randomness—we are not omniscient—that must be analyzed in probabilistic terms which are quite distinct from those of systems of equational description. When known, such equational descriptions should always constitute, thanks to appropriate computations, the primary instrument of scientific prediction and of positive knowledge.

Well, they don't. Poincaré (1854–1912) demonstrated that it suffices to consider three celestial bodies, the Sun and two planets, say, under gravitational interaction for the system of equations describing the movement to become unable to predict the system's evolution. In the approach of this paper, we may say that the system of equations is epistemically "incomplete" with respect to knowledge as prediction of the physical process.

Where is the problem? Newton had already realized this: his law of gravitation is "universal", meaning that it applies to the interaction between any celestial bodies; it even applies to the interaction between planets themselves. Therefore, if one can deduce from his equations the Keplerian orbit of a planet around the Sun, two planets also exert attraction upon one another and reciprocally disturb each other's movements. With time, these small perturbations can cause important changes, "secular" changes as Laplace would say, also being aware of the problem. And Newton had proposed the only solution likely to guarantee the stability of the system "*in saecula saeculorum*": once in a while, skillful adjustments by God reestablish order.

Laplace, on the other hand, wanted to avoid any metaphysical hypothesis; he believed that a thorough mathematical analysis should demonstrate the stability of the system and its full predictability. It is thus that astronomers and mathematicians applied themselves during decades to resolving the equations of planetary movements; but when considering three bodies or more, they would encounter insurmountable difficulties.

In 1890, Poincaré noticed an error in his own demonstration of the convergence of Linsted's series. This series should have provided an analytical solution to the system of gravitational equations for three bodies (the "Three-Body Problem"). And, with all of his genius, he deduced from his own error the intrinsic impossibility of resolving this system. He demonstrated that almost everywhere, one obtains increasingly small divisors in the coefficients of the series, preventing convergence.

In an even more audacious and certainly innovating way, he gave a *physical sense* to this mathematical difficulty, to its "negative result", as he called it: radical changes for the evolution of three bodies can depend on very small (non-measurable) variations in the initial conditions—we will later speak of "sensitivity to initial conditions".

Poincaré reaches this physical sense via geometry: he proves that in the "phase space" (of which the points are not only given by their position but also by the value of the momentum) trajectories present "bifurcations" points, while stable and unstable periodical trajectories intersect in an extremely complex way (in points that he calls "homoclines"). They indeed intersect infinitely often, in "infinitely tight" meshes and are also folded upon themselves "without ever intersecting themselves".[1] Poincaré presents here deter-

---

[1] Poincaré "sees" the geometry and the complexity of chaos, without even drawing it:

> To represent the figure formed by these two curves [the stable and unstable periodic "trajectories"] and their infinitely numerous intersections [homocline points], these intersections form a sort of lattice, of fabric, a sort of network made up of infinitely tight meshes; each of these curves must never intersect itself, but must fold upon itself in a very complex manner so as to intersect an infinite number of times with all other meshes in the network. One is struck by the complexity of this figure, which I will not even attempt to draw. Nothing is more apt for giving an idea of the complexity of the three-body problem and in general of all problems of dynamics where there is no uniform integral (Poincaré 1892).

ministic chaos for the first time. He deduced then, as early as 1892, and later in a more developed way, that

> prediction becomes impossible [...] and we have random phenomena (Poincaré 1902).

It is often hard to give physical meaning to mathematical *solutions* of systems of equations; by first seeing it geometrically, Poincaré gave physical meaning to the *absence* of integral solutions.

To conclude, equational determination, here being relatively simple—only three bodies—does not imply the predictability of the system. More precisely, the geometry of its evolutions *enables to demonstrate* this unpredictability as a consequence of its complexity. The homocline points, the bifurcations ... produce sensitivity to the system's initial conditions: fluctuations (internal) and/or perturbations (external) below observability can cause the system to have very different trajectories over time, see (Barrow-Green 1997) for more details.

This work of Poincaré, which leads him to invalidate a program of knowledge, marks the beginning of the "geometry of dynamical systems" and of the qualitative analysis of unpredictable deterministic systems. It is mostly a topological analysis of global fluxes, of evolutions and limits, including quantitative ones, and of the limits of predictability (Charpentier et al. 2006). This will lead to the computation of the time necessary for a system, the solar system in particular, to become unpredictable, to be discussed next.

From an epistemological standpoint, the innovation of Poincaré's approach is to understand that random evolutions can even be found in systems of which the determination is relatively simple and that classical randomness can be understood as an unpredictable determination. A key fact, largely ignored by common sense: determinism, or the possibility to fully "determine", by equations typically, may co-exist, it is not the opposite of randomness. A dice, a double pendulum or even a planetary system ... all of these are deterministic but chaotic systems as will be later asserted (see Laskar 1989, 1994 for the solar system). To describe a throw of dice, it would require numerous equations and it is not even worth it to attempt to write them: high sensitivity to initial or border conditions makes its movement unpredictable. But in no way does this change the fact that a thrown dice will follow a trajectory that is perfectly determined by the least action principle, a physical geodesic (a trajectory which minimizes the variation of energy over time), although it is unpredictable. Only two equations determine the movement of the double pendulum, but its evolution is quick to become chaotic and, therefore, is also unpredictable.[2] In what concerns the solar system, its time ofunpredictability

---

[2] A pendulum can be conceived as a bar connected to a pivot. If we attach another bar to the bottom of this first bar, one that is also free to rotate, what we have is a double pendulum. A recent and amusing theorem (Béguin 2006) demonstrated the following: if we choose a sequence of integer numbers $a_1, a_2, a_3, ...$, we can put the double pendulum in an initial position so that the second limb will make at least a_1 clockwise turns, and then change direction to make at least a_2 counterclockwise turns, and then at least a_3

has been recently computed (Laskar 1989, 1994). If we associate, to the detailed analysis of (non-linear) movement equations, a lower bound for the best measurement possible, we obtain, by non-obvious computations, an upper bound for predictability. This bound is relatively modest in astronomical terms (a few tens of millions of years, depending on the planet).

So, for modern dynamics, from Poincaré onwards, the solar system is chaotic. Let's note however that there are those who, having understood this a little late, even felt compelled to apologize on behalf of a whole scientific community. This was done, in a very British way, in a famous and mathematically very interesting article, but without referring to the previous century's illustrious French mathematician otherwise than as the source of an error (Lighthill 1986).[3]

As a consequence, we insist, classical randomness, seen as the unpredictability of a physical process, is a specific case of determination: that which governs a chaotic deterministic system. One could bet on the odds that Earth will still be in orbit around the Sun 100 million years from now: it is roughly as unpredictable as a throw of dice relative to its own time scale. One will note that "chaoticity" is a precise *mathematical* property, defined in general by means of three well formalized properties that have been clarified with rigor and with full generality after 1970 (sensitivity to the initial conditions, existence of dense orbits, density of periodic points).[4]

---

clockwise turns, etc. If we choose a random sequence a_1, a_2, a_3, ... (see Sect. 6), this purely mathematical result makes chaos and unpredictability "understandable" (but does not demonstrate it) in one of the simplest deterministic systems possible. We can also see this by observing an actual physical double pendulum or a computer simulation (such simulations can be found on the Web, we will return to this).

[3] It is interesting to compare explicit theorization and geometric analysis of the unpredictability of a Newtonian system in Poincaré (1892) and the references in Lighthill (1986), (as well as the title: "The recently recognized failure of predictability in Newtonian dynamics" ... "recently"?).
Indeed, two schools particularly distinguished themselves in the twentieth century regarding the theory of dynamic systems: the French school (Hadamard, Leray, Lévy, Ruelle, Yoccoz ...) and the Russian school (Lyapunov, Pontryagin, Landau, Kolmogorov, Arnold ...). To these, we must add, namely, the Americans Birkhoff and Lorentz. But works on the subject by Hadamard, Lyapunov and Birkhoff have long remained isolated and seldom quoted. Up until the results by Kolmogorov and Lorentz in the 1950s and 1960s, and well beyond, Classical Rational Mechanics—of which Lighthill presided the international association in 1986—has dominated the mathematical analysis of physical dynamics (as well as the way the subject matter was taught to the author of this article, prior to such apologies, alas).

[4] Ruelle and Takens (1971a,b) faced many difficulties for getting published. As has been said above and as will be said again as for genocentric approaches in Biology, the Laplacian mentality (but Laplace, two centuries ago, was a great mathematician) is still present in many minds of today, although outside of the sphere of Mathematical Physics. And, in general, "negative results" are the most difficult to accept. Thus, they are the most difficult ones to finance, even if they are most likely to open up new horizons. And this is exactly what the institutional administrators of research steered towards *positive projects* and towards patents will succeed in hindering even more definitively, bolstered by their

In most cases, from a mathematical point of view, chaos appears when the system's evolution equations or evolution function are non-linear. It is the typical mathematical means of expressing interactions and effects of attraction/repulsion or of resonance (technically, two planets enter gravitational "resonance" when they are aligned with the Sun; it is a situation where there are great mutual perturbations). Unpredictability is, for its part, a problem at the interface between the mathematical system and the physical process, via *measurement*: if the mathematical description of a physical process (the equations or a function which describes its evolution) verifies the formal conditions of chaos, it is the *physical process* which becomes unpredictable by this mathematical system.

A measurement, in Classical (and of course Relativistic) Physics, is indeed always an interval, it is always an approximation. Because of this, *non-measurable* fluctuations or perturbations (within the best measurement's interval of approximation) can entail, over time, changes which are quite observable, but which are unpredictable. In other words, in order to predict or to demonstrate that it is impossible to predict, it is necessary to *view* a *physical process* mathematically. If the determination produced by the mathematical approach is "sensitive to initial or border conditions" (a crucial mathematical property of chaotic systems) and if the measurement is approximate, as always in Physics, then unpredictability appears.

We will later indicate how it is possible to relate, from both an epistemological and technical point of view, the unpredictability of deterministic systems to the Gödelian undecidability of logico-formal systems. From a historical standpoint, it is easy to see a first analogy (we will see others). The new conception of the physico-mathematical "determination" which stems from Poincaré's negative result, this limit to equational knowledge, as well as its qualitative geometrical analysis, have paved the way for the geometry of modern dynamic systems.

Analogously, Gödel's theorem, setting a limit to formal knowledge, marks the beginning of contemporary Mathematical Logic (Computability Theory, Model Theory and Proof Theory). The *epistemological fracture*—as Bachelard puts it—of great importance and very surprising at the time (and often still so today), caused by each of these great negative results, was extraordinarily fruitful in science.[5]

---

bibliometric indices: the role of critical thinking and of intrinsically "anti-orthodox" innovation that are characteristic of scientific research, see MSCS Ed. Board (2009) and Longo (2014, 2018b).

[5] Besides those already mentioned, there are numerous other highly important "negative results", particularly in Physics (not to mention that in Mathematics, by a skillful use of double negations, any result can be presented as "negative"). The results of which it is question here are among those which contradicted major projects of knowledge, or theories that marked the history of science and which sometimes continue to guide common sense. They are also results that are linked to the negation of an assumed completeness (in its various forms) of these theoretical propositions.

3 From Geometry to Logic

The program for the progressive and full occupation of reality using formal writing has an epistemological parallel in the formalistic "creationist" view and an illustrious predecessor in the person of George Berkeley. The English bishop was particularly impressed by the invention of complex numbers, and by this imaginary "i". Such an audacious linguistic and symbolic notation, enabling to resolve an equation without "real" solutions, led him to conceive of the mathematician as a creator of formal instruments for comprehension serving to gradually construct knowledge.

According to Peano (1889), first came $\sqrt{2}$, which goes beyond the ratios between integers, introduced for the purpose of understanding the diagonal of a square, and then came π for understanding the circle: these formal inventions allowed new mathematics and new understanding or computations. And so the algebraically *complete field* of complex numbers was progressively reached, gloriously culminating by the invention of the imaginary "i": *any* algebraic equation has a solution in it.

Hilbert made use of these considerations in the context of a deep analysis of the foundations of Mathematics. He sought formal systems which are demonstrably consistent and *complete*, which he will designate in the 1920s as constituting a "definitive solution" to the foundational problem that was such a humongous issue. Thankfully, in science, there is no such thing as a definitive/final solution.

But which foundational problem? It is certainly not a question of these antinomies of the beginning of the century concerning a barber who shaves all those who do not shave themselves (must the barber shave himself?), Sunday amusements and contradictions at the barber's shop that are (and were) easily resolved. The mathematical practice (or "doxa") is "typed": in mathematics, we do not generally authorize the barber to shave himself, no more than we allow functions to apply to themselves.

We start by defining functions over natural and real numbers, having values within these (or other) "types" of numbers; then we define functionals over functions, for example, the integral, and we continue in a hierarchical manner. A formalization which frees itself from meaningful precautions, failing to take them into account, easily leads to contradictions. This happened several times: such attempts are a part of research.[6] These antinomies (for-

---

[6] We can, for example, recall the first formalization of one of these fundamental systems of computability, Church's untyped lambda-calculus (1932). The ensuing "Curry paradox", an antinomy similar to that of the barber's, will first entail a refinement of the calculus (1936) and the invention of another one, with "types" (1940). The first formal system of types by Martin-Löf will also be contradictory (1970). The formalizations, which loose "meaning" along the way (we will return to this), easily tend to produce contradictions: "Logic is not sterile", said Poincaré (1906), "it has created contradictions". Given the innovations they brought and the responses they were quick to receive, it must be noted that these formal theories, in spite of these errors of syntax due to the lack of a mathematical interpretation,

mal contradictions) do not deserve however the designation of "paradox" (against the "doxa", seen as common "mathematical" knowledge), a designation rich in history since Ancient Greece, and even less so when the doxa already contains the solution. One only needs to think about Zeno's *paradox*, which constitutes a true challenge to the doxa and which opened century-long discussions.

The true problem of the foundation of mathematics was rather to be found in the collapse of the Euclidean intuition of space, the Newtonian absolute in Cartesian coordinates. For over 2000 years, Euclid's "Elements" had provided the link between the geometric constructions in sensible space and physical space on any scale. The theorems which were constructed in such space with the intuition of our world of the senses and of action produced both the geometric relationships between the stars and between Democritus's atoms. For Kepler, Newton and Kant, sensible intuition like mathematical intuition was at the roots of this geometric reconstruction of the universe. And this universe was described, as Galileo had said, using the language of Euclidean circles, triangles and straight lines.

Well, such is not the case, proved the geometers of the nineteenth century: the interesting space manifolds *are not* "closed under homotheties". What does this mean? Riemann, in his habilitation of 1854, proposed a general framework for what we call non-Euclidean geometries. In short, by following the algebraic treatment by Klein's Erlangen Program (1872), we can observe that one of their crucial properties is the fact that the group of automorphisms (internal transformations or symmetries) does not contain homotheties—that is, arbitrary changes of size.

In Riemann's geometry, it is possible that the theorem regarding the sum of a triangle's internal angles—which is equivalent to Euclid's axiom of parallels—gives more than 180° when the triangle is expanded to a stellar order of magnitude. Moreover, Riemann conjectured that the "forces of cohesion between bodies are linked to the metric of space" when he demonstrated the general theorem of the metric tensor which in turn links the metric to the curvature of space. Einstein will give specific physical meaning to this audacious negation of the Euclidean universe, by the role of the distribution of energy and matter in structuring (the curvature of) relativistic spaces. Locally, in "tangent" planes of null curvature, Euclidean geometry provides a good approximation; but on a global scale, on the scale of the Universe, it is precisely the non-null curvature which enables to unify gravitation and inertia, the keystone of Einstein's relativity.

The non-null curvature of space, its metric structure ... a revolutionary geometrization of physics, originated from the "negation" of the Euclidean doxa. "A delirium", Frege will say in 1884, which made Riemann and his followers to renounce to the Cartesian intuition in Euclidean spaces. As fora

---

were at the origin of very interesting ideas and systems and not of a major crisis, except among logicists (Longo 1996).

conceptual and intuitive analysis, those spaces were the only possible ones for Frege, to a point where he will continue even after 1920 to think of the foundations of geometry in Euclidean terms. But prior to that, at the very end of the nineteenth century, in reaction to the non-Euclidean frenzy which marked the genuine crisis of all certitudes in mathematical intuition, he established the basis of a new mathematical discipline, one that is important and rigorous: modern Mathematical Logic.

Of course, numerous other people participated in this work, among whom Peano. But Frege is the first for whom the foundational preoccupation radically emancipated itself from the relationship to sensible and intuitive space, in order to focus on logico-deductive analysis, constructed using the founding (and absolute) concept of integer number and a rigorous treatment of "quantification" (for all ... there exists ...). Arithmetics is logical; the principle of induction (or of recurrence), formalized by Dedekind and Peano, is a logical principle which fully captures, and even identifies itself to the conceptual structure of integers.

The extraordinary developments of this foundational approach are before our eyes: logical arithmetic machines change the world. They are the direct products of a mathematical and philosophical work initiated by English algebraists such as Peacock and Boole, and reaching Frege, Peano, Hilbert, Gödel and finally, Turing, all focusing on (formal) Arithmetics and Logic.

4 From Hilbert to Gödel

Hilbert as well was foremost preoccupied by the loss of certitude due to the non-Euclidean shift in Geometry. His most important foundational text, the *Foundations of Geometry* (Hilbert 1899), sets the basis for an original approach to the question, well beyond the algebraic unification of geometries proposed by Klein. Perfectly abstract axiomatics must formally capture the various systems, while ensuring the mathematical transformations of each of them and while revealing the "foundational" properties from which to derive all theorems of each system. It is a calculus of signs, freed from the incertitude of the intuition of space, based on axioms and rules of deduction of which we could "potentially" mechanize the application. It is the locus of mathematical certitude, precisely because it is devoid of meaning, of spatial signification, which is a source of ambiguity and the locus of space intuition, which had turned out to be rather unreliable.

During the following years, "formalists" will insist that certitude resides in the formal manipulation of finite sequences of signs, on account of rules that are also described by finite sequences of signs without any semantic and intuitive reference. Given a "sequence-of-signs rule", such as "*from A and A → B follows B* ", formally deduce B from A. That is, by applying the rule, if the first A is (constituted by) a sequence of signs *identical* to those in the second A, write/deduce B. What is the meaning of $\rightarrow$, the arrow? It doesn't matter: a machine must be able to apply this formal schema of deduction.

So, the existence of mathematical objects is not, for Hilbert, an ontological question: it is ensured by the sole *consistency* of the axiomatic system within which they are defined, that is, by the impossibility of deducing a contradiction from the axioms by using the system's rules of deduction. In other words, if (it is possible to demonstrate that) a system is non-contradictory (consistent), then hypotheses and existence proofs, even proofs by contradiction, are the guarantee of existence specific to the objects of Mathematics. It is a strong and bold choice, a veritable turning point by its rigor and its clarity of exposition with respect to the ancient and ontological myths of ideal triangles and circles that "exist" because they are present in the mind of God. For Frege, instead and in spite of his error in formalizing Set Theory, mathematical signs and properties must make sense, must evoke in language meaningful or derivable concepts. And it is he who will oppose, in a polemic manner, an axiomatic theory of the properties of God that is non-contradictory; he ill then will observe that he has thus proven God's existence. This is not what Hilbert had in mind, infuriated by Frege's observation. He was talking about a formal deductive practice, one that is purely linguistic and specific to Mathematics and to its objects, with no ontological content in Frege's sense.

So, how then may the consistency (the non-contradiction) of axiomatic theories be demonstrated? Hilbert, in his book, translates or "interprets" the various geometric axiomatics, including Riemannian axiomatics, within the continuum of Analysis, which can be constructed from Arithmetics, following Cantor-Dedekind's method, yet another fantastic achievement of late nineteenth century. He thus observes that if it is possible to prove the consistency of Arithmetics, the analytic interpretation guarantees the consistency of all axiomatized geometries. This is why he posed, in a very famous conference presented in Paris the following year (1900), the question of the consistency of Arithmetics (and of Analysis) as being among the great problems for twentieth century Mathematics. And he was right, given the consequences that this problem will have and the importance of the formal mathematical framework proposed. His work on the formalization of geometries, highly rigorous, marks indeed the birth of the Axiomatic Method, one of the most fruitful methodological turning points of the twentieth century.

4.1 ... Through Poincaré and Weyl

Poincaré reacted strongly to Hilbert's bias and wrote a lengthy review of Hilbert's 1899 book. He indeed appreciated its technical novelty and profoundness, but not the foundational vision of Mathematics as a mechanical issue and practice devoid of reference to signification.

It is viewed, he noted, as a "mechanical logic piano" which produces theorems in a purely formal way:

> [...] a machine might be imagined where the assumptions were put in at one end, while the theorems came out at the other, like the legendary Chicago machine where the pigs go in alive and come out, transformed into hams and sausages

he would write in Poincaré (1908). And the rift between the two visions will widen over time. Hilbert, as we said earlier, will develop—in his own way, given his scientific originality—this formal linguistic tradition which relied on the set of signs, be they new or devoid of meaning (like the imaginary "i"), to practice and broaden Mathematics. On the one hand, the potentially mechanical manipulations of signs should be the locus of the certitude of Mathematics. On the other hand, as we will clarify later on, the completeness of sound formalisms will guarantee the possibility of reducing all Mathematics to the formal method. And we see once more the trace of the old positivist program. The formal systems of Laplace equations should cover the world, should fully explain its determination and predict its evolution ("by allowing to deduce all astronomical facts", said-he). Thus, any question regarding future astronomical evolution, in deterministic systems such as the solar system, must have an answer. In the same way, any mathematical problem should, for Hilbert, have a solution, an answer: yes or no. In particular, the formal system of Arithmetics should be complete: each of its assertions, as long as it is well-formulated, should be decidable.

Of course, among possible answers, we could have impossibility results. The Greeks were able to provide such results regarding the impossibility of expressing $\sqrt{2}$ as a ratio of integers; the transcendence of π (the impossibility of an algebraic representation) had been recently demonstrated. But Hilbert's conceptual reference, the theoretical example towards which he aspired, was the same as for his predecessors, particularly Peano: it was the complete field of complex numbers. The audacious formal maneuver, the invention of an "i", was devoid of meaning and represents, for this school of thought (the formalist school), the paradigm of the practice and creativity of Mathematics.

If we extend the field of real numbers using this meaningless sign, then, within the field of complex numbers it generates, we obtain algebraic closure or "completeness": any algebraic equation will admit a solution, as we said. "*Non ignorabimus*" in Mathematics, as Hilbert stressed at the Paris conference of 1900. At most, it will be a question of extending the chosen formal system, in a consistent way, with new notions and *principles of proof* that are well formed: this will allow to answer any purely mathematical question.

Poincaré will disagree with Hilbert's approach in several texts: unsolvable problems, those which are demonstrated to be unsolvable, exist and are the most interesting because they open up new avenues. We can add that there does not exist any formal extension of the Newton-Laplace equations which enables to predict the evolution of three bodies. Of course, Poincaré could not use this argument against Hilbert, because deterministic unpredictability is born at the interface between an equations' system and a physical process which the system describes, via measurement. This is not a "purely mathematical" problem, as are the existence of solutions, the equality between two formally defined functions or an assertion of Arithmetics ... Hilbert believes that these problems must always find a solution, even if it is possible to obtain impossibility theorems, such as the irrationality of $\sqrt{2}$. The latter are only stages for attaining a complete theory which would answer any well-formalized mathematical question.

Of course, such a theory must be consistent: in a contradictory system, anything and its opposite can be demonstrated. And since certitude resides in *finitude*, it resides for Hilbert only in the formal play of signs, in a combinatorial calculus on finite sequences which we could mechanize, it is also necessary for the proof of consistency to be finite. In other terms, a rigorous demonstration is composed of finite deductions, of finite sequences of signs, of formal statements, line by line. By their mechanical character, they remain far removed from the ambiguities of meaning. So for the first theory of Mathematics, Arithmetics or Formal Number Theory, to which Hilbert had reduced the various geometric axiomatics, a proof of consistency must also be obtained by means of a finite formalism, that is of a finite analysis of sequences of signs, line by line. It would thus have been possible to ensure at once the non-contradictory character of Number Theory and of Euclidean and non-Euclidean geometries.

During a 1904 symposium, Hilbert proposed a program for the proof of this consistency, a schema based on an inductive analysis of the formal proofs of Arithmetics. Not without irony, Poincaré observed in 1906 that "Monsieur Hilbert" thought to demonstrate by induction the consistency of Arithmetics, of which the main axiom is induction! For more than 10 years, Hilbert will be less interested in the problem of foundations, to the greatest benefit of Mathematical Physics and Analysis to which he will contribute very significantly.

Besides, the "best among his students", the great geometer, physicist and mathematician Hermann Weyl, will also during these years distance himself from Hilbert's foundational philosophy. In his book *The Continuum* (1917), Weyl explains several times that Mathematics is rendered trivial by the idea of its potential mechanization and of its decidability, by all demonstrations made "with fixed deductive techniques and in a finite number of steps". And above all, in a way that is uncertain, confused, and arguably hesitant (how would one dare thinking in opposite of his great professor?) he conjectures the incompleteness of Arithmetics (1917, end of Sect. 4). He will later on define himself as a "lone wolf".

Hilbert's steadiness regarding his program is exemplary indeed. In the beginning of the 1920s, he returns to his proof by induction of the consistency of Arithmetics using another framework: "meta-mathematical" induction. Throughout all of these years, he stresses an important distinction: a formal system is a very specific fragment of mathematics whereas the mathematical work one can do upon it is *meta-mathematical*. In other words, Meta-Mathematics has for object of study the formal axiomatic systems, particularly insofar as they can be examined as finite sequences of signs.

One must note once more the originality of Hilbert's vision: from 1900 to 1922 (and even in 1928, during a famous Mathematics symposium in Bologna), he proposed a *meta-mathematical* analysis of mathematical deduction, described as an algebraic-combinatorial calculus. This approach to the question of the foundations of Mathematics is truly innovating. In what concerns his proof of consistency, however, it is Weyl (Poincaré being deceased) who will point out to him that his proof by meta-mathematical induction nevertheless remains a proof by arithmetic induction. It can therefore not serve as "foundation" for a theory of which the axiomatic core is induction. Wittgenstein will insist in 1929:

> Hilbert's Metamathematics must necessarily turn out to be Mathematics in disguise.

And because a meta-mathematical proof should be

> [...] based on entirely different principles w.r.t. those of the proof of a proposition [...] in no essential way there may exist a meta-mathematics.

Therefore:

> I may play chess according to certain rules. But I may also invent a game where I play with the rules themselves. The pieces of the game are then the rules of chess and the rules of the game are, say, the rules of logic. *In this case, I have yet another game, not a metagame.* (Wittgenstein 1975, §153 and p. 315).

As we will see, Gödel will shatter, both mathematically and within Formal Number Theory, the foundational role of this distinction between theory and meta-theory, by encoding the latter as part of the former. This distinction is useful from a technical standpoint, but it is artificial, or at least it tends to exclude from the Hilbertian framework the epistemological aspects of the foundations of Mathematics; we will return to this.

4.2 Arithmetics, an Absolute

Arithmetics, as a (formal) Theory of Numbers, is very important in Mathematics and occupies, no less according to Hilbert than to Frege, a central place in the search for foundations. However, the gnoseological frameworks used by these two founding fathers are completely different. For Frege, ultimate certitude resides in the signification of natural numbers understood as *concepts*, as logical and ontological absolutes. For Hilbert, conversely, it resides in Arithmetics as the locus of the finite, which can be counted or written using a finite set of signs and which has finitude as its very object of study.

Both start with the problem of space, of the crumbling of Euclidean certitudes. But Hilbert, being one of the great mathematicians and geometers of his time, wishes to save non-Euclidean geometries. It is his main objective, as shown by his 1899 book, contrarily to Frege. Hilbert, by the relative consistency proof passing by Arithmetic, refers to finitistic/mechanical formalisms as a tool for solving once and for all the problem of foundations and, then, at last, working freely and safely in "Cantor's paradise of infinities". Both authors nevertheless propose a new absolute reference for foundational analyses: Arithmetics.

Indeed, the consistency of Mathematics itself would have been guaranteed if Hilbert's program had succeeded. In order to be certain, a proof of consistency of Arithmetics had to be itself formal, finite and therefore arithmetizable. Now, Arithmetics—the Theory of Integer Numbers enables the encoding of everything which is finite, as Gödel will formally demonstrate. Arithmetics would then have removed itself from the intuition of counting and ordering in time and in space, thanks to finite (arithmetic) computations using pure formal signs; it would have elevated itself over the world by itself, by pulling "on itself", just as the Baron of Münchausen would lift himself up by pulling on his own hair. It would have become the place of ultimate certitude, without recourse to meaning.

This perfectly formal and closed locus of deductive certitude, capable of self-demonstrating its own consistency, would have been an absolute that was both different and parallel to the ontological absolute of Frege's concepts and numbers. Hilbert actually speaks of an *absolute* formal proof of the consistency of Mathematics as part of the definitive solution to the problem of foundations. For those who consider Mathematics to be central from an epistemological and cognitive stand-point, this program proposes the definitive foundation of all knowledge,

to make a protocol of the rules according to which our thinking actually proceeds,

Hilbert asserts in the *The Foundations of Mathematics* published in 1927.

5 The Theorem

Well no, it doesn't work, even not for thinking (proving) properties of integer numbers. If Arithmetics (Formal Number Theory) is consistent, i.e. it does not prove a contradiction, then not only is it incomplete—meaning that there exists in its language assertions which are undecidable, that is that can not be proved and of which the negation is also unprovable—but it is also *impossible to complete*: it has no consistent and complete formal extension—Arithmetics is "non-completable". The analogy with the algebraically complete field of complex numbers does not work: it is impossible to add signs or formal axioms to define a complete (or maximal) theory which contains Arithmetics.

But there is more: the consistency of Arithmetics, when it is formalized, in the manner of Hilbert one might say, can not be demonstrated within Arithmetics. In other words, there is no finitary proof of consistency for Arithmetics. That is, in a few words, Gödel's results represented a true cold shower on formalist ambitions, which some will still try to save by introducing different variations and modulations to the notion of "finitary proof".

We will indeed see how it is possible to "lengthen" finite induction along infinite ordinals in order to improve the situation and, in a technically interesting way, to set a hierarchy between theories and to shift the problem of consistency from theory to theory. It nevertheless remains that "non-completability" is provable and is intrinsic to the system. This signals the death of the possibility for an ultimate foundation of Mathematics on an absence of meaning, on a potentially automatable computation of signs. The "non-completability" result is a difficult and shocking fact, still to be digested by many.

Let's examine a few technical points of Gödel's proof, without going into the details of the proof of the first theorem, which is a formal masterpiece. But before this, one remark must be made. Gödel never used, neither in his statements nor in his proofs, the notion of "truth", which is not a formal concept. It is necessary to stress this point, because in current readings of this theorem, it is often too hastily said that it shows the existence of "statements that are true but unprovable" in Arithmetics. "True" statements? But where, how, according to which notion of truth? This is a delicate question to which we will return, avoiding Platonizing flights of fancy postulating a list of true statements that already exist in the mind of God, but among which some are "unprovable". Such ramblings have nothing to do with Gödel's proof. The strength of his work is, to the contrary, of shattering the formalist program from the inside, using formal tools. He uses pure computations of signs without meaning and therefore does not invoke "transcendental truths"; he presents his argument by purely formal means. We can see a first analogy with Poincaré's Three Body Theorem, a result which demolished the myth of an equational determination capable of fully predicting the evolution of the world, and this was also done from "within", by means of a purely mathematical analysis of the equations, that is of their non-integrability, only later followed by an original geometric and physical interpretation of this. Of course, also Gödel's theorem needs to be (correctly) interpreted.

The first among Gödel's great ideas was to encode, using numbers, all propositions of any formal system given by a finite number of finite sequences of signs, in the form of axioms and of rules of deduction. In particular, by numbering each sign and each letter of the language of Arithmetics, Gödel bijectively associated a natural number-code to each statement of Arithmetics as formalized by Dedekind, Peano, Frege and Russell (which we will call PA, for Peano's Arithmetics).

We do not need to go into the details of this formalization which rigorously describes the well-known axioms,[7] and even less so into the details of its encoding (which we call "Gödelization" or Gödel Numbering, see Kreisel (1984) for a discussion on a category-theoretic understanding of this fundamental notion and more references). Today, these numerical encodings of letters are, indeed, everywhere.

By the Gödelization of propositions, of sentences, but also of music and images, logico-arithmetic machines enrich and transform our existence. All the sentences that you read from your computer screen are encoded using binary integers, just as Gödel proposed to do for the assertions of any formal language. We will then designate here as $\underline{A}$ the Gödel number of the proposition A.

For example $2 = 1 + 1$ is a proposition, whereas $\underline{2 = 1 + 1}$ is its Gödel number, let's say 65,1847, or the number which digitally encodes in this author's computer memory this proposition as displayed on its screen. Gödel will thus be able to mathematically address the until then informal notion of "effective" or potentially automatable *deduction*. The deduction of formulas from other formulas, such as of $2 = 1 + 1$ from the axioms of PA, written as "$PA \vdash 2 = 1 + 1$", will be treated as a function associating numbers to numbers (the Gödel Numbers of such formulas).

It is therefore a *calculus of formal signs*. To do this, he describes a class of functions defined by the computations one can finitely and effectively describe in PA if one considers, as did Hilbert, that PA, formalized Arithmetics, is the locus of finitist effectivity. These functions use for basis the constant function 0, the successor function "Succ", and almost nothing else. From here, one defines by *induction* the operations of sum and product, as well as a huge class of arithmetic functions, the *computable* or (primitive) "recursive" functions. There already existed definitions of such functions, but Gödel completed and stabilized their definition with great rigor.

So, we write "$PA \vdash B$" to say that proposition B is *deduced from the axioms* of PA, that is, that B is a *theorem* of PA. Gödel then constructs, by induction over the structure of formulas, functions and predicates in PA which encode the formation and deduction of formulas from PA. For example, he defines the primitive recursive functions $\text{neg}(x)$ or $\text{imp}(x, y)$, which represent in PA the negation of a formula or the implication between two formulas through the Gödelization of these formulas. In other terms, $\text{neg}(x)$ and $\text{imp}(x, y)$ are functions, written in the language of PA, such that:

$$PA \vdash \text{neg}(\underline{A}) = \underline{\neg A} \quad \text{and} \quad PA \vdash \text{imp}(\underline{A}, \underline{B}) = \underline{A \to B}.$$

---

[7] The properties of 0 and of the successor symbol (0 is not a successor and the successor operation is bijective) and, especially, of induction: suppose $A(0)$ and that from $A(n)$ one is able to deduce $A(n + 1)$, then deduce A for all integers, i.e. $A(m)$ for all m, see (Gödel 1986–2003) for details.

Thus, Gödel encodes the operations of construction and of deduction of the formulas of PA until reaching a predicate of PA, which is written $Prov(x,y)$, such as $Prov(\underline{A}, n)$ represents or encodes the fact that the formula A is provable, from the axioms of PA, using the finite sequence of formulas represented by their Gödel number n.

The reader will notice the gradual emergence of a huge tide of circularity. Indeed, we have just quickly seen how to define *in* PA deductions *over* PA. So we see how to write a predicate $Theor(\underline{A}) = \exists y \, Prov(\underline{A}, y)$ which encodes the fact that "there exists" as $\exists$, in PA, a proof of A, that is the (Gödel) number of a proof "y" of A, or that A is a theorem of PA. This predicate is a predicate on numbers, because it is numbers that are the objects of PA.

More formally, Gödel's great feat in terms of encoding and of computation enables him to write an arithmetic predicate $Theor$ and to demonstrate that:

$$\text{If } PA \vdash B, \text{ then } PA \vdash Theor(\underline{B}). \tag{1}$$

$$\text{If } PA \vdash Theor(\underline{B}), \text{ then } PA \vdash B. \tag{2}$$

In other words, point (1) states that if B is a theorem of PA, this very fact can be stated and proved within PA, in the sense that also $Theor(\underline{B})$ is a theorem—the meta-theory, i.e. the provability of B, gets into the theory. Point (2)[8] says the opposite: if one can prove within PA that B is a theorem, i.e. if $Theor(\underline{B})$ is proved, then B is indeed a theorem of PA—the coding of the meta-theory in the theory is sound. Another formulation: $Theor(\underline{B})$ is nothing else than the writing in PA of the function that computes the "effective deduction" of B from the formulae-axioms of PA.

Another step, and we will have closed the loop of circularity. We write as $\neg B$ (not-B) for the negation of B in PA. Then, all we need to write, thanks to non-obvious ingenuities of computation, deduction and recursion, i.e. fix points, is a formula G such as:

$$PA \vdash (G \leftrightarrow \neg Theor(\underline{G})). \tag{3}$$

Let's now suppose that PA is consistent, i.e. that it does not generate any contradiction (it is impossible, for any A, to prove both A and $\neg A$). We then demonstrate that G is not provable in PA. If it was, that is, if $PA \vdash G$, point (3) would imply that $\neg Theor(\underline{G})$ is provable (that is, $PA \vdash \neg Theor(\underline{G})$). Now point (1) states that, from $PA \vdash G$, one can also derive $PA \vdash Theor(\underline{G})$. Contradiction.

But we can show that $\neg G$ as well is not provable. One just needs to use the rule of "contraposition" that is formalized, for any theory T, by

---

[8] In Gödel's proof, point (2) requires a hypothesis only slightly stronger than consistency: ω-consistency. This is a technical, yet very reasonable—and natural—hypothesis: the natural numbers are a model of PA—it would be "unnatural" to assume less. This hypothesis was later weakened to consistency, see (Smorynski 1977).

$$(\text{Contrap}): \quad T \vdash (A \to B) \quad \text{implies} \quad T \vdash (\neg B \to \neg A),$$

and the result of classical logic $PA \vdash (\neg\neg A \to A)$. Point (3) can then be rewritten as $PA \vdash (\neg G \leftrightarrow Theor(\underline{G}))$. So a proof of $\neg G$ gives a proof of $Theor(\underline{G})$, and therefore a proof of G by point (2). Contradiction.

In conclusion, we constructed a proposition of PA which is undecidable: not provable itself, and its negation not being provable either. So if PA is consistent, it is incomplete.

Formula G is a sequence of signs, which we produced rigorously using a pure computation on signs; that should suffice for us. But it is possible to give it an intuitive "meaning". The reader, whose brain is not a Turing Machine, will thus be able to informally "understand", attribute a suggestive "meaning" by a formally *inessential abuse*, to this proof, whose formal construction we rapidly presented. By (3), formula G "says" that "G is not provable". That is, PA proves that G and its unprovability, formalized by $\neg Theor(\underline{G})$, are equivalent. We are thus forcing meaning where there is none—and where, formally, there is no need to be any: (one proves that) G is (equivalent to) the sentence "G is not provable".

The analogy with the liar paradox ("this sentence is false") is obvious: just replace "false" by "unprovable". In his paper's introduction, Gödel also acknowledges this brilliant invention of Greek culture as one of his sources of inspiration. But to obtain this contradiction, one must not refer to meaning (true/false) as in the liar paradox. To the contrary, it is necessary to remain within the formal theory of proofs and challenge the (provable/unprovable) contraposition. This is what Gödel does with great rigor. Now "this sentence is false" is neither true nor false, and there lies its great force and its paradoxical nature. Likewise, G will not be provable, nor will its negation be, if we suppose that PA is consistent.

But what have we used from PA? Only its capacity to encode propositions and formal proofs. So any sufficiently expressive formal theory T, that is, one that can be axiomatized (thus encoded), and which contains PA, enables to construct an assertion GT that is independent from T, if T is consistent. Thus, for any consistent extension T of PA, undecidable propositions exist for T. PA is therefore *impossible to complete*: there is no "field" (consistent formal theory) which is complete (maximal) and which contains Arithmetics, to use once more the inspiring analogy with the algebraically closed (complete) field of complex numbers. As regards mathematical theories which do not contain PA, they do not know how to count using integers: generally, we can't do much with them (induction pops out everywhere in real mathematics).

During this brief overview of encodings and contradictions, we had to omit details that are essential (and that are sometimes, but not always, mathematically difficult). The very technical aspect of the First Theorem, encodings and formal deductions, span several pages, does not afford us, in a text such as this one, the possibility of delving further into it. But we are not however done with our ponderings: there is a Second Theorem of incompleteness. What does it mean in formal terms that PA is consistent?

As we have seen, a (mathematical) theory is consistent if it does not generate contradictions: it is impossible for any A to prove both A and $\neg A$. Indeed, only one contradiction is enough to deduce everything from it: "*ex falso quodlibet*", as my Roman ancestors used to say. PA is therefore already contradictory if we only demonstrate $0 = 1$, which negates one of the axioms. It is then easy to state in PA that PA is consistent: all one needs to do is to write $\neg Theor(\underline{0 = 1})$. So the proposition $Cons \equiv \neg Theor(\underline{0 = 1})$ is a formula of PA which states the consistency of PA, by saying formally that PA does not demonstrate $0 = 1$.

One must note the strength of the formalism: $Cons$ soundly describes in the theory PA the eminently *meta-theoretical* (meta-mathematical) assertion "PA is consistent", or PA does not entail a contradiction. If we prove that $Cons$ is not provable in PA, then we will have proven the impossibility of demonstrating the consistency of PA using methods that are formal and finite, and therefore encodable in PA.[9]

Building upon all the techniques produced while addressing the first theorem, Gödel proved the second in a few lines. These lines are however extremely dense and abstract. There is no longer any need (nor even any mathematical opportunity) for interpretation. All is based on the syntactic proximity between the $\neg Theor(\underline{G})$ and $\neg Theor(0 = 1)$ formulas. And as a result of a few very formal (and meaningless) lines, the second incompleteness theorem demonstrates:

$$PA \vdash (Cons \leftrightarrow G). \tag{4}$$

In other words, in PA, $Cons$ and G are proved to be equivalent. Of course, the implication that interests us most here is: $PA \vdash (Cons \rightarrow G)$. That is, in PA, one may formally deduce G from $Cons$. As G is not provable, $Cons$ is not provable either.

Let's pause a moment on the extraordinary pun that has been constructed. Let's write, for short, "$(PA, A) \vdash B$" to say that B is a consequence of the axioms of PA with the *additional hypothesis* A. So $(PA, Cons)$ designates the axioms of PA to which we have added the *formal consistency hypothesis*, $Cons$. Let's now observe that $PA \vdash (Cons \rightarrow G)$ and $(PA, Cons) \vdash G$ are equivalent (it is an obvious result of propositional calculus). We also use the abbreviation $PA \nvdash B$ to state that PA *does not prove* B. We can synthetically rewrite the first and second theorems (the left-right implication of the second), respectively:

$$\text{If } PA \text{ is consistent, } PA \nvdash G \text{ and } PA \nvdash \neg G. \tag{5}$$

$$(PA, Cons) \vdash G. \tag{6}$$

---

[9] There has been some debate on the actual meaning of $Cons$: does "$Cons$" really expresses consistency? Piazza and Pulcini (2016) rigorously confirm the soundness of the approach we informally followed here and clarify the issue by a close proof-theoretic analysis.

The passing from point (5) to point (6) is most important and is seldom highlighted. Under the *meta-theoretical hypothesis of consistency*, point (5) says that *PA* does not prove *G* nor its negation. On the other hand, if we formalize consistency in the theory, by *Cons*, and if we add it to *PA* as a hypothesis, we can then formally deduce *G* in *PA* (!). In both cases, be it an issue of *proving the undecidability* of *G* or of *proving G*, the hypothesis of consistency is essential and gives different results. More precisely, after having encoded the meta-theory within the theory, by means of Gödelization and by the construction of *Theor* (points (1) and (2)), now points (5) and (6) prove that the theory is in a way "stronger". Indeed, with the hypothesis of consistency, encoded and added, *PA* does prove an assertion which is formally *unprovable* if we suppose consistency only at the meta-theoretical level.[10]

This is a definitive stop blow to Hilbert's vision. As Weyl and Wittgenstein thought, Meta-Mathematics, when rigorous, is part of Mathematics. Once again, Gödel proves this with points (1) and (2): *Theor* encodes in theory *PA* the meta-theoretical character of demonstrability. Using points (5) and (6) he also shows that the theory is even more expressive than the meta-theory (or, as we will see better in 5.1, the meta-theoretical deduction of *G* from consistency follows from the deduction in the theory). In particular, a finitistic meta-induction does not exist: it is a form of induction, which can be perfectly encoded by theoretical induction.

The use of the terms meta-theoretical or meta-mathematical can be practical, namely from a didactic standpoint, for instance to distinguish between the "consistency of *PA*" and *Cons*. But it is not "fundamental": one cannot found Mathematics, nor any other form of knowledge, by having recourse to its own meta-knowledge, which still has a mathematical (or that knowledge's) form, as Wittgenstein had observed

there is no game which is a meta-game: it is yet another game.

No meta-language can found language:

we are locked in the prison-house of language,

also wrote Wittgenstein. As for Arithmetic and its mathematical extensions, Gödel's coding locks up the prison by a powerful circularity.

As a philosophical conclusion beyond Wittgenstein, let's observe that only a "genealogy of concepts", said Riemann—that we must entrench, with language of course, but *beyond* language, *before* language, *beneath* language, in action in space (Poincaré), such as ordering, or in time (Brouwer: the discrete flow of time)—can propose an epistemological analysis of Mathematics,

---

[10] Piazza and Pulcini (2016) prove the truth of *Cons* in the natural or standard model of *PA*, by applying Herbrand's notion of "prototype proof"—a proof of a "for all" statement, by using a "generic" element of the intended domain, instead of induction. This is a key notion also for the analysis of true and interesting (non-diagonal, like *G*) but unprovable propositions of *PA*, see below and Longo (2011). Formal induction is not the bottom line of the foundation of mathematics, even not for the (meta-)theory of Arithmetic.

as *meaningful* knowledge construction within the world, between us and the world, to organize and understand the world.

Both space and time active experiences are needed in order to propose the conceptual invariant, in language, the notion of integer number, which then becomes independent from each one of these actions, yet preserves its structural, geometric meaning. Mathematics is grounded in the "primary gestures" of knowledge, such as pursuing or tracing a trajectory, drawing borders and then posing "lines with no thickness" (Euclid's definition beta) (Longo 2005, 2016). Language and writing stabilize geometric practices in space and time by the intersubjectivity of our human communicating community (Husserl 1933). Thus, the philosophy of mathematics should not be just an annex of a philosophy of language, as it has been "*From Frege to Gödel*" (the title of a famous book) and till now, but a component of a philosophy of nature and its sciences, see Weyl (1949) and Bailly and Longo (2011).

5.1 And What About "Truth"?

Easy and customary popularizations of Gödel's theorems continue to be proposed—including by illustrious colleagues, in terms of Platonizing "ontologism", often to impress the reader by ontological miracles in Mathematics. Such popularizations still adhere to the rapid and ontologically naïve reading of Gödel's theorem. There are many who invoke, making big gestures and gazing towards the sky, the stupefying existence of true but non-provable assertions, such as G (an ontological or quantum miracle: and what if the understanding of G's unprovable truth was due to quantum processes in the brain?). In the face of such assertions, one must always ask how can we state that G is "true"? Besides, how can we state, in Mathematics, that an assertion is true without demonstrating it (or taking it for hypothesis)?

The interlocutor must then produce a *proof* convincing us of the "(unprovable) truth" of G. The hypothesis of consistency, he/she points out, *implies* that G is unprovable (first theorem). And since G "asserts" that it is not provable (it is equivalent in PA to $\neg Theor(G)$), then it is true. This reasoning based on the "meaning" of G is informal, vague and unwritten, as we observed. But once formalized, it is a semantic version of the rigorous formal implication $PA \vdash (Cons \rightarrow G)$ that constitutes the core of the second theorem.

As a matter of fact, the latter formally deduces G from $Cons$, and therefore proves G, *once Cons is assumed*. So, once we give ourselves an interpretation of PA in the model of standard integers (which gives consistency, that is the truth of $Cons$), G is "evidently" true, because it is a provable consequence of $Cons$. Ultimately, we *prove* the truth of G, and any Platonist will also be forced to do so, at least by handwaving. And we do prove it, even easily, in PA and from $Cons$: that is the second theorem.

As we were saying, we return to the extraordinary finesse of Gödel's result, to the subtle interplay between points (5) and (6). And there is no need for a miracle of ontology or of quantum effects, but just classical logic and a reference to the second theorem when trying the unneeded exercise of interpreting the first—a perfect formal diamond, independently of any interpretation.

We will later mention the "concrete" results of incompleteness, that is, combinatorial assertions of Number Theory (of the type "for any x there exists y ... and a complicated numeric expression in x and y"), which are interesting and which are not provable in PA—not even from $Cons$—but only by infinitary extensions of induction or by "prototype proofs" (see the previous note for references). Given their combinatorial complexity, no one dares say of these that they are "evidently" true by invoking ontological or quantum miracles. We are reduced to demonstrating them, of course outside of PA, as we will explain.[11] What is Gödel's responsibility in this?

The 1931 article is perfect: there is not a single assertion, not a single proof, nor a single argument which calls to "truth" or which refers to an *interpretation* of the formal game. Only in the introduction does Gödel want to informally explain the meaning of the First Theorem and he notes that G, the statement which will be unprovable, is *sound*. But he *immediately* adds that the specific analysis of the meta-theoretical reasoning which proves it— and which we outlined—will lead to "surprising results" from the standpoint of the "proofs of consistency of formal systems" (the Second Theorem).

Of course, the ontological vision can still be salvaged: the proof is only an access to a pre-existing reality which may sometimes be more than only formal. More specifically, we can give a good notion of *relative truth* to the relationship between a formal system and a given mathematical structure. For example, imagine the sequence of integers with the properties learned in elementary school. You know how to say that $4 + 3 = 7$ is true, or that $667 \times 52 = 34{,}084$ is false, or that $7 < 8$ ... The formal theory (PA) makes it possible to demonstrate it automatically (and a machine does this far better than we do). It is though possible to consider these properties as "true" or "false" by associating to the signs of the theory the concrete and meaningful numbers from one's school-age experience. In general, we will say that a formal theory is "sound" if it proves *only* true assertions in the associated (or standard) model and, following Hilbert, that it is "complete" if it proves *all* true assertions in this model.

---

[11] In the ontological search for an unprovable mathematical truth, sometimes the "fact" that G must either be true or false is used. Or—this amounting to the same thing—that either G or $\neg G$ must be true, *without saying which* because it would be necessary to prove it. This "weak ontology of truth" comes from a classical and legitimate hypothesis (the excluded middle) but one which is very strong and semantically unsatisfactory (or highly distasteful—and this is important in Mathematics) when it is question of discussing the *truth* of assertion G: this is a formal rewriting of the liar paradox which is precisely neither true nor false. Gödel also uses the excluded middle (G is independent from classical PA) but precisely to give us, in *Proof Theory*, the "middle": the undecidable.

Alfred Tarski indeed proposed in the 1930s a general theory of truth (Tarskian Semantics), the foundation of the new and very relevant logico-mathematical Theory of Models. It associates to each formal sign the corresponding "object" in the associated structure (the model): "0", as a sign, corresponds to the first element of the well ordered structure of integers; the function sign *Succ* will be the passage to the next one, following the order etc.. The formal description adapts to the underlying structure and, from it, everything will be derived. Developing a general theory of the truth of the linguistic and scientific expression as "*adaequatio intellectus et rei*" is a very delicate endeavor.

The misuses, inspired by the works of Tarski with no reference to its technical depth, were numerous. Some extended Tarskian semantics for example to historical languages and observed, say, that "snow is white" is true when snow is white (brilliant!). So "grass is green" is almost always true, whereas "grass is blue" is almost always false. Yet, this would have been a difficult observation for the Ancient Greeks, who had only a single word for designating both colors green and blue. And we would have trouble refereeing a dispute between two Eskimos where one would be saying "today, the snow is *white5*" and the other would be saying "the snow is *white7*" (it seems that Eskimos have over 20 different names for designating the whiteness of snow). Color is not a precise and defined wavelength, but a human act tracing delimitations within a quasi-continuum of wavelengths, an act that is rich in intersubjectivity and history. And the whole construction of objectivity goes likewise. But for Arithmetics, in a first approximation, such semantics may suffice and the reader can be satisfied with what he or she understood at school: one needs only to associate the formal signs to the elementary-school comprehension of integers. But we will see how the notion of truth (or of "element of reality" as Einstein will say), becomes an enormous challenge in Quantum Mechanics; we will return to this while examining its alleged "incompleteness".

As for now, we can summarize Gödel's proof of the existence of a statement—and he actually constructs one—that is true in Tarski's sense and that is not formally provable. We could explain that it is true because PA *proves* it from the formal hypothesis of consistency: that is, if we suppose that $Cons$ is true, since $PA \vdash (Cons \rightarrow G)$ and PA is sound, G is also true.[12]

The historical importance of Gödel's article must now be clear, not only regarding the foundations of Mathematics, but also for the *techniques* invented throughout the First Theorem's proof. Gödelization and the definition of the class of recursive functions will pave the way for the Theory of Computability

---

[12] Let's recall for the reader who may be somewhat numbed by this wonderful pun that the question resides in the difference between the meta-theoretical hypothesis of consistency and $Cons$, the theoretical hypothesis of consistency which encodes consistency in PA. When this extra assumption is added to PA, then, we insist, Gödel could formally derive G, within PA, thus its truth in the standard model, which realizes $Cons$, if PA is consistent (see the work in Piazza and Pulcini (2016) and the previous notes as for the meaning of $Cons$).

and hence for the works of Church, Kleene and Turing during the 1930s. These thinkers, especially Turing, will in turn establish the foundations of modern Computer Science by starting off with—and we insist on this point—entirely logico-mathematical problems: the question of undecidability and the definition of the computable real numbers (that is, those that are effectively generated by means of an algorithm).

It is interesting to note how Gödel and Turing (5 years later) invented the rigorous notion of computability or of effective decidability within the framework of formal languages and systems. By this, they also definitely stabilized the notion of Hilbertian formal system (Poincaré's "sausage machine"). Yet, they aimed to demonstrate that it is possible to exhibit undecidable propositions and uncomputable processes (which can not be automatically generated, like sausages, without using stronger hypotheses). To say no, it is necessary to define exactly that to which one says no. And then, if it is interesting, it can be made even more usable, ultimately taking the form of the digital machine (an arithmetic or Turing machine) which is in the process of changing our world.

As with Poincaré's three-body theorem, the negative result is the starting point of a new science due to its content and to the *methods* it proposes. It must be noted that in 1931, the scope of the analysis of computability proposed by Gödel was not obvious. Gödel, who was aware of this, wrote in the end of his article that his result did not necessarily contradict Hilbert's formalist point of view. One could possibly find other formalizations for the informal notion of effective deduction which would not necessarily be encodable using his recursive functions. It is only with the equivalence results of all formal systems for computability proved by Turing and Kleene in 1935 and 1937 that we will have a proof of the generality of Gödel's method.

Church's thesis will propose the mathematical invariance of all notions of computability and of the notion of effective deduction or of acceptable deduction as regards finitism. And in a 1963 note, Gödel will recognize the full generality of his theorem: it bases itself on a "sure, precise and adapted" notion of formal system and contradicts the decidability, completeness and (formally) provable consistency "of any consistent formal system which contains a sufficiently expressive finitary theory of numbers". And the search for extensions that are specific to (and consistent with) formal Arithmetics and Set Theory will mark the developments of Logic during the following decades.[13]

---

[13] We can mention Gentzen's ordinal analysis (1935). Larger infinities, as orders beyond integers or as cardinals beyond the countable, provide tools for the analysis of proof in order to fill the incompleteness of PA—or to postpone it to stronger theories. Set Theory, with an axiom of infinity, in its formal version (ZF or NBG) extends and proves consistency of PA, but it does not prove its own consistency—it is incomplete, of course, nor is it able to answer the questions for which it was created: the validity of the axiom of choice and of the continuum hypothesis. The respective independence results cast additional light on the expressivity and on the limits of formal systems (Kunnen 1980).

6 Poincaré vs. Gödel

We have attempted to explain how Poincaré's Three-Body Theorem can be seen as a "philosophical precedent" for Gödel's theorem. Unpredictability resembles undecidability, in time and space—in a sense, statements on future space configurations are undecidable. From a philosophical point of view, Poincaré always appreciated unsolvable problems, "negative results". But technically, both theorems can not be directly correlated; be it only because Laplacian predictability is a problem of the interface between the mathematical system and the physical process and not a purely mathematical question as is Hilbertian decidability. We can, however, establish a mathematical correlation between certain *consequences* of these two great theorems. We just give some hints here to an analysis more closely surveyed in Calude and Longo (2016a).

Poincaré's geometry of dynamical systems extends physico-mathematical determination and captures randomness, contrarily to Laplace's distinction. Classical randomness, as we have said, is *unpredictable determinism*—a fundamental insight by Poincaré to be recalled, as too often randomness is still opposed to determinism. Now, this randomness can also be given by purely mathematical means without reference to physical processes. Birkhoff provided a definition of it in the 1930s, following one of his important results. In very informal terms, if we give ourselves an observable in a particular dynamic (the speed or momentum in each point, for example), a point is said to be *random* if the average of the temporal evolution of the observable for the point coincides *at infinity* with the average of the observable on the full space (the temporal average coincides asymptotically with the spatial average). Think of a particle in an isolated volume of an ideal gas: its average speed over time will be equal to the average speed of all of the particles making up the gas. If we push this asymptotic analysis of an average to the actual limit, as the coincidence of two integrals, one integral expressing the average over time the other over space, we obtain a mathematical means of defining a random movement, and even a *random point*—the origin of the trajectory (Petersen 1983).

Let's return to Gödel. Martin-Löf (1966) proposed a notion of randomness for infinite sequences of numbers (for example of 0s and 1s) that is based on Gödel's (in-)computability. The idea, proposed in a doctoral thesis directed in part by Kolmogorov, was then further developed by G. Chaitin (Calude 2002). In short, the notion of "effective statistical test" is defined in terms of computable functions; informally, the possible regularities or computable segments in a sequence are effectively checked. A random sequence must not have any effectively recognizable regularity which is repeated indefinitely. Then all possible effective tests are enumerated and an infinite sequence which passes "all effective tests" is qualified as (*ML*)-*random* (for Martin-Löf random): that is, randomness for an infinite sequence is defined by the property of "passing all effective tests for regularities" or no "regularity" can be effectively detected.

Note that this asymptotic construction is necessary to deal with randomness in full generality. Kolmogorov had conjectured that incompressibility for finite strings could characterize randomness. Martin-Löf showed that any infinite sequences possesses finite *compressible* initial segments. Even more strongly, any sufficiently long finite sequence is compressible, by Van der Waarden theorem, see Calude and Longo (2016b).

It is easy to prove that an (ML)-random sequence is strongly undecidable in Gödel's sense: it is not only undecidable and even impossible to effectively generate (it is not semi-decidable), but, especially, no infinite sub-sequence of it can be effectively generated (it contains no infinite recursively enumerable sub-sequence). The interesting fact here is that asymptotic dynamic randomness, a la Birkhoff, and the "Gödelian" ML-randomness are equivalent. And, indeed, if one gives a structure of effectivity (effective metric spaces etc) to a vast and interesting class of physico-mathematical dynamics, from weakly chaotic (mixing) dynamics to full chaoticity, one can demonstrate the coincidence of Poincaré-Birkhoff randomness and Martin-Löf gödelian randomness, see Gács et al. (2009) or Calude and Longo (2016a) for a survey.

Let's be clear, Poincaré's theorem cannot be deduced from Gödel's (nor can the opposite be done). However, as we have said, the approaches proposed by the one and the other, and more specifically by reinforcing their negative results, allow to give purely mathematical limit notions of dynamical or algorithmic randomness. And these notions may be brought to coincide. Let's observe that the introduction of classical randomness in deterministic systems, i.e. considering it as unpredictable determination, is a very important element of the new vision of dynamic systems proposed by Poincaré. In the same way, undecidability is at the center of Gödel's theorem—and algorithmic randomness is a (strong) extension of it.

Let's finally note that to prove the equivalence, asymptotically, of algorithmic randomness and of the randomness of physical dynamics does not signify at all that the "the Universe is a (great?) algorithm". To the contrary, we have demonstrated that in a deterministic framework, asymptotically and under certain hypotheses, dynamic randomness or unpredictability coincides with algorithmic randomness, which is a (strong) form of undecidability. So, by contraposing (the Contrap rule stated above) of this equivalence, an algorithmic procedure, a method of semi-decision or a computable (recursive) function only generates *predictable* deterministic processes. Now, not only the dice of course, but also the solar system (or just three celestial bodies) and almost all which surrounds us is a fabric of correlations and, therefore, forms a "system". This makes most physical processes better described by non-linear mathematical systems, as interactions yield non-linearity. In view also of physical measurement, which is always an interval, and to sensitivity to initial conditions, they are therefore *unpredictable* and, *a fortiori*, non-computable. In this frame, measurement by interval, that is this thinking in terms of interval or "natural" topologies over *continuous* mathematical structures, is crucial.

In summary, the comprehension of the world provided by continuous Mathematics and the one provided by discrete Mathematics differ: the world is not the sum of little squares or of little points, as are Seurat's paintings, whose access (measurement) is exact. As soon one deals with a dynamics, not just the approximation of a static image by pixels, continuous and discrete space trajectories differ, approximation becomes a major challenge. This has been discussed by many authors, its dramatic consequences in the understanding of biology are hinted in Longo (2018a).

However, randomness, as unpredictability in the intended physical theory, may be brought to coincide, asymptotically, with algorithmic randomness, a theory grounded on discrete data types: "negative results" in a sense "converge" at the infinite limit. This is at the root of very interesting further work relating algorithmic randomness both to classical dynamics, as mentioned above, and to statistical physics (thus thermodynamics), see Baez and Stay (2012) for example. It may be so anytime randomness and limit processes play a role in the intelligibility of physical phenomena. Also Turing, during the last few years of his short life, dealt with continuous vs. discrete dynamics (Turing 1952), the fundamental aporia of Mathematics, as observed by Réné Thom.

6.1 Turing: From Formal Systems to Continuous Dynamics

"The Discrete State Machine", wrote Turing in (Turing 1950) concerning the Logic Machine he invented in 1936, the prototype of the digital computer, "is Laplacian": unpredictability can only be in practice (due to a long and complicated program), and does not exist in principle, he insists, as it does in the physics of "continuous systems". Thus he defines the systems he will study in his fundamental article of 1952 dedicated to morphogenesis (the continuous dynamics of forms). In his 1952 non-linear continuous systems of action/reaction/diffusion equations, the evolution of forms—for instance color patterns on an animal's fur—is sensitive to the initial conditions: it is subject to "exponential drifts" or to "catastrophic instability", says Turing. Imperceptible changes, over time, to physical measurement *and therefore to any discretization*, can cause great differences over time. Turing completely shifts his area of research and perspective.

He discusses the problem and works in the wake of Poincaré, all the while limiting the analysis of solutions to linear approximations—as he focuses at length on the non-linear case. Continuous dynamics replace his first Machine's sequence of discrete states.

The computation is no longer based on a fundamental distinction, which he invented, between software and hardware (the first made of finite strings of signs, the second a material support of 0s and 1s), but is rather a continuum of deformations, a *continuous genesis* of forms of the physical matter only. Turing's morphogenesis is a purely hardware/material dynamics of forms.

Let's very briefly attempt to grasp the meaning of Turing's reflection. Due to this change in point of view, we will understand why the correlation result between dynamic randomness and algorithmic randomness contributes in turn to the formal negation of the myth of a universe completely accessible to numerical computations. By approximation, these computations transfer equational determinations to discrete data bases; here the access to the data is exact, contrarily to physical measurement which is always an interval. Moreover, due to successive rounding-off, the orbits of chaotic dynamics, when they are computed by a machine, quickly differ from the physical orbits described in continuous space-time. So the sensitivity to initial conditions can be hidden in a theory of algorithms, one which is necessarily arithmetic, and discretization imposes evolutions which are different than those we are able to describe in the mathematical continuum.

Take the best computer simulation of the double pendulum (it's easy, there are only two equations; such simulations can be found on the Internet). If you launch the pendulum once and again using the same initial values, the algorithm will cause the simulated pendulum to take the exact same trajectory, be it one thousand or ten thousand times. But this does not make any physical sense. Indeed, the unpredictability of the (random) evolution of the actual physical device is very simple to show and is precisely characterized by the fact that, launched again using the same initial conditions (in the physical world where measurement is not exact and is by nature an interval), it generally does not follow the same trajectory. Due to the sensitivity to the initial conditions, after a few oscillations and from the very interval of the best possible physical measurement, it follows different orbits. Continuous Mathematics tell us this a priori. And some call "random" a physical process precisely when, repeated under the "same" initial conditions (in the physical sense), it does not follow the same evolution.

This is foreign to Algorithm Theory, and it is only artificially that one who has understood can imitate the physical unpredictable dynamics. One can, for instance, add to the time of each new launch, a one-number shift to the left or to the right according to a random number taken from the Internet (for instance, is there an odd or an even number of people using Skype at this very moment?).

But this is an "imitation" and not a "modelization" of the physical phenomenon. In this respect, Turing makes a very subtle distinction between *imitation* (the *game* described in the 1950 article) and *model* (1952). The latter does not seek to deceive the observer, as does imitation but rather to make intelligible the examined physical process (morphogenesis) and to propose a structure of determination for it, the equations.

For example, the sensitivity of the double pendulum to fluctuations in temperature is not made intelligible, from the point of view of "causality", by the recourse to randomness taken from the network in a discrete state machine. It is just (but effectively) imitated. The differential equations of its movement, a mathematical model, provide on the other hand its formal determination; they make it intelligible, by highlighting the forces at play and enable to analyze the divergence of trajectories (the "exponential drift", says Turing, the so called Lyapunov exponents). Turing elegantly contributed to the debate by teaching us, both as for discrete state machines and, later, for continuous dynamics (morphogenesis), how "to be within phenomena" (Longo 2018c).

We are not saying that the world is continuous rather than discrete. It is what it is. We are only saying that Continuous Mathematics, since Newton, enables to understand the properties of physical dynamics which elude Discrete Mathematics. The unavoidable interval of classical physical measurement, with the possible fluctuations/perturbations within, is better grasped by continuity. In a theory of the digital, nothing can happen *below* the proposed discretization, in principle, but also in actual applications: the repetition works—woe if there lacked a comma in a file, in the result of a program that has been relaunched a thousand times! Nevertheless, Discrete Mathematics, in turn, once implemented in extremely powerful machines, allows to analyze processes, chaotic ones in particular, that mathematical conceptual analysis can absolutely not reveal. Hence they provide us also with another type of intelligibility, one which is just as important.

In short, from the physical standpoint, a theory of algorithms does not produce an accurate model of the whole world, but of a small set of deterministic systems: predictable systems. And once transferred to the realm of the discrete, all deterministic systems become predictable, even if they are the implementation of non-linear equations or functions. It is possible to perfectly repeat, against Physics, even the wildest of turbulences. And it is not true that the discrete is an approximation of the continuous.

Numerical analysts very well know that difficult "shadowing" theorems (Pilyugin 1999) are required to prove that, in the numerical implementation of certain chaotic dynamics, the continuous trajectories "shadow" the discrete ones (and not the opposite). *In general, the discrete is not an approximation of the continuous. It is, at best, the opposite: a given discrete trajectory can be approximated by a continuous one.* So the images, displayed on a computer screen, of a chaotic evolution give qualitatively important information regarding continuous trajectories: they provide very useful imitations that are now indispensable for science and its applications.

And the richness of science and technology, the variety in history, from Lorenz onwards and especially since the 1970s, is so that we appreciate chaos on the screens of digital machines more than anywhere else. The meteorologist can look at turbulence and hurricane simulations over and over, and can repeat them identically if desired. He/she can thus have a better grasp on what appears to be the most interesting aspects and, based on experience, can make increasingly reliable predictions.

In a very specific sense then, any algorithmic theory of the physical universe is mathematically "incomplete" with respect to continuous descriptions. And the aforementioned theorems, which link classical and algorithmic randomness, demonstrate it again, by duality (or by contraposition, as we said).

If Gödel's theorem sets limits to any attempt at a mechanical reduction of mathematical deduction, its consequences (we will see other ones) also obliterate the algorithmic visions of an inert universe—and let's not even mention the living state of matter, the brain for example—because, as mentioned above, there are problems even in the algorithmic simulation of the double pendulum: no program follows the physical dynamics long enough. And when limits are better understood, it becomes possible to use our tools at their best and improve them, from Proof Theory to digital simulation; the latter being science's main instrument today.

As concerns the continuous/discrete dichotomy, even within Theory of Computation, the proofs of abstract properties of discrete structures (see Kreisel (1984) for a discussion and references) or the analysis of today's computer networks may require a difficult use of geometric tools, in the continuous realm. The latter are indeed immersed in a relativistic space-time, which we better understand using continuity (cf. for example Goubault (2000) for a relevant use of homotopy theory in Concurrency Theory in computer networks).

As for the discreteness of Quantum Mechanics—which some could invoke as an ultimate discretization of the world—the phenomena of entanglement or of non-separability are at the opposite of the topological separations specific to discrete databases in which each point is isolated, well-separated from all others. As a matter of fact, the understanding of these phenomena generated yet a further possible meaning of "incompleteness". Its consequences are opening today the way to new forms of computations and actual machines: Quantum Computing.

7 Einstein and the Thesis of the Incompleteness of Quantum Mechanics

Einstein was certainly no stranger to the debate concerning the foundations of Mathematics, firstly through his active collaboration in Zurich with Weyl who published a book in 1918 on the foundations of Mathematics (*The Continuum*) as well as another one, a veritable mainstay, on the mathematical foundations of Relativity (*Space, Time, Matter*). Einstein would later meet again with Weyl in Germany, as well as with Hilbert. He will also witness from afar the foundational contention between Hilbert and Brouwer, the founding father of intuitionism, a dispute that will result in the exclusion of Brouwer from the editorial board of the very prestigious mathematical journal directed by Hilbert. Such a preposterous outcome will be rather appalling to Einstein (he would call it a *"batrachomyomachia"* referring to ancient Greek comedy).

Einstein also had the opportunity to discuss with Von Neumann, also in exile from Nazi Germany at the Institute for Advanced Studies, where Weyl will also move to in 1933. Von Neumann had a good knowledge of Gödel's theorem. It is even said that when Gödel, at the age of 24, presented his result before a meeting in 1930, Von Neumann was the only person to grasp the scope of it. The result actually had a shattering effect on the staunch formalist who was Von Neumann; he had worked on Hilbert's program, as did Ackermann and so many others, and at a point he had been briefly convinced of having obtained an acceptable proof of the consistency of PA.

The great mathematician was rather swift: after having heard the first theorem, he drew from it the purely formal proof of the second one. However, by the time he informed Gödel of this, the latter's article was already under print with the two theorems. Thus, not only did Von Neumann know Gödel's theorems, but he had even worked on them. He then presented them to Princeton mathematicians and physicists at one of his first seminars during the fall of 1931 (Sieg 1994). Later, Gödel himself would also temporarily move to the Institute, in 1933–1934.[14] During the following years, Von Neumann developed his hyper-formalist approach in several fields ranging from the axiomatization of Quantum Mechanics to Probability Calculus and Game Theory, formal games of economics and war.

In 1935, Einstein will write an article with Podolski and Rosen that will be known under the initials "EPR" in which they will examine the problem of the "sound" and "complete" character of Quantum Mechanics (QM). These terms are specific to Mathematical Logic (we have used them) and are not common in Physics, especially in what concerns the term "completeness". It is therefore more than likely that it is no coincidence that the authors used the term "completeness" to criticize the descriptions of physical reality proposed by quantum formalism: they most probably imagined they would be dealing another blow like Gödel's against Hilbert. The Gödelian paradigm will in any case serve as a tool for comprehension: almost surely so for them, most definitely so for us.

We use the term "paradox" when referring to EPR, as it is often done, thus reminding of the employment of proof by contradiction as used by Gödel, as well as emphasizing the "paradoxical" aspect of QM (it is indeed a theory which is often positioned against the classical "doxa", in physics).

EPR begins by stating with great clarity the ontological hypotheses of the whole reflection: even in microphysics, there must exist a physical reality that is independent of measurement and of theory. At most, the measurement can "disturb" the measured physical quantity. As regards the theory, it must, of

---

[14] Gödel moved there permanently in 1939, after a spectacular escape from Nazi-occupied Austria.

course, be sound: a "satisfactory" theory must only lead to true assertions. Then, in order for it to be *complete*, "every element of the physical reality must have a counterpart in the physical theory", meaning that it must be described or deduced within the theory. One will recognize a requirement of "semantic" soundness and completeness, as we mentioned concerning Logic, as well as an ontological reading of these properties.

The classical semantic interpretation of Gödel's theorem tells us precisely that, in hypothesis of consistency of PA, the assertion G is valid over the natural numbers (of course, we observed, since $Cons$ implies G), but that the formal theory, PA, is not able to deduce it. EPR seeks a *complete* theory regarding a physical reality whose objects of knowledge, even in microphysics, must unambiguously be accessible (well-separated in space) by measurement and separated as well from the knowing subject. And it demonstrates, under this ontological hypothesis, that current QM does not constitute one. The arguments used by EPR are based on various fundamental aspects of QM, among which those we know under the names of "indetermination" and of "entanglement".

Quantum *indetermination* may be described as the non-commutativity of the measurement of the position and momentum of a particle. According to the theory, the values obtained *depend on the order* in which these measurements are made and therefore, as it is stated in EPR,

> we can no longer speak of the physical quantity A [or B] having a particular value.

Also,

> if the operators corresponding to two physical quantities, say A and B, do not commute, that is, if $AB \neq BA$ then the precise knowledge of one of them precludes such a knowledge of the other.

And EPR continues: the two physical quantities of position and momentum therefore "cannot have simultaneous reality" and at least one element of reality will not be described.

> If then the wave function provided such a complete description of reality, it would contain these values; these would then be predictable.

As regards *entanglement*, EPR deduces it from an observation which will become fundamental. From quantum formalism (Schrödinger's equation in particular), it is shown that if two systems interacted at time $t = 0$ and then were separated without any further interaction until time $T > 0$, it would be possible to know the value of a measurement over one of the systems at time T by performing this measurement on the other system. Two "entangled" particles, as we are saying, allow for an instantaneous knowledge of the value of a measurement made on the one because of the measurement made on the other. If the first has an "up" spin, for example, the result of measurement of the other spin will be "down". By repeating the same process, we can obtain the "down" spin for the first; the other will then have an "up" spin, if measured. Is this an instantaneous propagation of information, one which happens faster than the speed of light? That would be impossible, it would contradict Relativity. The theoretical explanation by QM is either inconsistent or incomplete, says EPR.

To summarize, EPR points out the incompleteness entailed by a fundamental property of the gap between theory and measurement in QM, if consistent: *that which is computed*, with the wave function (Schrödinger's equation), *is not what is measured*. In Classical and Relativistic Physics, computations are made over real numbers taken from measurements. These computations, in turn, produce real numbers which are verified by means of other measurements.

In QM, computations are made over complex numbers in Hilbert spaces that are very abstract, possibly having an infinite number of dimensions, and which are therefore outside of usual space-time. Then, real numbers are produced as projections (modules) of complex numbers obtained by means of the computations. These values are the probability of getting certain results in the process of measurement and, when verified by means of measurement, they are, on the one hand, dependent of the order in which the measurements are made (non-commutativity) and, on the other hand, they can be correlated if the particles, which are measured, are entangled (or which are in an "EPR" state, as physicists would still put it today).

And even recently, "hidden-variable" theories have tried to fill (to complete) these gaps in QM, its incompleteness. However, it is the "standard interpretation" which prevails, emphasizing the originality of the construction of knowledge in QM. Measurement is consubstantial with the physical object: there is not already a particle traveling along with its properties and states "already given" and which is to be, at most, disturbed by the measurement.

If we launch a "photon" against a double slit and if we measure, using an interferometer, the result on a wall beyond the slits, we will observe interference, a typical wave-like behavior. If, conversely, we put a particle counter behind each of the slits, we will "observe" a particle passing 50% of the time on one side and 50% of the time on the other. The action of measurement, the consequence of a whole theoretical framework, gives the specification of the object. The *scientific concept* of photon *isolates* a fragment of the universe which is specified in the theoretical and practical act of *its own production* and of *measurement*: a wave or a particle.

Likewise, Schrödinger's equation enables to calculate the evolution of a system of entangled particles and provides "correlated" values of probabilities for eventual measurements. In short, if we throw two classical coins into the air which then interact (for example, if they collide), and then take their own distinct trajectory without any further interaction, the two probabilistic analyses of the heads or tails values taken by the two coins will be independent. On the other hand, the Bell equations (Bell 1964) and the Aspect experiments (Aspect et al. 1982) demonstrated that the measurements (probability values) for two entangled quanta (having interacted) are correlated, not independent. If we know the one, we know the other, even at a great distance; this confirmed EPR's theoretical deduction. No "information" passes between the two distant events: it is necessary to make, a posteriori, a phone call in order to verify that the two measurements are indeed entangled.

This fact, undoubtedly extraordinary ("paradoxical") and now empirically verified several times over, is at the origin of very interesting theoretical reflections of which the practical consequences could be significant: Quantum Computing. Such a "calculus" could revolutionize actual computing: in the very least, computations that are impossible to perform because they are too complex would become quite feasible because entanglement is a form (a very original one) of "parallel computing". But what is being computed? It is not numerical information as we usually understand it, but the evolution of a system, which is global: the two particles are not separable by measurement and a variable associated to the object would not be local (it would not depend on the evolution of a "single point"). These are absurdities, from the standpoint of classical and relativistic physics, which EPR deduces from the theory and which have been verified empirically.

As we were saying, the world is not made up of little dots or of little squares, of classical bits and bytes that are well separable by the unique way we have of accessing them: the active constitution of scientific objectivity and of objects of knowledge, in the friction between ourselves and the world, which is measurement (sense or instrument-based).

Let's finally note that we have not said here that QM is *complete*, but that the proof given by EPR of its incompleteness is neither theoretically valid nor empirically corroborated: entanglement is there, it does not contradict physical evidence. EPR argument for incompleteness is founded upon topological (and ontological) hypotheses, the well separated locality of measured observables, that are inadequate as regards microphysics. EPR thus declared the impossibility of a situation that has been empirically shown to be possible (and very interesting). Einstein was wrong, but when he observed that

> QM is incomplete because entanglement is deduced from its theoretical and mathematical structure,

he first paved the way for research and experiments, and then for possible machines which may become of great importance.[15]

---

[15] Deduction in EPR may remind of another, from Aristotle:

> the void is impossible, because in it, all objects would fall at the same speed (*La Physique*, vol. 4, chap. 8).

Great theoretical minds, even when they are mistaken, propose very interesting ideas indeed.

8 The Concrete/Mathematical Incompleteness of Formal Theories

Following Gödel's theorem, the opposition between various schools of thought regarding the foundations of Mathematics deepened. Federigo Enriques said so with great lucidity in Paris, 1935:

> [...] if we avoid the Scylla of ontologism, we fall into the Charybdis of nominalism: could an empty and tautological system of signs satisfy our scientific reason? From both sides, I see the emergence of the ghost of a new scholastic (Enriques 1983).

One the one hand, the invocation of the eternal and pre-existing "truth", certain because absolute, that

> the mathematician discovers by looking over God's shoulder (John D. Barrow).

On the other hand, the insistence on the mathematical certitude founded upon the absence of ambiguities of meaning, on the mechanical nature of deduction and, why not, of all reasoning. Then some claimed that our humanity, could be fully transferred to a Logico-Mathematical machine, eventually producing the so-called super-brains foreseen by Artificial Intelligence in the 1960s and 1970s. Indeed, the formalists (nominalists) will say for many years, Gödel's theorem demonstrates the independence of a meaningless diagonal assertion. It is an astute, rather farfetched paradox; it is of no importance as regards interesting mathematical deduction and even less in what concerns human reasoning.

On the contrary, Gödel's theorem is only the starting point of an avalanche of formally unprovable assertions, among which some are very interesting. They are assertions of *Formal Number Theory* with a mathematical sense and of mathematical interest and which can only be demonstrated by means of more powerful arguments than those provided by formal finitism. To pass from one line of such arguments to the next, it is necessary at some point to invoke such a thing as "meaning" or "infinity". Let's try to explain this non-obvious matter, very briefly (more may be found in Longo 2011).

To remain within the sphere of Logic, let's recall that, in 1935, Gentzen (1969), gave a proof of the coherence of Arithmetics by using transfinite induction; a result which will inaugurate modern Proof Theory, in the form of "Ordinal Analysis". In short, he demonstrated the consistency of PA by transfinite induction over a restricted class of formulas (roughly: induction with an infinity of hypotheses, reaching the ordinal ϵ_0, an infinity which is "small", but which is large enough to resolve the equation $x = \omega^x$, where ω is the infinity of integers). The restriction to a certain type of formula and the rigor of proof, in an original framework called "natural deduction", will make the proof convincing, but it is obviously not formalizable in PA. In 1958, Gödel himself will give a proof in a "stratified" system (numbers, then functions over numbers, etc. ... the typed λ-calculus).

Let's note that this proof will be extended, in a non-trivial manner, by Girard (1971) to a system based on second-order quantification, that is, on "for all ... " or "there exists ... " also over sets or types (PA is a first order theory: only number variables are quantified). Girard's system turned out to be mathematically challenging and highly successful in Computer Science, for introducing a strong form of modularity in programming (Girard et al. 1990; Asperti and Longo 1991). Of course, here also, the effectivity of the calculus cohabits with the formal unprovability of its consistency, of which the proof is only formalizable in third-order Arithmetics (sets of sets) and which implies the consistency of PA. Thus, while with Gentzen begins the use of larger and larger ordinals in order to give infinitary proofs of the consistency of increasingly expressive theories; with Gödel or Girard, we pass onto higher orders, as quantification over infinite sets or types.

So, in order to salvage the paradigmatic theory of the finite, PA, it is necessary to have recourse to forms of infinity; in Mathematics, infinity is a difficult but omnipresent concept. We need only to think about the birth of infinitesimal calculus and the associated notions of instantaneous speed and acceleration, indispensable to Physics after Newton and obtained as limits to the infinity of finite approximations. Or to Projective Geometry, born in fifteenth century Italian painting, and in particular in the Annunciations, where a symbolic form of divine infinity, the vanishing point at the back of the painting, made the space on the finite plane more human. *In Mathematics, the infinite helps to better understand, describe, organize the finite.* And Set Theory demonstrates this: to formalize the concept of "finite", it is necessary to have an axiom of the existence of infinity (in PA, it is impossible to formally isolate standard, finite numbers, and hence to define the "finite"). Whether finitist formalists like it or not, the mathematical concepts of finite and of infinite are formally "entangled", inseparable: if we wish to formally capture the finite, it is necessary to work with the concept of infinity.

Paris and Harrington (1978), published a combinatorial assertion, PH say, also inspired by Logic, but one which was not artificial, a rather "meaningful" mathematical statement (Gödel's G is not so, according to many), formalizable in PA and without any apparent relationship to consistency (formalized by $Cons \equiv \neg Theor(0 = 1)$ in Sect. 5). From this statement, it is possible to deduce $Cons$ in PA, so PH is therefore unprovable. But it is possible to prove it, outside of PA, with Gentzen-like transfinite induction. We will mention another result, one which is similar but even more interesting. As a matter of fact, both proofs are similar and what we are about to say applies, implicitly, to the proof of Paris-Harrington's statement.

In a 1981 unpublished note, H. Friedman gave a finite version, formalizable in Arithmetics, of a famous theorem on finite trees. The trees, which in Mathematics, grow from the top towards the bottom of the page or the blackboard, are familiar and useful structures, with numerous applications. In particular, Kruskal's theorem (Kruskal 1960), which Friedman "miniaturized", proves a property which is widely used, especially in Mathematical Computer Science (for halting problems in formal systems of calculus, or of "rewriting", Bezem et al. 2013). We informally hint here to a result that has been discussed in several books and papers, see Harrington et al. (1985), Gallier (1991) among others. An analysis of its (un-)provability is also in Longo (2011).

It is easy to imagine how to say that a tree is included in another, that is to give a partial order between trees. Then the theorem says that no infinite sequence of trees can be completely disordered, i.e. there always exist comparable trees, the first included in the second, in the order—thus there are no infinite decreasing sub-sequences—and this has very interesting applications also in computing (term rewriting).

Friedman's Finite Form (FFF) "renders in the finite" the infinitary statement of Kruskal (which concerns infinite sequences of finite trees). FFF, for any n, gives the length m of the finite sequence in which we find two comparable trees. FFF is formalizable in PA: it is a "for any n there exists an m such as (...)" statement, where "(...)" is a property which is encodable in PA (finite trees are easily Gödelizable) and which is decidable (once n and m are fixed). Now, the function which associates n to m is computable, but it increases so fast that it definitively majorates any recursive function provably total in PA (and also in strong extensions of it). This is a way to prove the unprovability of FFF in PA.

Friedman, for his earlier proof, immerses trees in transfinite ordinals and, thanks to the absence of infinite decreasing sequences (that is, by transfinite induction), demonstrates that FFF implies $Cons$ in PA. And so, by this very difficult tour de force, he demonstrates that FFF is formally unprovable, a consequence of Gödel's Second Incompleteness Theorem ($Cons$ is unprovable). With one or the other technique, the proof of unprovability constitutes a surprising logical and mathematical feat to which a whole book was devoted shortly following the dissemination of the 1981 note (Harrington et al. 1985).

The observation to be made is that many of the applications of Kruskal's theorem are also obtainable from Friedman's arithmetic form. It is therefore clearly something other than an artificial/logical trick: it is Mathematics. And yet FFF, as well as its negation, are formally unprovable: this is the reason why we have called this section *The Concrete/Mathematical Incompleteness of Formalisms*, something we would have been unable to do if we were only thinking of Gödel's "logico-antinomical" statement G which is not very "mathematical"—nor concretely talks of numbers nor trees of/orders on numbers.

Now, in order to demonstrate the undemonstrability of FFF's *negation*, we can only prove something stronger: that FFF is true for integer numbers (or in any model of PA). There is no way of kidding here and claim that its truth is God given or due to quantum effects in the brain—all remarks based on a superficial reading of Gödel's theorems—with no reference to its *actual* proof. One only has to prove the statement. It is indeed Mathematics, not just Logic. So how is it possible to show that FFF is true (holds) in this structure?

Of course, we cannot make a finite formal induction, an induction in PA, due to its undemonstrability. The proofs given by Friedman and in the book we mentioned use induction in a way that is quite usual for mathematicians who do not work on foundations.

8.1 Towards the Cognitive Foundations of Induction

To explain and possibly justify such a use of induction, we will adopt a strong epistemological position, one which develops Riemann's reference to the foundations of Mathematics as a "genealogy of concepts", Poincaré's reflections on the role of action in space for the constitution of mathematical concepts, those of Enriques—sometimes vague, but often very stimulating—on the various forms of sensorial access to space, and the unity of Weyl's thought regarding symmetries as principles of conceptual construction, in Mathematics and in Physics (Weyl 1952). These great geometers, opposed to formalism, opened up, in a very incomplete and informal manner, avenues for foundational reflections of a strictly epistemological nature. They are sometimes revisited today in terms that are cognitive, relatively general and scientific, and beyond introspection, which was the only means of investigation at the time. We refer to the books by (Berthoz 1997) and (Dehaene 1997) and to previous reflections by this author (Longo 2005).

It is then possible to understand the incompleteness of formalisms as an insufficiency of the "principles of proofs" (of which informal induction is the paradigm) for capturing the "principles of construction" (firstly, well-ordering and symmetries)—the latter are increasingly shared with theoretical construction in Physics, whereas their principles of proof differ, cf. Bailly and Longo (2011).

The mathematician says and writes the following every day: if a set of integers—regardless of how it is defined—is non-empty, then it has a smallest element. You, the reader, see (we hope) the sequence of integers, well-ordered from left to right (for those of us who write in this direction, for those who write Arabic, it is the opposite). Look at it carefully, in your mind, as an infinite sequence of well separated numbers growing towards the horizon rather than on paper: 1, 2, 3, 4 ... If we isolate in the sequence, conceptually, an ordered set of integers numbers containing at least one element, we may observe that this set contains a smallest element—at worst it will be 0: the set is discrete and without infinite decreasing sequences—technically, such a set is said to be *well-ordered*.

This is a common practice in numerical intuition, one which is prohibited to the formalist because it is geometric and because it evokes "meaning", meaning as the act of counting or ordering. It is an act that is rich in signification—of writing, of ordering in space, of making this repeated movement towards the horizon. It originates in a human gesture (or maybe even in the pre-human one, in what concerns small numbers (Dehaene 1997)), of ordering (small) countable quantities together. It also refers to the "sense" of the discrete flow of time, in Brouwer's approach. Meaning is thus rooted in ancient gestures that are, in that, extremely strong. Language and writing gave them the objectivity of intersubjectivity, the stability of common notation, and independence as regards the objects denumerated.

The number and its order are first *practical* then *conceptual invariants* that make sense thanks to the independence they acquired with respect to a plurality of uses and acts of life, in space and in time. By repetition in space, by means of language and writing, we construct this discrete and increasing sequence to which the mathematician easily applies the abstract principle of "well order" thanks to its rich geometric meaning: a non-empty set of integers has a smallest element.

The mathematician uses such a signifying structure, one which evokes order in space and time, everyday and even also to construct a formal axiomatic, as did Peano and Hilbert, as a last stage in the construction of invariance or independence. But this last step, formalization, does not enable to completely separate the proof and its theory from meaning, of space and time and in space and time, which is constituted in this genealogy of concepts which is behind all of mathematical construction.[16] This is what the *mathematical* incompleteness of formal systems means: the principles of (formal) proof do not have the expressivity of principles of construction (order in space or time and symmetries) having produced the conceptual structures of Mathematics—they are incomplete w.r.to our active, concrete and meaningful mathematical structuring of the world.[17]

It is thus that even mathematicians who philosophically support or who are close to formalism demonstrate, in the 1985 book onwards, the validity of Friedman's statement by invoking, in a repeated but highly visible way, the principle of "well-order". With calm certitude, they pass at some point from the argument of one line to the next by observing that a non-empty set of integers, defined in the demonstration, has a smallest element. Such proof is perfectly rigorous and is founded upon a most solid cognitive practice: the invariance and conceptual stability of well-order specific to the "gestalt", rich in meaning, of the sequence of integers. Contrarily to so many formalizations, it does not entail contradictions. This is how the formally unprovable proof works.

---

[16] Husserl (1933):

> Original certainty can not be confused with the certainty of axioms, because axioms are already the result of the formation of meaning and always have such formation of meaning as a backdrop.

[17] For technical details regarding order and symmetries in the demonstrations we refer to, see Longo (2011).

Of course, some mathematicians later produced a detailed analysis of the proof, since it is not formalizable in PA. They demonstrated that the well-ordered set eludes finitist formalization, since it implicitly uses infinite quantification (over sets, Σ_1^1 technically, cf. Rathjen and Weiermann 1993). They thus prove Friedman's statement using induction over a huge transfinite ordinal, one which is far greater than that proposed by Gentzen and which is definable by means of a very difficult construction. Some justify the infinitary audacity by observing that the set involved in the well-ordering statement is non-empty only by hypothesis of a *"reductio ad absurdum"*. It will then disappear precisely because it gives rise to absurdity ... it will therefore be empty. And yet, this detour by infinity is necessary, because of the proof that the assertion is unprovable in a finitary way (its unprovability in PA).

But has then all of this work been useless? Even the Greeks could believe in the consistency of Arithmetics, they who "saw" the potentially infinite and well-ordered sequence of integers with, scattered in their midst, the prime numbers. Gödel's theorem is a pun with no "mathematical" meaning; the mathematical statements which do have meaning are demonstrated using presumed infinite sets which are not definable in PA, sets that, in the end, turn out to be empty ... Is then this detour useless? Not at all, this path is extremely rich, in itself and because of its spin-offs. Simply, in what concerns the play between the finitude and infinitude of numbers, of space, it traverses all Mathematics. It began with the use of potential infinity with Euclid, as *apeiron* (limitless). Then, it was clarified by Aristotle, and refined by the Thomist school, which was used to working with the difficult and controversial infinity of the Christian God: thanks to its contribution, we clearly established the distinction between potential and actual infinity, specific to God. Then came projective geometry, as we were saying, a first mathematical consequence of the practice of actual infinity, followed by infinitesimal calculus, both having entailed huge developments and applications.

It was then necessary to clarify *how* demonstrations were made, particularly when using this limit concept and in particular following the brilliant congestion of nineteenth century Mathematics; *how* rigorous definitions are produced, after a century that was so prolific mathematically, although its mathematics often lacked rigor. The formal systems turned out to be incomplete, but far from useless: they taught us how to produce good definitions, how to rigorously generalize, how to unify methods and proofs using the axiomatic method ... . The mistake was rather to think that it was possible to work without meaning in order to prove consistently, mechanically, rigorously; to be able to avoid any reference to action in space and in time, which are the loci of the constitution of Mathematics, even that of integers. However, as we have mentioned, it was necessary—in order to demonstrate that there are undecidable statements—to specify what is meant by decidable or computable in a mechanically certain way; and so were set the Mathematical bases, with Gödel and Turing, of Computer Science. And in the end, we are brought back, but with a whole set of tools, to this sense of space and of action within, to its

> [...] geometry, generated in our human space from a human activity (Husserl 1933).[18]

9 Information and Encoding in the Cell

> In calling the structure of the chromosome fibers a code-script we mean that the all-penetrating mind, once conceived by Laplace, [...] could tell from their structure whether the egg would develop, under suitable conditions, into a black cock or into a speckled hen, into a fly or a maize plant [...]

wrote Schrödinger in his 1944 book during his exile in Ireland from Germany. The immense figure of Laplace remains in the backdrop of the whole history we have examined, beginning with Poincaré's work. Some of the thinkers we mentioned here explicitly recognized his mark upon their own scientific analysis: in Sect. 6.1, we quoted Turing's remarks on the "laplacian" nature of his Discrete States Machine.

Schrödinger proposed in 1944 the idea of seeing, halfway between metaphor and science, chromosomes as a "code-script", encoding hereditary information. And from his perspective as a physicist, he understood its implicitly Laplacian nature—and he provided prudent and plausible examples. In the second part of the book, though, Schrödinger hints

> [...] to the possible meaning of the principle of entropy at the *global scale* of a living organism, while forgetting for the time being all what we know on chromosomes.

---

[18] The date at which Husserl's manuscript was written reminds us that almost all of the story we have told took place during the first and dramatic half of the twentieth century, 1933 being a pivotal year, with the rise of Nazism and the flight from Germany of so many people we have met in these pages. During that year, Husserl, who was 74 years old at the time, was prohibited from publishing and even from accessing the University Library. And this frequent appearance of some illustrious names reminds of another important/small academic/political story. In 1923, Einstein, having recently been awarded the Nobel prize, thought about returning to Italy, maybe for a long period, after a short stay in Bologna. He had a very good knowledge of the results by Levi-Civita and was in contact with several colleagues, among whom Volterra and Enriques. The latter, in the previous years had become familiar in dealing with the governments, managed to obtain a meeting with the new prime minister, formally not yet dictator, Benito Mussolini: he hoped to obtain exceptional financing for the guest. This was in early 1924. The Duce's response was: "Italy has no need for foreign geniuses" — this reminds by contraposition of the great Princes of the Renaissance or of Princeton in the 1930s (and afterwards). And so Einstein did not return to Italy. In 1929, Marconi added to a list of his colleagues drafted for Mussolini a little e. (for Jew — "*ebreo*" in Italian) in front of the names of the three aforementioned Italian mathematicians, the greatest of their time. The Duce, nine years prior to his racial laws, excluded them from the Academy of Italy (Faracovi et al. 1998).

In particular, Shrödinger investigates the possible role of "negative entropy" as a form of Gibbs' free energy (available energy for work). He opened by this yet another possible path for reflection in organismal biology (see Bailly and Longo (2009) and Longo and Montévil (2014, chap. 9) for more in this direction hinted by Schrödinger).

What can we find in common between these various forms of determination which involve predictability and therefore a full understanding of the world based on a few equations and a few signs? The expressive completeness of writing, more specifically of alphabetic writing, can provide a key for interpreting the omnipresence of this way we have of doing science.

The Laplace equations are of course formal or formalizable writing which were believed, up until Poincaré, to be a complete determination, i.e. able to predict the possible evolutions of the physical universe—with beside it randomness that was supposed to be distinct from equational determination. We have also recalled in the first section how, for Laplace, the fundamental level is always in the *elementary*, in the *simple* particles of which it is necessary to isolate, describe, and then integrate the movement into systems by the progressive sum of individual behaviors.

Hilbert, in turn, will make explicit the discrete nature of mathematical formalisms, as a sequence of *simple* and *elementary* material, alphabetic signs: sequences of signs, the axioms, were meant to completely allow to deduce, "determine", the properties of the intended mathematical structures. He paved the way for Turing's digital machine, once letters and words had been Gödelized—encoded by numbers. Alpha-numeric formal systems should have told us everything about them. And for some, Turing's machine should have fully modeled at a point the functioning of the brain. If it remains each time audacious, the process of knowledge seems to increasingly narrow itself and deteriorate. It is original and justified in the case of Laplace and Hilbert and two immense theorems were required, by Poincaré and Gödel, to undo them—theorems which were made possible by the mathematical rigor of the original proposals. But this project hits rock bottom when reaching the 0 and 1 of a brain seen as a digital switchboard, in Classical Artificial Intelligence, or when reaching the four-letter alphabet of the nucleotide bases which compose DNA. The latter becomes

> the program of any individual's behavioral computer (Mayr 1961)

(Mayr later opposed the idea of a central role for genes in evolution). And the assumption of completeness assures that the

> DNA contains all information required for the reproduction of the cell,

and of the whole organism (Crick 1966).
So the

> one gene — one enzyme (Beadle and Tatum 1941)

hypothesis and then the

Central Dogma of molecular biology

(information passes linearly and unidirectionally from DNA to RNA, to proteins and then to the structure of the organism (Crick 1958)) are of a Laplacian nature as regards the structure of determination they suggest: DNA is complete, it contains the information for any phenotype and information propagates in a linear fashion and in a single direction from it ("one gene—one enzyme" and the "dogma"). The first hypothesis was considered to be valid for over 50 years before it was demonstrated to be false; as it concerns the dogma, it still permeates research in molecular biology (in cancer etiology, for example, see below), although it has recently been rejected by the majority, albeit not always aloud.

More or less implicitly the idea of completeness of DNA w.r. to all phenotypes is still prevailing, in spite of growing empirical evidence against those claims and various alternative proposals for organismal biology—some work is synthesized in Soto et al. (2016). Note, though, that the complete knowledge of the chemical structure of the DNA and the alleged identification of all human "genes", its "decoding", has been a major technological success, in 2001. Unfortunately, the number of genes keeps changing, from about 80,000 in 1999, still assumed by the head of the Human Genome Project launched in 1990, (Collins 1999), to 25,000 in 2001, down to about 20,000 today.[19]

It is not our aim here to develop such considerations further, see for example (Fox Keller 2000). Our goal is to compare the scientific practices, as for negative results, in physics and mathematics to those in a relatively young and very important field such as Molecular Biology. As for an analysis of the incompleteness of genocentric analyses in biology, let's first note that the hypotheses or dogmas which were at the basis of numerous works for such a long time, and which purported to be "physicalistic" or "materialistic", seem to not have taken enough into account what happened in physics (Longo and Tendero 2007; Longo 2018a).

Since Poincaré, we have understood that in the presence of simple interactions (only three celestial bodies), the initial measurable situation does not contain all "the information" (to use a rather unfitting expression) on future trajectories, if we mean by that the "complete determination" of the *system's evolution*. And we remain Laplacian when adding to what is "necessary" a fragment of randomness, as "noise", quite distinct from the former, as done in Monod (1970)—recall that Poincaré had integrated the two. Monod's necessity, because of its Laplacian nature, turns out to be programmable (the theory of the "genetic program").

What seems to be neglected, in the hypotheses and dogmas regarding sequential molecular cascades, is that even Physics of the twentieth century, after Relativity, sees the universe as a fabric of interactions: *if the interactions*

---

[19] In 1999, Collins was pleased to stress the difference between humans and *Caenorhabditis Elegans*, a one millimeter worm with less than 1000 cells, whose DNA had just been decoded: it has only 19,000 genes!

change, the fabric and its space are deformed; if we act upon the fabric and upon space, the interactions change. The Central Dogma is foreign to this vision of interactions as constitutive of a unity specific even to contemporary physics; and it concerns, let's recall, objects within a structure such as the cell and the organism, where almost everything is correlated to almost everything. Noise as well is a largely inadequate notion to understand the role of randomness as biological unpredictability, see Bravi and Longo (2015), as well as of stochasticity in genes' expression (Elowitz et al. 2002). In general, the focus on discrete structures sets a bias on the analysis of determination and randomness, see Longo (2018a).

DNA is, of course, a most important component of the cell, but the analyses of life phenomena, which base themselves solely upon it and upon the molecular cascades that follow, are incomplete, in a sense which is indeed impossible to specify within a theorem but which is suggested by Physics itself. When we see that it is described as "the book in which the essence of life is written", we realize that the alphabetic myth still governs a part of science: it is a myth in the Greek sense, a positive myth which is a powerful constructor of knowledge, but which needs to be continuously reviewed and to have its own limits brought to light. From Democritus—who fragmented the world into atoms and who associated them to the letters of the alphabet—to Descartes—for whom certitude is obtained by decomposing reasoning into elementary and simple components—and to Laplace and Hilbert, certainty in understanding must always refer to the elementary and simple, to the atomic and alphabetic.

The model of alphabetic reconstruction, discrete and elementary, of the continuous song of language has been presiding our sciences for millennia, with extraordinary productivity: we wish to understand everything in this manner. So, we have believed for centuries that, like language with the alphabetic structure, we can reconstruct, for all of knowledge, the world by projecting letters onto it and that these completely determine the intended structures, in Physics, in Logic, in Biology (atoms, the sequences of signs of a formalism, the letters of DNA). In other words, discrete signs and letters make it possible to express all which is sayable and *therefore* all which is thinkable.[20] So, from Mathematics, to Physics and Biology, the signs and

---

[20] Jacob (1965):

> The surprise is that genetic specificity is written not with ideograms, like in Chinese, but with an alphabet.

In this perspective, see Jacob (1974) for more, also the philosophy of biology, reduced to Molecular Biology, is transformed in an annex of a philosophy of (alphabetic) language, cf. Sect. 5 above. As a matter of fact, Molecular Biology deals with information, programs, expressions, signals ... since "life is fully coded" in chromosomes, following Schrödinger (1944), *thus* in discrete sequences of meaningless signs, as theorized also by Maynard-Smith (1999), Gouyon et al. (2002) and many others.

discrete sequences of signs (formal encodings) contain the full determination of all possible evolutions, at all levels of phenomena.

Now, it is necessary to highlight the strength and limits of this vision of knowledge, its incompleteness to put it shortly. Indeed, even the image of the language thus proposed, as an instrument of human communication, is quite incomplete. We forget that the "compiler" or "interpreter" of alphabetic languages is the production of sound, by a composition of phonemes: meaning is in the spoken-sung and in its expressivity. It is necessary to read, to produce a sound—even silently in one's head—to find meaning, in the same way that a musician *hears* music, "interprets it", when reading a score, which is another form of alphabetical writing of the musical continuum, (but a two-dimensional writing which is enriched with symbols and signs of continuity). So the context, sometimes linguistic and written, and the tone, the gesture or the drawing contribute in an essential way to expression and to comprehension ... in sum, to meaning.

Furthermore, a pout, a smile, a punch, making love, all of these enable to say something else, and contribute to human expressivity, to what is thinkable, in a essential way, *beyond* and *with* the sequences of alphabetic signs. In the same way, the meaning in space of the well-order of integers is part of mathematical proof and, for the fervent anti-formalists we are, of its foundations, in the epistemological sense, *with* but *beyond* formal systems, demonstrated to be incomplete.

So, to return to Biology, we are slowly steering away from the alphabetic myth, which is unfortunately still the priority in what concerns financing, and which claims that the stability and organization of DNA, and of molecular cascades stemming from it, fully determine the stability and organization of the cell and of the organism. This myth is false, because the physical and biological stability and organization of the cell and of the organism contribute causally to the stability and to the organization of DNA and of the molecular cascades which follow from it. Isn't this circular? We are used to such challenges: let's recall what Gödel did with a very subtle circularity, far away from logicist fears.

The problem of "how it started", the origin of life, remains in any case enormous. With no membrane, without a cell, an organisms, no significant metabolic cycle is created, even less is it maintained over time. Similarly to what Gödelian incompleteness led us to understand, for Mathematics with respect to formal systems, "strict" extensions (in the logical sense) of molecular theories seem necessary in order to say something more about the physical singularity of the living state of matter, see Bailly and Longo (2011) and, for recent advances on an organismal perspective (Soto et al. 2016).[21]

---

[21] The empirical evidence on the incompleteness of the genocentric approach as for a dramatic phenotype, cancer, and some general consequences on the understanding of causality, in particular on the etiology of that disease, are discussed in Longo (2018a).

Let's conclude these considerations with some questions and by identifying general challenges. Why should the fundamental always be the "elementary"? Galileo's theories of gravitation and inertia, which are fundamental theories, tell us nothing about Democritus's atoms which did however constitute his masses. Einstein unified inertia and gravitation; he proposed another theory, the relativistic field, also fundamental, without saying anything about quanta. Of course, the problem of unification with the quantum field is an issue. However, physicists will say *unification*, and not reduction: it is a question of putting fundamental theories into perspective, of modifying them both in view of a synthesis to invent. The greatest progresses are possibly achieved today by reconstructing, from quantum measurement, the geometry of space and time, see A. Connes's "non-commutative geometry".

And finally, why should the elementary always be simple, as if we were transposing the alphabetic and Cartesian method to the phenomena at hand? Two frontiers of contemporary knowledge, microphysics and the analysis of life phenomena, seem to call for another vision: their elementary components, the quantas and the cell (which is elementary, atomic, since it is no longer alive if we split it in two) are very complex. Their comprehension requires approaches that are "non-local", to use quantum terminology, or systemic analyses, as many increasingly put it in Biology (see Soto et al. (2016) for work and references), well beyond the supposed causal completeness of DNA and well beyond the myths, from Mathematics and Physics to the analyses of human cognition, of the completeness of alphabetic formalisms.

Appendix: On Gödelitis in Biology

In an attempt to bypass the mechanistic-formal approach, enriched by some noise, Danchin (2003, 2009) tried to bring Gödel's theorem into the genocentric view of biology. Within the formal-linguistic approach to biology, Gödel's incompleteness would prove the "creativity" of biological dynamics by recursion and diagonalizing on the programs for life: in short, the DNA would generate unpredictable novelty by a creative encoding of phenotypes, a la Gödel. A remarkable attempt for a leading biologist, as these issues in Logic are far from common sense, as we hinted above.

Indeed, (Rogers 1967), a classic in Computability Theory, calls "creative" the set of (encoded) theorems of arithmetic, i.e. the formal-mechanical consequences of its axioms. As we know, by Gödel's first theorem, this set is not computable (not decidable)—and, to the biologist, its evocative name may recall Bergson's Creative Evolution. However, this set is *semi-computable* (semi-decidable), meaning that it may be effectively generated and, as such, is far from "unpredictable", since an algorithm produces all and exactly all its infinite elements—the set of encoded theorems. Moreover, the generation of Gödel's undecidable formula is effective as well: it is an incredibly smart recursive and "diagonal" construction (it recursively uses the encoding of logical negation), as we have seen, which allows to *construct* a formula not derivable from the axioms. This procedure may be indefinitely and effectively iterated.

In short, Gödel's undecidabile sentence is effectively produced by an effective encoding of the *metatheory* into the formal theory and it does not finitely "create" any "unpredictable" information: the diagonal formula may be constructed, even though it is not derivable from the axioms. In summary, on one side, formal derivability is not decidability (Gödel's first theorem), as the "information" in the axioms does not allow to decide all formulae, typically Gödel's diagonal formula. Yet it still yields *semi*-computability or *semi*-decidability: the theorems can be effectively generated, by passing through the encoded metatheory (what would be the metatheory in evolutionary biology?). On the other side, the construction of the sentence that escapes the given axioms is also effective (semi-computable), as we have seen.

Theoretical unpredictability, instead, that is the least property one expects for "creativity" in nature, is at least (algorithmic) randomness, for infinite sequences (Sect. 6). This yields a very strong form of incomputability, far from semi-computa-bility. As observed in Sect. 6, a random set of numbers and its complement cannot even contain an infinite semi-computable subset. This form of randomness may be soundly compared, asymptotically, to unpredictability in physics, as we observed (note that biological unpredictability includes both classical and quantum randomness (Buiatti and Longo 2013; Calude and Longo 2016a)).

We also observed that finite incompressibility does not soundly relate to randomness in nature: an incompressible sequence may be programmable—by a program of its length or just one bit longer; moreover, there are no sufficiently long incompressible sequences, Calude and Longo (2016b)—except by a restriction on the allowed machines, a la Chaitin (Calude 2002).

In summary, physical/biological randomness is unpredictability relative to the intended theory (Calude and Longo 2016a), and a time related issue: it concerns the future and is associated to time irreversibility (Longo and Montévil 2014: chap. 7). It relates only asymptotically to algorithmic randomness; it is necessary, but insufficient, for analysing evolutionary changes. It goes well beyond Gödel's constructive diagonal craftiness.

The merit of Danchin's remarks, though, is that they are based on precise mathematical notions, thus they may be proved to be wrong. This is in contrast to the commonsensical abuses of vague notions of information and program, as mostly used in Molecular Biology, from which strong consequences have been too often derived, see Longo (2018a).

References

Aspect, A., Grangier, P., & Roger, G. (1982). Experimental realization of the Einstein-Podolsky-Rosen-Bohm Gedankenexperiment: A new violation of Bell's inequalities. *Physical Review Letters, 49*, 91–94.

Asperti, A., & Longo, G. (1991). *Categories, types and structures: An introduction to category theory for the working computer scientist.* Cambridge, MA: The MIT Press.

Baez, J., & Stay, M. (2012). Algorithmic thermodynamics. *Mathematical Structures in Computer Science, 22*(5), 771–787.

Bailly, F., & Longo, G. (2009). Biological organization and anti-entropy. *Journal of Biological Systems, 17*(1), 63–96.

Bailly, F., & Longo, G. (2011). *Mathematics and natural sciences. The physical singularity of life.* London: Imperial College Press. (original version in French, Hermann, Paris, 2006).

Barrow-Green, J. (1997). *Poincaré and the three-body problem.* Providence/London: American Mathematical Society/London Mathematical Society.

Beadle, G. W., & Tatum, E. L. (1941). Genetic control of developmental reactions. *American Nauturalist, 75*, 107–116.

Béguin, F. (2006). Le mémoire de Poincaré pour le prix du roi Oscar. In E. Charpentier, E. Ghys, & A. Lesne (Eds.), *L'héritage scientifique de Poincaré.* Paris: Belin.

Bell, J. S. (1964). On the Einstein-Podolsky-Rosen paradox. *Physics, 1*(3), 195–200.

Berthoz, A. (1997). *Le sens du mouvement.* Paris: Odile Jacob. (English transl.: The Brain Sens of Movement, Harvard University Press, 2000).

Bezem, M., Klop, J. W., & Roelde Vrijer, R. (2013). *Term rewriting systems.* Cambridge: Cambridge University Press.

Bravi, B., & Longo, G. (2015). The unconventionality of nature: Biology, from noise to functional randomness. In C. S. Calude & M. J. Dinneen (Eds.), *Unconventional computation and natural computation* (pp. 3–34). Cham: Springer.

Buiatti, M., & Longo, G. (2013). Randomness and multi-level interactions in biology. *Theory in Biosciences, 132*(3), 139–158.

Calude, C. (2002). *Information and randomness* (2nd ed.). Berlin: Springer.

Calude, C., & Longo, G. (2016a). Classical, quantum and biological randomness as relative unpredictability. *Natural Computing, 15*(2), 263–278.

Calude, C., & Longo, G. (2016b). The deluge of spurious correlations in big data. *Foundations of Science* (March), 1–18. https://doi.org/10.1007/s10699-016-9489-4

Charpentier, E., Ghys, E., & Lesne, A. (Eds.). (2006). *L'héritage scientifique de Poincaré.* Paris: Belin.

Crick, F. H. C. (1958). Central dogma of molecular biology. *Nature, 227*(8), 561–563.

Crick, F. H. C. (1966). *Of molecules and man.* Seattle, WA: Washington University Press.

Collins, F. (1999). Medical and societal consequences of the Human Genome Project. *The New England Journal of Medicine.*

Danchin, A. (2003). *The delphic boat. What genomes tell us.* Cambridge, MA: Harvard University Press.

Danchin, A. (2009). Bacteria as computers making computers. *Microbiology Reviews, 33,* 3–26.

Dehaene, S. (1997). *La bosse des Maths.* Paris: Odile Jacob. (English transl., Oxford University Press, 1998).

Einstein, A., Podolsky, B., & Rosen, N. (1935). Can quantum-mechanical description of physical reality be considered complete? *Physics Review, 47*(10), 777–780.

Elowitz, M. B., Levine, A. J., Siggia, E., & Swain, P. S. (2002). Stochastic gene expression in a single cell. *Science, 297*(5584), 1183–1186.

Enriques, F. (1983). Filosofia scientifica. *Dimensioni, 8*(28/29), 46–50. (Special Issue: "La Filosofia Scientifica a Congresso, Parigi, 1935", G. Polizzi (Ed.)).

Faracovi, O., Speranza, F., & Enriques, F. (1998). *Filosofia e storia del pensiero scientifico.* Livorno: Belforte.

Fox Keller, E. (2000). *The century of the gene.* Cambridge, MA: Harvard University Press.

Frege, G. (1884). *The foundations of arithmetic.* Evanston, IL: Northwestern University Press. (original German ed.: 1884).

Gács, P., Hoyrup, M., & Rojas, C. (2009). Randomness on computable metric spaces: A dynamical point of view. In *26th International Symposium on Theoretical Aspects of Computer Science, (STACS 2009)* (pp. 469–480).

Gallier, J. (1991). What is so special about Kruskal's theorem and the ordinal Γ_0? *Annals of Pure and Applied Logic, 53,* 199–260.

Gentzen, G. (1969). In M. E. Szabo (Ed.), *The collected papers of Gerard Gentzen.* Amsterdam: North-Holland.

Girard, J.-Y. (1971). Une Extension de l'Interpretation de Gödel à l'Analyse, et son Application à l'Élimination des Coupures dans l'Analyse et la Théorie des Types. In *Proceedings of the Second Scandinavian Logic Symposium,* Amsterdam (pp. 63–92).

Girard, J. Y., Lafont, Y., & Taylor, P. (1990). *Proofs and types.* Cambridge, MA: Cambridge University Press.

Gödel, K. (1986–2003). In S. Feferman et al. (Eds.), *Collected works* (5 volumes). Oxford, NY: Oxford University Press.

Goubault, E. (2000). Geometry and concurrency: A user's guide. *Mathematical Structures in Computer Science, 10*(4), 411–425.

Gouyon, P.-H., Henry, J.-P., & Arnoud, J. (2002). *Gene avatars, The Neo-Darwinian theory of evolution.* New York, NY: Kluwer Academic Publishers.

Harrington, L. A., Morley, M. D., Sčědrov, A., & Simpson, S. G. (Eds.). (1985). *H. Friedman's research on the foundations of mathematics.* Amsterdam: North-Holland.

Hilbert, D. (1899). *Grundlagen der Geometrie.* Leipzig: Teubner. (English ed.: "Foundations of Geometry". La Salle, IL: Open Court,1971).

Husserl, E (1933). The origin of geometry. In J. Derrida (1989), *Edmund Husserl's origin of geometry. An introduction* (pp. 157–180). Lincoln, NE: University of Nebraska Press.

Jacob, F. (1965). *Génétique cellulaire.* (Leçon inaugurale prononcée le vendredi 7 mai 1965). Paris: Collège de France.

Jacob, F. (1974). Le modèle linguistique en biologie. *Critique, 30*(322), 197–205.

Kreisel, G. (1984). *Four lettters to G. Longo.* (http://www.di.ens.fr/users/longo/files/FourLettersKreisel.pdf).

Kruskal, J. (1960). Well-quasi-ordering and the tree theorem. *Transactions of the American Mathematical Society, 95,* 210–225.

Kunnen, K. (1980). *Set theory: An introduction to independence proofs.* Amsterdam: North-Holland.

Laskar, J. (1989). A numerical experiment on the chaotic behaviour of the Solar system. *Nature, 338,* 237–238.

Laskar, J. (1994). Large scale chaos in the Solar system. *Astronomy and Astrophysics, 287,* L9–L12.

Lighthill, M. J. (1986). The recently recognized failure of predictability in Newtonian dynamics. *Proceedings of the Royal Society of London A, 407,* 35–50.

Longo, G. (1996). *The Lambda-Calculus: Connections to higher type recursion theory, proof-theory, category theory.* A short (advanced) course on lambda-calculus and its mathematics, Spring 1996. (Revision of "On Church's Formal Theory of Functions and Functionals". *Annals Pure Appl. Logic, 40*(2), 93–133, 1988).

Longo, G. (2005). The cognitive foundations of mathematics: Human gestures in proofs. In P. Grialou, G. Longo & M. Okada (Eds.), *Images and reasoning* (pp. 105–134). Tokio: Keio University Press.

Longo, G. (2011). Reflections on concrete incompleteness. *Philosophia Mathematica, 19*(3), 255–280.

Longo, G. (2014). Science, Problem Solving and Bibliometrics. In W. Blockmans, L. Engwall & D. Weaire (Eds.), *Bibliometrics: Use and abuse in the review of research performance* (pp. 9–15). London: Portland Press.

Longo, G. (2016). The consequences of philosophy. *Glass-Bead* (web-journal). (http://www.glass-bead.org/article/the-consequences-of-philosophy/?lang=enview).

Longo, G. (2018a). Information and causality: Mathematical reflections on cancer biology. *Organisms Journal of Biological Sciences, 2*(1), 83–103.

Longo, G. (2018b). Complexity, information and diversity, in science and in democracy. In V. Bühlmann, M. R. Doyle & S. Savic (Eds.), *The Ghost of Transparency: An Architectonics of Communication*. New York, NY: Springer.

Longo, G. (2018c). Letter to Alan Turing. *Theory, Culture and Society*, (Special Issue on "Transversal Posthumanities"). https://doi.org/10.1177/0263276418769733.

Longo, G., & Montévil, M. (2014). *Perspectives on organisms: Biological time, symmetries and singularities*. Berlin: Springer.

Longo, G., & Tendero, P.-E. (2007). The differential method and the causal incompleteness of programming theory in molecular biology. *Foundations of Science, 12*, 337–366.

Martin-Löf, P. (1966). The definition of random sequences. *Information and Control, 9*, 602–619.

Maynard-Smith, J. (1999). The idea of information in biology. *The Quarter Review of Biology, 74*, 495–400.

Mayr, E. (1961). Cause and effect in biology. *Science, 134*(348), 1501–1506.

Monod, J. (1970). *Le Hasard et la Nécessité*. Paris: Éditions du Seuil.

MSCS Editorial Board. (2009). Editors' note: Bibliometrics and the curators of orthodoxy. *Mathematical Structures in Computer Science, 19*(1), 1–4.

Paris, J., & Harrington, L. (1978). A mathematical incompleteness in Peano Arithmetic. In J. Barwise (Ed.), *Handbook of mathematical logic* (pp. 1133–1142). Amsterdam: North-Holland.

Peano G. (1889). *Arithmetices principia, nova methodo exposita*. Torino: Bocca.

Petersen, K. (1983). *Ergodic theory*. Cambridge: Cambridge University Press.

Piazza, M., & Pulcini G. (2016). What's so special about the Gödel sentence G? In F. Boccuni & A. Sereni (Eds.), *Objectivity, realism, and proof: FilMat studies in the philosophy of mathematics* (pp. 245–263). Cham: Springer.

Pilyugin, S. Yu. (1999). *Shadowing in dynamical systems*. Berlin/New York: Springer.

Poincaré, H. (1892). *Les Méthodes Nouvelles de la Mécanique Céleste*. Paris: Gauthier-Villars.

Poincaré, H. (1902). *La Science et l'Hypothèse*. Paris: Flammarion.

Poincaré, H. (1906). Les mathématiques et la logique. *Revue de Métaphysique et de Morale, 14*(3), 294–317.

Poincaré, H. (1908). *Science et Méthode*. Paris: Flammarion.

Rathjen, M., & Weiermann, A. (1993). Proof-theoretic investigations on Kruskal's theorem. *Annals of Pure and Applied Logic, 60*, 49–88.

Rogers, H. (1967). *Theory of recursive functions and effective computability*. New York: McGraw Hill.

Ruelle, D., & Takens, F. (1971a). On the nature of turbulence. *Communications in Mathematical Physics, 20*, 167–192.

Ruelle, D., & Takens, F. (1971b). Note concerning our paper "On the nature of turbulence". *Communications in Mathematical Physics, 23*, 343–344.

Schrödinger, E. (1944). *What is life? The physical aspect of the living cell.* Cambridge: Cambridge University Press.

Sieg, W. (1994). Mechanical procedures and mathematical experience. In A. George (Ed.), *Mathematics and mind* (pp. 71–117). New York, NY: Oxford University Press.

Smorynski, C. (1977). The incompleteness theorems. In J. Barwise & H. J. Keisler (Eds.), *Handbook of mathematical logic* (pp. 821–866). Amsterdam: North-Holland.

Soto, A. M., Longo, G., & Noble, D. (Eds.). (2016). From the century of the genome to the century of the organism: New theoretical approaches. (Special issue), *Progress in Biophysics and Molecular Biology, 122*(1), 1–82. (https://doi.org/10.1016/j.pbiomolbio.2016.06.006). (Epub 2016 Jul 2. Review. PubMed PMID: 27381480, Oct., 2016).

Turing, A. M. (1950). Computing machinery and intelligence. *Mind, 49*(236), 433–460.

Turing, A. M. (1952). The chemical basis of morphogenesis. *Philosophical Transactions of the Royal Society B, 237*(641), 37–72.

Weyl, H. (1949). *Philosophy of mathematics and natural sciences.* Princeton, NJ: Princeton University Press. (original German ed.: 1927).

Weyl, H. (1952). *Symmetry.* Princeton, NJ: Princeton University Press.

Wittgenstein, L. (1975). In R. Rhees (Ed.), *Philosophical remarks.* Chicago, IL: The University of Chicago Press. (original German ed.: 1964).

First Steps Toward a Systemic Ontology

Lucia Urbani Ulivi

1 Introduction

Looking at the objects of the world as systems is a perspective point used in the disciplines we usually consider as "scientific", as well as in those qualified as "humanistic". Systemic literature, both scientific and humanistic, is a river in spate, continually being enriched with contributions and also attracting to itself new disciplines, drawn by the possibility of shattering interpretative cages, now worn and inadequate, thanks to a tool for understanding that improves our comprehension of certain aspects of the world.[1]

If it is the ambition of the systemic approach to become, for all effects, a new paradigm, able to replace, or integrate, within a broader perspective, the analytical one—which is still in many ways dominant and, certainly, by far the most widespread—it is necessary for the so-called "systemic approach" to be structured also as "systemic thinking" and to show that it is capable of elaborating a philosophical perspective that tests its scope in philosophical contexts and problems that somehow are "classical": ontology, anthropology, philosophy of mind, epistemology, ethics, aesthetics, the philosophy of language, social philosophy, the history of philosophy, etc. For many of these domains there already are contributions of great interest,[2] while for

L. Urbani Ulivi (✉)
Department of Philosophy, Catholic University of the Sacred Heart, Milano, Italy
e-mail: lucia.ulivi@unicatt.it

---

[1] Minati et al. (2016). The authors offer an updated bibliography of recent works in the systemic world.

[2] Among many, there are the Proceedings of Systemic Conferences organized by AIRS from 1998: Minati (1998), Minati and Pessa (2002), Minati et al. (2006, 2009, 2012, 2016). Three books of essays: Urbani Ulivi (2010, 2013, 2015). The tenth volume of the series "*Handbook of the Philosophy of Science*": Hooker (2011). The *Rivista di Filosofia Neo-Scolastica* has

© Springer Nature Switzerland AG 2019
G. Minati et al. (eds.), *Systemics of Incompleteness and Quasi-Systems*,
Contemporary Systems Thinking,
https://doi.org/10.1007/978-3-030-15277-2_2

others—and I think of ethics—we can glimpse a great potential still waiting
to be exploited.

The aim of this work is to take the first steps in the elaboration of a
systemic ontology, testing its possibilities, its theoretical efficacy, its limits.

2 Conceptual Genealogy: Aristotle

2.1 What Ontology Is

Ontology is, with metaphysics, the branch of philosophy that strives to answer
the most general and universal question about the world: What there is?
The question thus expressed corresponds fully to the etymological meaning
of the term "ontology": reasoned discourse (*logos*) on what is there (*on*).
Although without assuming an ontological engagement of universal scope,
systemic thought expresses itself regarding what systems are, outlining the
traits that characterize them as systems. With this clarification, we can say
that systemic thinking engages itself in an ontology of systems.

2.2 Aristotelian Genealogy of Systemic Thought

The most ancient and most authoritative antecedent of systemic ontology is
Aristotle, and it is worthwhile to selectively retrace some concepts of his philosophy, not so much to ennoble the genealogical tree of systemics with such
an illustrious ancestor, but because Aristotle's indications, often neglected or
misunderstood in many successive moments, are full of suggestions and ideas
that can still guide systemic research today. We need only think of the concept of substance—closely related to that of the system—forged by Aristotle
for understanding the entities of the world while safeguarding their unity,
without breaking them into parts or reducing them to disincarnate and formal abstractions. Aristotle introduces the concept of substance in response to
the question: What causes a congeries of separate parts to become a unitary
object? The answer is that we must hypothesize the existence of a principle
of unification and of activity, which cannot be directly observed, but which
accounts for the identity and unity we observe in some entities present in the
empirical reality. As some later critics have not failed to observe, the concept
of Aristotelian substance conserves a good dose of vagueness, because it is not
explained how the transition occurs from separate parts to unitary object.

published a section of "System researches in philosophy, the sciences and the arts" in the
following numbers: (vol. CII (2), 2010; CIII (4), 2011; CIV (4), 2012; CVI, (3), 2014; CVII
(1–2), 2015; CVIII, (2), 2016; CIX, (2), 2017).

Centuries later, systemic thinking would be capable of specifying that what it is that gives unity to an entity are the relationships of interaction and interference connecting the parts. The Aristotelian substance is expressed as organization, and the ancient dictate acquires theoretical breadth and precision.

2.3 What Exists?

The Aristotelian answer to the ontological question par excellence "What exists?" is ecumenical, and could be formulated like this: everything exists. Aristotle eliminates nothing, pushing without hesitation to include also non-being among entities. The Stagirite unequivocally says, in *Metaphysics*,

> we even say of non-being that it *is* non-being (Aristotle, *Metaphysics*, IV, 2, 1003 b, 11).

A notoriously difficult step, which can only be understood if it is brought back within a pluralistic ontological vision: everything exists, but not everything that exists, exists in the same way. Non-being also exists, because it enters as subject into the predication "non-being is different from being", and that is sufficient to differentiate it from other entities and to position it within the ontological plane of entities whose being consists in being words endowed with significance; different ontological statutes will be attributed to other entities. Every entity's being something constitutes the minimal and universal reference of ontological unification: all entities are something; nothing and no one is excluded. But this does not in any way lead Aristotle to ontological monism. On the contrary: entities must be placed under different and various ontological qualifications, and next to unification, diversification is positioned immediately adjacent: not all entities are in the same way (just consider that, besides empirical entities, there are others, of thought, still others that are represented, possible, ideal, propositional, semantic, etc.). If Aristotelian ontology were to limit itself to identifying that which entities have in common, this would allow the possibility of distinguishing the traits rendering each entity different from all the others to evaporate, and a uniform and "gray" metaphysical vision would result from that, with neither words nor concepts for distinguishing between one thing and another. We would be in a Parmenidean, or in a Cartesian, position, in which the rich variety of entities could only be devalued to *doxa*, or explained as an ephemeral and apparent secondary quality.

To avoid the opposite risk, that of having to deal with heterogeneous singularities, intractable to the thought, Aristotle introduces categories as principles of classification for bringing order into things. The first of the categories, substance, is based on the *noetic* apprehension of separate and determinate entities with which the world is immediately grasped by human beings: it constitutes the fundamental, ontological, and logical reference of reality and of knowledge. The *noetic* determination, pre-categorical support of the subsequent *dianoia*, is shared by systemic thinking, aimed at describing and conceptualizing unitary entities, systems, or substances, as you will, that populate the world.

2.4 Substance

We use the term substance for a unitary entity that can stand of itself and that, even while composed of parts, is not a multiplicity, but rather a unity. Aristotle immediately asks the question:

> for clearly the thing is one, but in virtue of *what* is the thing one, although it has parts? (Aristotle, *Metaphysics*, VII, 2, 1037 a, 19–20).

And a bit farther along:

> the differentiae present in man are many, e.g., endowed with feet, two-footed, featherless. Why are these a one and not many? (Aristotle, *Metaphysics*, VII, 2, 1037 b, 22–24).

A first hint of a response can be found in the famous passage:

> Since that which is compounded out of something so that the whole is one, not like a heap, but like a syllable — now the syllable is not its elements, ba is not the same as b and a, nor is flesh fire and earth (for when these are separated the wholes, i.e., the flesh and the syllable, no longer exist, but the elements of the syllable exist, and so do fire and earth); the syllable, then, is something — not only its elements (the vowel and the consonant) but also something else [...] But it would seem that this "other" is something, and not an element, and it is the cause which makes *this* thing flesh and *that* a syllable. And similarly in all other cases. And this is the *substance* of each thing (for it is the primary cause of its being); and since, while some things are not substances, as many as are substances are formed in accordance with a nature of their own by a process of nature, their substance would seem to be this kind of "nature", which is not an element but a principle.
> (Aristotle, *Metaphysics*, VII, 17, 1041 b, 12–31 and passim).

Aristotle introduces a principle that acts as a cause of unification of the elements, but he does not explain how the principle operates. Systemic thinking, which, as has been said above, pinpoints the foundation of the unity of a system in the organization, that is to say, in the bonds of interactions and interferences that connect and constrain the parts, performs this further passage.

Successively, Aristotle observes that those unitary entities to which the name "substances" applies maintain their identity even in the flow of changes to which they are subject. How can we give reasons regarding becoming, an apparently contradictory phenomenon, for which neither common sense nor the sciences are helpful in understanding? The philosopher must introduce new concepts: in this way potency and act make their entry into philosophy. To each substance pertains a constellation of potentials, which, passing to the act, allow the mutations governed by the principle proper to that substance. Aristotle does not explain how the potency-act dynamic is realized, and on this point systemic thinking develops the Aristotelian intuition, by specifying that a system can substitute its parts, while the organization is maintained stable. The organization, a dense nucleus of relations giving unity to the system, is in act, while the elements, substitutable in response to internal or external perturbations, constitute the basin of potentials that an entity can realize.

Even if in passing, it is necessary at least to point out the greater extension of the concept of system with respect to the Aristotelian concept of substance. While, for Aristotle, only natural living substances are substances in the full sense of the term, systemic thinking identifies as systems all entities exhibiting emergent properties, including collective artifacts and entities, which generate and sustain systemic properties thanks to the maintaining of coherent states.

2.5 The Pluralism of Entities

In the opening of *Metaphysics*, IV, we read the well renowned affirmation:

> There are many senses in which a thing may be said to be.
> (Aristotle, *Metaphysics*, IV, 2, 1003 b, 33–34).

From the statement that entities have different ontological statutes it follows that the methods and tools for knowing them must also be different. From ontological pluralism descends epistemological pluralism, because in order to know structurally different entities, different scientific domains must be constituted, appropriate to the specificities and characteristics of different objects. While pluralism is explicitly assumed by Aristotle in ontology, epistemological and methodological pluralism, which so clearly derive from it, are not explicitly theorized; the foundation of the sciences—in the plural— by Aristotle indirectly confirms his pluralistic position also in epistemology, but the task of a theoretical justification remains open.

Systemic thinking carries forward the research in the Aristotelian line, and starting from the observation that the world's entities/systems are different, argues in favor of the irreducibility of the sciences to a single one, because the different objects of the sciences require different tools for observation and different methods of knowing and verification, adapted to their different characteristics and peculiarities.

Pluralism, which may seem obvious to those familiar with systemic thinking, is in reality an important, and not yet pacific, achievement in contemporary epistemology, which has not yet been freed from the ideal based on univocity and monism suggested and supported by Platonic inspiration. I

L. Urbani Ulivi

propose a brief, critical *focus* without pretense of historical precision: in the Platonic or Platonic-leaning view, the visible and experimentable world is a mobile and smoky veil placed on the fundamental structure of the world, which is mathematical. Knowing means conquering the *episteme*, i.e., the knowledge without errors of the reality that perpetually underlies the transient world of phenomena.[3] This perspective is based, radically, in univocity: the world consists of rules, and knowing means extracting them with the tools of reason. The Platonic ideal has captivated a long line of successors to this day, who have continued to seek "The" method, "The" true knowledge of reality, "The" ultimate and universal constitution of all things in the world. Aristotle struggled against Platonic reduction and has safeguarded the plurality of the different phenomena and objects of the world by founding different sciences, each appropriate for describing a certain type of objects.

Adopting the Aristotelian perspective means questioning many prejudices and assumptions widely shared even today by scientists, philosophers, and widespread also in common sense, who believe more or less explicitly and consciously that only one type of causes (of the physical sort) acts upon the world, that only one type of laws (once again, physical) is sufficient for describing any phenomenon completely, that the non-microphysical aspects of phenomena are nothing other than epiphenomena.

Against this closed, univocal, and reductionist worldview, the voices of some philosophers of science—the so-called Stanford group—have risen, along with that of John Dupré,[4] who, in particular, has effectively argued against any form of reductionism in favor of pluralism of objects and of sciences. Once again, the revolt against the reductionist and physicalist paradigm came from the domain of biology, with the voice of Von Bertalanffy, recognized as the founding father of systemic thought, speaking in the 1960s (see Von Bertalanffy 1969). Today the researches of Bertolaso (2016), Giuliani (2016), Longo (2011), and many others, on the complexity and cellular plasticity of the living with its laws, the different levels intertwined in different causal modes, the adaptive link with the environment, allow the scientist and the philosopher to glimpse a new and boundless world to be conquered by knowledge, irreducible to the laws of physics.

As Agazzi (2015) has well seen, with his knowledge of the sciences as a physicist and his knowledge of philosophy as a logician, science has always been systemic, and, I would add, also pluralistic: if no scientist starts looking at the stars using the microscope, nor at bacteria using a telescope, this is because when he performs research, he is at least implicitly pluralistic:

---

[3] Plato exposes his theory of knowledge in *Theaetetus*, *Meno*, and *Republic*.

[4] Dupré (1995, 2003). In both books Dupré maintains an ontological pluralism: things are different and heterogeneous. Ontological pluralism implies epistemological pluralism: there is no general scientific method, process, or attitude valid for every domain.

he designs the instruments of observation according to the characteristics of what he is observing and according to his objectives.[5]

2.6 Distinction of Levels

Ontology's program of research can be read in Book IV of Aristotle's *Metaphysics* where he says that the object of his investigation is

> being as being, and the attributes which belong to this in virtue of its own nature (Aristotle, *Metaphysics*, IV, I, 1003 a, 20–22).

With these words, Aristotle attributes to ontology an object—entities considered for the fact that they are entities—and immediately adds:

> Now this is not the same as any of the so-called special sciences; for none of these others treats universally of being as being (Aristotle, *Metaphysics*, IV, I, 1003 a, 22–24).

Aristotle's first concern is to avoid having ontology be referred to as simply one science among the others. Why? For the reason that being, which is the object of ontology, is at a different level than the object of the sciences: the sciences have as their object the empirical entities, at the basic level, while ontology considers entities from a meta-level. The distinction among levels, even while remaining implicit, is an inalienable characteristic of Aristotle's philosophical and scientific project: understanding the world requires scientific knowledge, which traces the "second", empirical, causes of phenomena, but from this level it is necessary to reach the next, which seeks the knowledge of the "first" causes, beyond the empirical, caught by ontology and metaphysics.

Systemic thinking, open to multiple levels of knowledge and to both human and natural sciences,[6] traces different levels also in the empirical world and recognizes that systems, regardless of the nature of the elementary constituents, are articulated into subsystems and supersystems. Systemic hierarchy allows detection of a special class of errors, errors of level. Here, too, it was biology that served as teacher, in particular, through the studies of Bertolaso (2016) and Giuliani (2016): in cancer research, scientists' attention is focused on the cell; there are convincing motivations in biology to hypothesize that cancer, instead, lies at the level of tissues, as a disorder of tissue

---

[5] A Neo-Aristotelian Renaissance has been flourishing in recent times also in the direction of ontological pluralism. See Koslicki (2008), Turner (2010), Tahko (2012) and Novotny and Novàk (2014).

[6] The theologian Camillo Card. Ruini introduced systemic thinking in Catholic theology in order to better understand and prove the possibility that the human soul survives death. See Ruini (2017).

information (Bertolaso 2016). Systemic thinking teaches that placing a problem at the appropriate level is an indispensable condition for knowledge to progress effectively and to generate effective interactions with the world.

2.7 Points of Contact and Differences Between Aristotelian Ontology and Systemic Ontology

It seems clear from what has been said, that despite the depth and richness of the Aristotelian worldview, we cannot simply be Aristotelian or neo-Aristotelian. If we were, we would not have made Aristotle's deepest lesson our own: all knowledge starts from the observations we can make here-and-now, and even the highest, most sophisticated, and most abstract levels of thought are hinged on that basis. But it is precisely that basis which today is completely changed with respect to what was available to Aristotle: we have an extraordinary heritage of scientific knowledge, which, utilized as a source of a philosophical vision capable of being contemporary to its time, will induce profound changes in philosophy. If one wants to be faithful to the spirit of Aristotle, keeping his most precious indications firm, it will often be necessary to be unfaithful to its literal expressions, that rely on historical conditions that are now completely outdated.

What are the living traits of Aristotelian ontology to be taken up again and developed? What form, today, does an ontology inspired by Aristotle take? It will be, like Aristotelian ontology, not eliminativistic, nor even reductionist, pluralist, sorted into hierarchical levels, not formal, realist, focused on substance. It will have the task of developing and deepening many concepts already introduced by Aristotle, including abduction (Urbani Ulivi 2016), analogy, substance/system, hierarchy, causes; to introduce others: environment, the observer-observed relationship, collective entities, the relations of interaction and interference, constraints; to correct or simply abandon others, dependent on scientific knowledge that is almost nil or unreliable. The profound convergence of systemic thinking with Aristotelian thinking is in the auroral gaze, which Calogero[7] with Aristotle, would have called noetic, with which human beings structure the world into determinate, unitary, dynamic entities, leaving to the next moment, the *dianoia*, the task of explanation, of theory, of argumentation.

---

[7] Calogero (1927). In this book Calogero exposes his interpretation of human knowledge in Aristotle's philosophy. Human knowledge has two levels: the first and fundamental one, called *noetic*, consists in the immediate apprehension of things as separate and determined by which human beings are in direct contact within the world. The second level, called *dianoia*, is the realm of judgment, where the unity of *noesis* is broken into subject-object propositional language and ideas are expressed and connected in a logic form.

3 Systemic Ontology: First Steps

The sense of this work could be summarized in one question: what worldview is gained if we use the concept of system in philosophy? In other words: what ontological engagement assumes systemic thinking—if it assumes one?

3.1 Theoretical Nucleus

Every ontology is characterized by a nucleus, which is subsequently developed to the point of reaching a theoretical view that can even be quite general.

Systemic ontology also makes use of a characteristic and inalienable theoretical core, which can be summarized in a few sentences:

A system is an organization of parts connected by relationships, which has properties that its parts do not have (called emergent, or second-level, or systemic properties). The system is able to maintain identity while replacing parts and relationships. Systems are related to the environment in which they are immersed. The concept of system is independent of the nature of its elementary constituents.

Even so barely stated, the concept of system is of great interest for those dealing with ontology, because it clarifies two crucial ontological questions in an elegant and immediate way. The first one: how do we explain that the things of the world maintain identity? The response to this ancient and *vexata quaestio* is implicit in the definition of system: systems remain identifiable as long as the variable and fluctuating parameters of their internal relations support the specific and characteristic properties of that system. The second: how is change occurring in systems explained? The change of a system depends on the ability of the system to replace parts and relationships as an adaptive response to the perturbations and to the stresses imposed by the environment to which the system is connected.

3.2 Philosophical Consequences of the Systemic Approach

We have some important consequences for epistemology and for ontology:

1. Since the system properties cannot be attributed to the parts and cannot be deduced from observation of the parts, it follows that knowledge of the system and its properties cannot be obtained a priori, but must be acquired a posteriori, through observation.
2. The concept of system is a cognitive instrument that guides and orients our observation and description of some (not all) objects of the world, and which is enriched through experience.

3. The objects of the world are different, irreducible, and heterogeneous.
4. There are no isolated systems, but rather each of them is immersed in at least one environment. And, I would willingly add,
5. The predicate of existence is not univocal: objects have different modes of existence, or, otherwise stated, have different ontological statutes.

3.3 Does Systemic Thinking Assume an Ontological Position?

Certainly, because it proposes a theory that applies to all objects that satisfy the systemic description, regardless of the nature of their elementary constituents.

3.4 What Kind of Ontology Is It? What Engagements Does It Assume?

It is a pluralist, realist, and a posteriori, regional (or special) ontology. Let's briefly examine each property of the systemic ontology.

1. **Pluralism.** Systemic pluralism takes up and develops Aristotelian pluralism: objects are different and require different instruments appropriate to their specificity and characteristics.
   Pluralism seems far too easy a conquest, but it is still a countercurrent position, and not shared in ontology. I shall explain why briefly.
   The pluralism to which Aristotle explicitly committed, when he affirmed that

   > There are many senses in which a thing may be said to "be".
   > (Aristotle, *Metaphysics*, IV, 2, 1002 a, 33–34),

   seemed a weak position to many successors who preferred to follow Plato in the enterprise of giving a solid foundation to knowledge (preferably mathematical) equal for any object in the world. Thus began, starting from Avicenna's Commentary on Aristotle's *Metaphysics*,[8] a genealogy of philosophers who cultivate the ideal of a formal and a priori knowledge having nothing to learn from the results and progress of sciences, animated by the desire to achieve incontrovertible knowledge (the Platonic *episteme*), in which errors and corrections have no place, that cannot admit limits and doubts, certain only of the logical construction of concepts

---

[8] Morewedge (1973). Good guides to understand the historical and philosophical context of Avicenna's thinking are Bertolacci (2003, 2006), Hasse and Bertolacci (2011) and Gutas (2014).

and objects, completely devoid of a content of reality, for the reason that reality is judged uncertain, therefore inadequate.

To pursue ideals and anxieties of certainty, ontology slowly and continuously withdraws from reality, which is present to it only in its conceptual or formal form: a univocal and monistic, disincarnate reality echoes a mentalist and/or formal epistemology. Today, the de-realization continues as a reduction of the real to its computational simulacrum, which replaces it without residues and without appeal. A potent and effective ideal, that inspired *Brain Project* and *Human Brain Project*,[9] constructed and largely funded on the hypothesis that, in order to know the brain, it is sufficient to launch a computational program starting from the data collected in the neurosciences.

It is the ideal that can be traced, with some variations, up to contemporary formal ontology and to mereology[10] that, intolerant of ontological diversification, considers objects as the simple sum of their parts. Quite differently systemic thinking recognizes plurality of objects and differences among them, recovering to rational comprehension every aspect of objects.

2. **Realism.** If by realism—through pruning the forest of meanings for a now philosophically fashionable term—we soberly mean that reality is other than the knowing subject, systemic thinking assumes a realistic engagement, not for philosophical, but for systemic reasons: the emerging properties of each system are grasped by the subject through observation of the behaviors of the system, which cannot be understood as the autonomous fruit of the subject, because they are unexpected and unpredictable. Unless one wants to support an extreme solipsism, that, since cannot offer any proof or reason to support its fundamental statement "I, and only I, exist", should be classified among the pathologies of thought and left to sick fantasies.

3. **A posteriori,** because it arises from observation of things of the world: a system and its properties are observed. Observation is no longer, as for Aristotle, powered solely by perceptual abilities, but is extended thanks

---

[9] Two "Big brain projects" have been launched independently in the United States and in the European Union, both in 2013, called, respectively, Brain Initiative (BI) and Human Brain Project (HBP). Human Brain Project explicitly aims to realize a computer simulation of the whole brain, to reach with the development of a new kind of Information and Communication Technologies (ICT). It puts overwhelming emphasis on technological development and on computing tools, leaving aside fundamental neurobiology and cognitive neuroscience. It is a "paradigm shift" that goes in the direction of substituting scientific research with computation.

[10] Generally speaking mereology is the theory of parthood relations, concerning both the relations of part to whole and the relations of part to part within a whole. A good presentation of mereology both in logic and in ontology can be found in Varzi (2016).

to the fine observational capacities guaranteed by the sciences. Systemic ontology makes reference to the results of the sciences to enrich its vision of objects.

4. **Regional (or special).** It is a regional or special ontology because it only applies to objects that can be described as systems, and not all entities satisfy this description: for example, time, numbers, space are not systems. Monism is not accepted and even the highly useful concept of the system is not universal: its use is limited to entities with certain characteristics.

3.5 What Systemic Ontology Is Not

Systemic ontology differs from most of the ontologies available (which, beyond the alleged differences, present a panorama of disheartening conformism) as:

1. it is not a formal ontology, because it is not neutral over the domain;
2. it is not an a priori ontology, because it derives its concepts (even those concerning systems as systems) from experience;
3. it is not an analytical ontology, because it does not reduce objects to the sum of their parts (nor it is a mereology, or a topo-mereology, which seek to reconstitute the unity of entities through identification of the boundaries between parts) and it does not concentrate on language, on meaning, on conditions of meaningfulness of propositions, nor even on the logical structure of the world;
4. it is not universal, because it is only valid for objects that satisfy the conditions of being a system.

4 Systemic Thinking: Developments and Implications

From systemic ontology derive numerous concepts, which, integrated into a general theoretical framework, open to the elaboration of a systemic paradigm. I shall point out the most relevant. In a logical context, systemic thinking emphasizes the importance of *abduction*, the inference through which, given a context of incomplete information, the subject introduces an element not present in the context, therefore "new", which explains the problematic data available. The abductive inference implies, to a certain extent, a creative activity, which goes beyond the information available.

In the epistemological context, systemic thinking suggests and carries forward a revision of the concept of cause. Starting from the post-Galilean period, cause had been understood exclusively as efficient cause, easily expressible in mathematical language. It was a conceptual impoverishment with respect to Aristotelian thought, which not only had identified four causes, but which moreover, by interpreting "cause" to be any explanation valid for a given circumstance, had also implicitly authorized the search for other types of causes.

In this, as well, systemic thinking has carried forward the suggestion of Aristotle and has acquired many types of causality, including final cause, intentional cause, cause by absence, top-down, bottom-up, meso-level, etc.

Of great philosophical importance is the definitive overcoming of the subject-object dualism. Systemic thinking, in line with the observations of neuroscientists such as Edelman (2006), highlights that there is an inseparable relation between the observer and the observed: the observer with his biological, cognitive constitution, his past, his culture, his objectives, selects aspects and characteristics of reality that are accessible and relevant to him, and, to put it in the words of Maturana and Varela (1992), "bring forth a world". From this consideration it follows that the separation between epistemology and ontology should be reconsidered, because, if reality is not given absolutely, but always through the mediation of a knowing subject, and if the subject can only exercise his cognitive activity towards something, subject and object are inextricably connected and interdependent. The *quoad nos* is intertwined with the *quoad se*, and considering them as two independent contexts can induce the dangerous illusion that reality can be grasped as it is, in itself, regardless of the knowing subject and, conversely, that the processes of knowledge can be described while neglecting that on which they practice.

The concept of potency introduced by Aristotle to explain the becoming, which in the Stagirite remains in many ways critical and sub-theoretical, acquires a more precise theoretical status in systemic thinking, at least for some aspects: the knowledge of a system is realized by observing his behavior (inasmuch as the entity is in act, Aristotle would say), but to explain its dynamics, that is, its transformations in time, it is necessary to attribute to it also hidden variables that define the framework of the possible variation that such system can acquire. From this follows a strong epistemological limitation, inasmuch as the knowledge of any observed system is structurally incomplete, because a system has a bundle of potencies that are not explicit, nor directly observable.

This observation opens up to *incompleteness*, another relevant theme of systemic thought in process of elaboration. There is an epistemological incompleteness (*quoad nos*), which, recognizing that the potencies are not all known, concludes that the dynamics of becoming cannot be completely modeled. The knowledge of a system is necessarily incomplete also for the limits of our capacity of observation, which can take into consideration only some aspects or features of an object, never entirely or completely.

Besides theoretical incompleteness, there is another type of incompleteness, which we can call intrinsic (*quoad se*, as the Medievals would say), which starting from the observation of the transformations of systems, and also of their disintegration, leads us to conclude that systems have a structural incompleteness that feeds their dynamics, connected to the environment in which the system is immersed. Some philosophers might object that if we identify a system as a sum of what it currently is and of the potencies that are attributed to it, then it can be said to be complete. The objection can be accepted on the speculative level, but it would have the result of freezing research: it would be totally useless in furthering our knowledge of a system that remains irremediably incomplete for us human observers.

As it is often observed, incompleteness is not only a limit, but also a resource, because it is from incompleteness itself that we derive the possibility of corrections, improvements, and progress in knowledge.

Systemic thinking emphasizes the theme of *environment*: there are no isolated bodies (Del Giudice 2010), nor are there even isolated systems. Systems are in continuous exchange with the environment through dissipation and the acquisition of elements.[11] A corollary of the interaction between systems and their environment is the concept of *logical openness*, introduced in literature by Von Foester (1984) and developed by Minati et al. (1998) and Licata (2008).

Systemic thinking knows very well that "The past helps shape the future", as G. Longo said lecturing in Milan (2017), recognizing the importance of *history* to describe the current state of a system; it introduces the concept of *equifinality*, that of *coherence* as an instrument for maintaining the properties of a system; it accepts Maturana's and Varela's (1980) concept of *autopoiesis* as the ability to maintain the organization by replacing the parts; it studies *emergence*, and recognizes several types of it, including syntactic (or necessary) emergence and radical (completely new) emergence (Pessa 2006).

Starting from the observation that the sciences are neither isolated nor complete, and that no science exhausts the knowledge of a system, and also that no science can exhaust the complex of phenomena and events that interfere with the system, systemic thinking sustains a position of *radical antireductionism*, opening to studies on complex systems, and *complexity* in general, understood as irreducibility to one single formal model, and thus to one single science.

5 From Systemic Ontology Toward a New Paradigm

The extraordinary ability of human beings to take themselves and the world as objects of knowledge is possessed in a natural way, but acquires a disciplinary status when it gains a unitary vision by means of concepts and arguments, becoming philosophical thought, capable of delineating an understanding of the world from a meta-level.

---

[11] Vitiello (2001). The author considers the brain as an open and dissipative system that through continuous interaction with its environment leaves in it a trace, or a copy, or a "Double", where world objectiveness and the brain's implicit subjectivity are conjugated.

The object-level to which philosophy refers, both explicitly as well as implicitly, is acquired through various cognitive contributions, ranging from common sense, to introspection, to formal reduction, to acquisition of the results of the sciences.

A philosophy inspired by systemic thinking, by its capacity to rethink in a new way numerous traditional themes and problems of thought, maintains close contact with the different branches of knowledge, from which it incorporates knowledge and concepts forged in different domains, and elaborates a unitary, solid, and, at the same time, open, philosophical perspective.

Since the sources of human action are mainly ideas, more or less receptae, more or less the result of a critical scrutiny and of personal rethinking, also the systemic vision as a complex set of ideas inspires and suggests behaviors. Considering the world as an open and mobile unity of ordered systems in intertwined hierarchies, of which we ourselves are part, fosters an attitude of cooperation and prudence toward all interactions, including those with the environment in which we are immersed. The principle of prudence is not motivated by an evaluation of interest, nor even by an ethical imperative, but by an epistemic-ontological consideration, in that we have a structurally limited and partial knowledge of structurally incomplete objects, and any intervention must take into account this twofold limitation, in order to avoid altering the equilibria whose deep connections are known only in the smallest part.

Systemic ontology, which guides us in the description and the theorizing of an important typology of objects, does not remain isolated, but opens to a general philosophical vision, which welcomes the contributions of other branches of knowledge, structuring them in an organic vision, which can rightfully aspire to the title of "systemic philosophy".

References

Agazzi, E. (2015). Presentazione. L'Orizzonte Sistemico. In L. Urbani Ulivi (Ed.), *Strutture di Mondo. Il Pensiero Sistemico come Specchio di una Realtà Complessa* (Volume terzo) (pp. 7–28). Bologna: Il Mulino.

Aristotle. *Metaphysics*. Edited with Introduction, commentary and translation by W. D. Ross (1924). Oxford: Clarendon Press.

Avicenna. *The "Metaphysica" of Avicenna (Ibn Sīnā). A critical translation-commentary and analysis of the fundamental Arguments in Avicenna's "Metaphysica" in the "Dānish Nāma-i 'alā'i" (The Book of Scientific Knowl edge)*. Edited by P. Morewedge (1973). London: Routledge and Kegan Paul.

Bailly, F., & Longo, G. (2011). *Mathematics and the natural sciences. The physical singularity of life*. London: Imperial College Press.

Bertolacci, A. (2003). Some texts of Aristotle's metaphysics in the Ilāhīyāt of Avicenna's Kitāb al-Šifā'. In D. C. Reisman & A. H. Al-Rahim (Eds.), *Before and after Avicenna. Proceedings of the First Conference of the Avicenna Study Group* (pp. 25–45). Leiden, Boston: Brill.

Bertolacci, A. (2006). *The reception of Aristotle's metaphysics in Avicenna's Kitāb al-Šifā': A milestone of western metaphysical thought.* Leiden, Boston: Brill.

Bertolaso, M. (2016). *Philosophy of cancer. A dynamic and relational view.* Berlin/New York: Springer.

Calogero, G. (1927). *I Fondamenti della Logica Aristotelica.* Florence: Le Monnier.

Del Giudice, E. (2010). Una Via Quantistica alla Teoria dei Sistemi. In L. Urbani Ulivi (Ed.), *Strutture di Mondo. Il Pensiero Sistemico come Specchio di una Realtà Complessa* (Volume Primo) (pp. 47–71). Bologna: Il Mulino.

Dupré, J. (1995). *The disorder of things. Metaphysical foundations of the disunity of science.* Cambridge: Harvard University Press.

Dupré, J. (2003). *Human nature and the limits of science.* Oxford: Oxford University Press.

Edelman, G. M. (2006). *Second nature. Brain science and human knowledge.* Yale: Yale University Press.

Giuliani, A. (2016). Complessità organizzata: Perché lo studio delle reti di interazione ci costringe a ripensare la biologia da cima a fondo. *Rivista di Filosofia Neo-Scolastica, CVIII*(2), 305–319.

Gutas, D. (2014). *Avicenna and the Aristotelian tradition: Introduction to reading Avicenna's philosophical works.* Leiden, Boston: Brill.

Hasse, D. N., & Bertolacci A. (Eds.). (2011). *The Arabic, Hebrew and Latin reception of Avicenna's metaphysics.* Berlin/Boston: Walter de Gruyter.

Hooker, C. (Ed.). (2011). *Philosophy of complex systems.* Handbook of the philosophy of science (Vol. 10), Amsterdam: North Holland.

Koslicki, K. (2008). *The structure of objects.* Oxford: Oxford University Press.

Licata, I. (2008). *La logica Aperta della Mente.* Torino: Codice Edizioni.

Longo, G. (2017). *Interfaces of incompleteness.* Opening Lecture of the Seventh National Conference on Systems Science (AIRS), Università Cattolica del Sacro Cuore, Milano, 16th–17th November 2017 (in this volume).

Maturana, H. R., & Varela, F. J. (1980). *Autopoiesis and cognition. The realization of the living.* Boston: Reidel.

Maturana, H. R., & Varela F. J. (1992). *The tree of knowledge. The biological roots of human understanding* (Rev. Edition). Boston: Shambhala.

Minati, G. (Ed.). (1998). *Proceedings of the first Italian conference on systemics.* Milano: Apogeo Scientifica.

Minati, G., & Pessa E. (Eds.). (2002). *Emergence in complex cognitive, social and biological systems.* New York: Kluwer Academic/Plenum Publishers.

Minati, G., Penna, M. P., & Pessa, E. (1998). Thermodynamic and logical openness in general systems. *Systems Research and Behavioral Science, 15*(3), 131–145.

Minati, G., Pessa, E., & Abram, M. R. (Eds.). (2006). *Systemics of emergence: Research and applications.* New York: Springer.

Minati, G., Abram, M. R., & Pessa, E. (Eds.). (2009). *Processes of emergence of systems and systemic properties. Towards a general theory of emergence.* Singapore: World Scientific.

Minati, G., Abram, M. R., & Pessa, E. (Eds.). (2012). *Methods, models, simulations and approaches. Towards a general theory of change.* Singapore: World Scientific.

Minati, G., Abram, M. R., & Pessa, E. (Eds.). (2016). *Towards a post-bertalanffy systemics.* Cham: Springer.

Novotny, D. D., & Novàk, L. (Eds.). (2014). *Neo-Aristotelian perspectives in metaphysics.* London: Routledge.

Pessa, E. (2006). Physical and biological emergence: Are they different? In G. Minati, E. Pessa & M. R. Abram (Eds.), *Systemic of emergence. Research and development.* New York: Springer.

Ruini, C. (2017). L'Anima e la sua Immortalità tra Teologia e Approccio Sistemico. *Rivista di Filosofia Neo-Scolastica, CIX*(2), 277–286.

Tahko, T. E. (Ed.). (2012). *Contemporary aristotelian metaphysics.* Cambridge: Cambridge University Press.

Turner, J. (2010). Ontological pluralism. *Journal of Philosophy, 107*(1): 5–34.

Urbani Ulivi, L. (Ed.). (2010). *Strutture di Mondo. Il Pensiero Sistemico come Specchio di una Realtà Complessa.* Bologna: Il Mulino.

Urbani Ulivi, L. (Ed.). (2013). *Strutture di Mondo. Il Pensiero Sistemico come Specchio di una Realtà Complessa. (Volume Secondo).* Bologna: Il Mulino.

Urbani Ulivi, L. (Ed.). (2015). *Strutture di Mondo. Il Pensiero Sistemico come Specchio di una Realtà Complessa. (Volume Terzo).* Bologna: Il Mulino.

Urbani Ulivi, L. (Ed.). (2016). L'abduzione come Momento della Scoperta in Contesti di Realtà. Abduction is the Inference That Discovers a Solution in Problematic Contexts. *Cassazione Penale, LVI*(5), 2240–2251.

Varzi, A. (2016). Mereology. In E. N. Zalta (Ed.), *The stanford encyclopedia of philosophy.* Palo Alto: Stanford University.

Vitiello, G. (2001). *My double unveiled. The dissipative quantum model of brain.* Amsterdam/Philadelphia: John Benjamins Publishing Company.

Von Bertalanffy, L. (1969). *General system theory.* New York: George Braziller.

Von Foester, H. (1984). *Observing systems.* Seaside: Intersystems Publications.

Part II
Models of Incompleteness and Quasiness

All the Shades of Incompleteness: The Interesting Case of Structure/Function Relations in Biochemistry

Alessandro Giuliani

1 Structure-Activity Relationships

The story of QSAR (Quantitative Structure Activity Relationships), the science dealing with the sketching of effective models to predict the biological activity of small organic molecules (drugs, toxicants, pesticides, ...) from their structural and chemico-physical properties, begun with the classic 1964 Hansch and Fujita paper appeared on the *Journal of American Chemical Society* (Hansch and Fujita 1964). The idea of correlating biological activity with chemical structure and physical properties of compounds is not new. Indeed, its origins can be traced back to the famous Meyer-Overton theory (Meyer 1899) where the bioactivity of anesthetics was correlated with their lipid/water partition coefficients.

The merit of Hansch and Fujita was twofold: they both turn a "vague intuition" of a necessary correlation between changes in the chemico-physical properties of a molecule and its biological activity into an operational quantitative approach and gave a sound theoretical basis of the phenomenon in the context of physical organic chemistry. To reach this goal, they relied on the so called Hammett equation (Hammett 1937), a linear free-energy relationship linking reaction rates and equilibrium constants for reactions involving aromatic compounds. The basic idea is that for any two reactions involving two aromatic reactants only differing in the type of substituent, the change in free energy of activation is proportional to the change in Gibbs free energy. The basic equation is:

$$\log\left(\frac{k}{k_0}\right) = \sigma\rho \tag{1}$$

A. Giuliani (✉)
Department of Environment and Health, Istituto Superiore di Sanità, Roma, Italy
e-mail: alessandro.giuliani@iss.it

© Springer Nature Switzerland AG 2019
G. Minati et al. (eds.), *Systemics of Incompleteness and Quasi-Systems*,
Contemporary Systems Thinking,
https://doi.org/10.1007/978-3-030-15277-2_3

where k is the equilibrium constant relative to a given reaction with a substituent R and k_0 is the reference constant when R is an hydrogen atom. The value of σ depends only by the particular substituent R and not by the particular reaction, while ρ is a reaction constant that depends only by the type of reaction but not on the substituent used.

This result opens the way to the so called "extra-thermodynamic relations" (linear free-energy relationships) that Hansch and Fuijta invoke (Hansch and Fujita 1964) to explain their success in modeling "biological reaction rates" (expressed in terms of biological endpoints like the antibiotic concentration needed for achieving the 50% of bacteria mortality in a plate) in terms of electronic (σ), hydrophobic (π) and steric (ρ) parameters along an homogeneous series of chemicals.

The success of QSAR approaches, that are routinely used in chemical industry for the refinement (we will go more in depth in the following on this crucial concept) of an already detected lead compound showing some biological activity of interest, can be considered as a paradigmatic case of conscious incompleteness of a model. The drastic reduction to an organic reaction mechanism of an extremely complex (and largely unknown) mechanism of action going from the administration of a chemical to the emerging of a biological effect, can work only by neglecting the by far major part of the features of the system at hand. This "extreme reduction" works because QSARs do not actually model the chemico-biological interaction as such, but only the "noise" added to an ideal "optimal drug" by relatively small modifications (addition of different substituents) to a "pharmacophore" (a molecular chassis endowed with biological activity).

In other words, a successful QSAR deals with the variations on a theme constituted by an "already there" lead compound. This is why QSAR approach needs the collection of an homogeneous series of organic molecules as such (Fig. 1).

The leading compounds (on the left of Fig. 1), embedding the supposed pharmacophore (middle panel) are expanded into a series of dozens of analogues differing among them only for R_1 (substituent) site.

If the leads already have a certain degree of activity, the generation (via least squares optimization) of an equation linking the electronic, hydrophobicity and steric features of the R_1 substituents to the biological activity of the corresponding molecules, will guide the synthesis of the most potent (and/or less toxic) molecule. This is a refinement and not an *ab initio* (still largely dependent on serendipity) discovery process, the (apparent) reduction of biology to organic chemistry is only a "gift of biology" that settles the "recognition scenery" in which we can only marginally act (even if these marginal actions can have huge practical consequences).

This is in some sense the most drastic acceptance of incompleteness: the modeler accepts the impossibility to enter into the actual functioning of the complex biological systems and, thanks to the agency of a "rate limiting step" involving an (often unknown) interaction between a small organic molecule

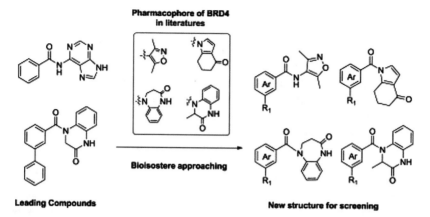

Fig. 1 The leading compounds correspond to molecules already showing (at least at some degree) the desired pharmacological activity. The "shared moieties" across the leading compounds (central panel) correspond to an empirical "pharmacophore". In the right panel different molecules are generated by different substitutions at R_1 sites. These molecules are in turn screened as for their biological activities in order to sketch a quantitative model linking chemico-physical properties of R_1 and molecules therapeutic index (ratio between desired pharmacological activity/unwanted toxicity)

(drug) and a biological macromolecule (receptor), focuses on the modulation exerted by small changes in the drug structure.

The above sketched general paradigm was applied across a huge number of cases encompassing the generation of "more sophisticated" chemico-physical and structural descriptors, the application of extremely smart and acute "deep learning" optimization techniques or the development of physically motivated "molecular dynamics" approaches to model drug-receptor interaction. The (often unsaid) goal of many of these applications, was to push the refinement boundaries and entering into the "real thing" i.e. the actual modeling of the system as such. This implies to abandon the "safe territory" of QSAR modeling: the homogeneous series. In the following we will analyze in depth what happens just outside these territories discussing successes and failures of different modeling strategies from the view point of their (perceived and/or actual) degree of "accepted incompleteness".

2 To (Think To) Know Too Much Is (Often) a Curse

In their 2005 paper (Bender and Glen 2005), the authors propose an extremely interesting challenge located at the very beginning of "biological complexity": the estimation of the ability of small organic molecules to bind to a biological receptor.

The authors analyzed two big data bases reporting structures and biological activities of a very large set of compounds. The first data base (A) contains 5 subsets: each element of the subset was assayed as for its ability to link to a specific pharmacological receptor (a receptor is a macromolecule, usually a protein). Each subset corresponds to a different receptor and contains a number of putative ligands (test compounds) varying from 49 to 134 (for a total of 957 molecules).

The second data base (B) contains 7 distinct subsets varying from 341 to 1236 chemicals, for a total of 8293 distinct structures.

The authors operate a virtual screening in which each molecule is compared with the leading compound (an already known ligand to each of the 11 (5+7) receptors) of its subset in order to predict its ability to interact with the correspondent receptor. The prediction was based on nine different structure-activity models, endowed with different degrees of sophistication. These models, together with a purely random selection, were evaluated in terms of their predictive ability. It is worth noting the chemicals relative to the eleven subsets are not homogeneous, i.e. they do not derive from small variations of the lead, thus preventing the "pure refinement" attitude of QSAR.

The results were summarized in terms of "Average hit rate" over the entire molecules population (Fig. 2).

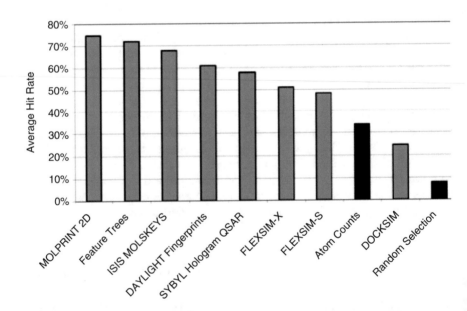

Fig. 2 The ten models are reported in decreasing order of accuracy. Atom Counts and Random Selection (dumb methods not considering in the actual molecular structure) have hit rates comparable with the by far most sophisticated one (DOCKSIM)

The results of the challenge are somewhat astonishing: the most sophisticated method (lower level of incompleteness) is DOCKSIM, that takes explicitly into account the 3D structure of both the ligand and the receptor, is halfway between two very "dumb" approaches like random selection and the simple count of atoms of the molecules expressed as a 12 component vector corresponding to the frequency of different atomic species.

Going up toward the "good performing end" we find FLEXSIM-X and FLEXSIM-S: these methods are based on the computation of the Euclidean distance between a molecule A and a query structure (here the leading compound) on the basis of the shared presence of a priori defined sub-structures weighted as for their probability to bind to the receptor. The weights derive by the analysis of huge data bases (Lessel and Briem 2000) and (at odds with classical QSAR methods) do not require homogeneous series sharing the same pharmacophore. It is worth noting both FLEXSIM approaches are optimized by sophisticated computational methods like genetic algorithms.

SYBYL Hologram QSAR (Heritage and Lowis 1999) has an at intermediate level of predictive power (Average Hit Rate = 50%), this corresponds to a more than fivefold enrichment with respect to random selection (hits, i.e. effective binders are around 10% of total molecules as evident by random selection results). This technique tries to enlarge the reach of classical QSAR extra-thermodynamic models to deal with non-homogeneous series of chemicals by projecting the molecules on a 3D-grid upon which the usual hydrophobicity, electronic and steric parameters are superimposed. The 3D-QSAR representation of each molecule is then aligned to the query compound and a distance is computed. The "hits" are then estimated in terms of similarity with the query. It is worth noting how the apparent "addition of realism" (molecules are effectively three- and not bi-dimensional as structural formulas appear) to the QSAR approach does not counterbalance the absence of an homogeneous series (classical QSAR on homogeneous series predictability ranges from 70 to 95%).

The top four methods (Daylight Fingerprints, ISIS Molskeys, Feature Trees, MOLPRINT 2D) display prediction accuracies ranging from 62 to 76% and are all based on 2D structural formula. These four methods build upon the representation of molecules by arrays having as components the sub-graphs of the structural formulas taking 1/0 values corresponding to presence/absence of the sub-graph in the molecule. These sub-graph are named "structural keys".

Figure 3 reports an example of structural keys extraction from 2D organic formulas.

Fingerprints are structural keys "self generated" by the molecule itself. The fingerprinting algorithm examines the molecule and generates the following: a pattern for each atom, a pattern representing each atom and its nearest neighbors (plus the bonds that join them), a pattern representing each group of atoms and bonds connected by paths up to 2 bonds long, ... atoms and

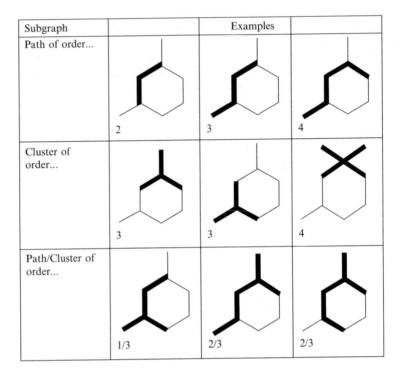

Subgraph		Examples	
Path of order...	2	3	4
Cluster of order...	3	3	4
Path/Cluster of order...	1/3	2/3	2/3

Fig. 3 Structural keys are paths of different length correspondent to paths along molecular graphs (2D structural formula)

bonds connected by paths up to 3 bonds long, ... continuing, with paths up to 4, 5, 6, and 7 bonds long (Butina 1999).

Given the number of possible paths grows exponentially with the complexity of the molecules, fingerprint approaches utilize different heuristics to limit the dimension of the Boolean vectors correspondent to the structural keys distribution.

MOLPRINT 2D (the best performing method) extracts the so-called "Circular fingerprints" (Glen et al. 2006). The process of generation of circular fingerprints is illustrated in Fig. 4: every heavy atom of a molecule is sequentially used as a starting point for the generation of descriptors and is assigned an atom type. The atom types for a number of layers, for example, two layers in addition to the central heavy atom, are then assigned to neighboring atoms. To calculate descriptor values, the number of atoms with each given atom type and at each distance from the central atom is recorded. This "count vector" then serves as a descriptor of the local, chemical environment of the central atom. To create fingerprints for an entire molecule, this process is repeated for each of its atoms. Because the growth of a descriptor is radial, this type of fingerprint is referred as circular.

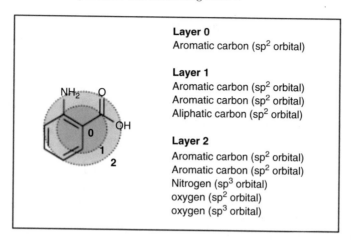

Fig. 4 Circular fingerprint extraction: the molecular graph is represented in terms of the "neighboorood" of different heavy atoms. The metrics on the resulting vectors (analogously to the Feature Trees method that scored second in the challenge) is hierarchical, taking into consideration the different "neighboorood layers"

From this exercise in model building we can derive some very useful general lessons about the conscious use of incompleteness. The main points to consider are:

1. 2D is better than 3D. In Fig. 2 the modeling methods that (more or less explicitly) take into account the three-dimensional character of the structures occupy the "low hit" (right) half of the distribution, while 2D methods are in the "high hit" zone.
2. The "most realistic" model explicitly accounting for the 3D structure of the receptor is the less efficient one.
3. The molecular graph (structural formula) embeds a latent information that goes much further a pure 2D projection of a molecule structure. The best performing models exploit such information by a largely redundant (albeit very incomplete) representation style (fingerprints).

3 Picture of Incompleteness

We are now in the position to sketch a general picture of the role played by incompleteness in modeling. The first lesson comes from the reason of the success of classic QSAR approach: limiting to small variations on a fixed molecular skeleton eliminates the contextual information (and consequently biology) from the picture. The "mutual recognition" between the complex biological device (receptor protein) and the simple small organic molecule (drug) happens for unknown reasons we decide not to investigate. We only focus on the modulation of this interaction by little chemical changes: we are totally inside the organic chemistry realm where extra-thermodynamic relations hold and support the success of the phenomenological *a-la-Hansch* structure-activity equations.

We do not know when and how our model will cease to function (when we do exit from the homogeneous series/same mechanism of action paradigm) and for sure the degradation of the model predictability will not be gradual but we can safely rely on chemical knowledge to minimize the risk of incorrect predictions.

Problems start when we make a "little step" into biological complexity abandoning the homogeneous series safe space. In this case the naïve notion of "realism" (and consequently the idea that the "more facts I take into consideration the more efficient my model will be") does not hold anymore. Here context dependence enters into play and has a strong selective effect in deciding of the success of our "incompleteness choices". It is certainly true that the modeled interaction involves both the receptor and the drug but trying to "put inside the model" a thorough structural description of the receptor interaction site (DOCKSIM approach) has the paradoxical effect to almost completely destroy the model efficiency. What happens is that we make the model much more demanding: our predictions are now driven by the very strong (and false) assumption that the drug-receptor interaction can be equated to a static key-lock fit while, on the contrary, protein molecules have a rich dynamics in solution and we have no idea of the way a drug approaches its binding site on the biological macromolecule. This provokes the uncertainties of our knowledge to exert a much greater effect than the (supposed) realism addition.

Similar considerations can be made as for the superiority of 2D approaches with respect to 3D ones. Structural formula (or molecular graph) are very peculiar objects that by no means can be equated to 2D pictures of molecules.

All the relevant chemico-physical properties of a given organic molecule (even molecules that still not exist but are only "well designed on the paper following the correct valence rules") can be derived from its structural formula (Katritzky et al. 2001; Todeschini and Consonni 2009). This comes from the fact molecular graphs topology stems from the optimization of the physical principles at the basis of molecule stability and formation. The edges between the atoms of a structural formula do not only mark "proximity in space" relations but point to a shared molecular orbital supporting the covalent bond. This implies the molecular graph wiring architecture is a symbolic summary of the energy landscape of the system from where it derives its reactive and/or interaction properties.

The word "symbol" derives from the Greek language and is made by two parts: "syn" ($\sigma\upsilon\nu$) that means "together" and "ballo" ($\beta\alpha\lambda\lambda\omega$) "to throw". This means that a symbol collects together (but in a difficult to deconvolve manner given they were "thrown" and not orderly positioned) different entities and/or concepts (that in our case derive from physics). Embedded into each organic formula there is a big part of modern physics, the problem is that while we can draw the molecular graph, we are not able to explicitly extract all the implicit information burden. We can try to derive "meaningful descriptions" by purely empirical means, exploiting the huge combinatorics of graph-like representations.

This is exactly the case of the most successful choices in drug-receptor interaction modeling. The bet on the possibility to extract a latent organization by the action of redundancy, is typical of multivariate statistics that, according to a classical definition of Guigou (1977)

[...] deals with variables which are numerous, approximate, not very significant (in the sense that each of them, singly, carries limited information), discrete or continuous, heterogeneous, qualitative or quantitative.

The emerging of a latent meaning from these variables is made possible by the exploitation of their mutual correlation (Giuliani 2017) on a purely data-driven way.

This style of reasoning that, notwithstanding its success in organic, medicinal (García-Domenech et al. 2008) and biological (Di Paola et al. 2012) chemistry, met a very strong opposition by a large part of scientific community (Randić 2001) implies a conscious acceptance of incompleteness in favor of "consistent" representations.

The criteria of deciding about the "consistency" of a representation traverse different fields of scientific enquiries and are the object of enquiry of statistical methodology. It is out of the scope of this chapter to go in depth into this issue, we can only drive the reader attention toward very general statistical indexes like: "non-trivial determinism" (Pascual and Levin 1999) or Akaike information (Ludden et al. 1994), but the essential issues are related to semantic aspects like the "stability of meaning" (e.g. an edge between any two nodes in a network must stem from the same metrics for all the nodes and must point to the same kind of process) dealing with the simplicity and clarity of the adopted operational models.

Incompleteness is an inescapable property of human approach to the world, thus it is neither an evil nor a good thing per se, that is why we must not look for the minimal incompleteness of our approaches. The exemplar case described in this chapter tells us how each problem asks for its own level of conscious incompleteness corresponding to the modeler's bet. In the case of drug-receptor interaction the best choice was to consider molecules as bi-dimensional graph discarding the possibility of further reduce the model incompleteness: it was the best bet.

References

Bender, A., & Glen, R. C. (2005). A discussion of measures of enrichment in virtual screening: Comparing the information content of descriptors with increasing levels of sophistication. *Journal of Chemical Information and Modeling, 45*(5), 1369–1375. https://doi.org/10.1021/ci0500177.

Butina, D. (1999). Unsupervised data base clustering based on day-light's fingerprint and Tanimoto similarity: A fast and automated way to cluster small and large data sets. *Journal of Chemical Information and Computer Sciences, 39*(4), 747–750. https://doi.org/10.1021/ci9803381CCC.

Di Paola, L., De Ruvo, M., Paci, P., Santoni, D., & Giuliani, A. (2012). Protein contact networks: An emerging paradigm in chemistry. *Chemical Reviews, 113*(3), 1598–1613. https://doi.org/10.1021/cr3002356.

García-Domenech, R., Gálvez, J., de Julián-Ortiz, J. V., & Pogliani, L. (2008). Some new trends in chemical graph theory. *Chemical Reviews, 108*(3), 1127–1169. https://doi.org/10.1021/cr0780006CCC.

Giuliani, A. (2017). The application of principal component analysis to drug discovery and biomedical data. *Drug Discovery Today, 22*(7), 1069–1076. https://doi.org/10.1016/j.drudis.2017.01.005.

Glen, R. C., Bender, A., Arnby, C. H., Carlsson, L., Boyer, S., & Smith, J. (2006). Circular fingerprints: Flexible molecular descriptors with applications from physical chemistry to ADME. *IDrugs, 9*(3), 199–208. ISSN 1369-7056.

Guigou, J. P. (1977). *Methodes Multidimensionelles*. Paris: Dunod.

Hammett, L. P. (1937). The effect of structure upon the reactions of organic compounds. Benzene derivatives. *Journal of the American Chemical Society, 59*(1), 96–103.

Hansch, C., & Fujita, T. (1964). p-σ-π Analysis. A method for the correlation of biological activity and chemical structure. *Journal of the American Chemical Society, 86*(8), 1616–1626.

Heritage, T. W., & Lowis, D. R. (1999). Molecular hologram QSAR. *Rational Drug Design, 14*, 212–225. https://doi.org/10.1021/bk-1999-0719.ch014.

Katritzky, A. R., Perumal, S., Petrukhin, R. & Kleinpeter, E. (2001). Codessa-based theoretical QSPR model for hydantoin HPLC-RT lipophilicities. *Journal of Chemical Information and Computer Sciences, 41*(3), 569–574. https://doi.org/10.121/ci000009t.

Lessel, U. F., & Briem, H. (2000). Flexsim-X: A method for the detection of molecules with similar biological activity. *Journal of Chemical Information and Computer Sciences, 40*(2), 246–253. https://doi.org/10.121/ci990439eCCC.

Ludden, T. M., Beal, S. L., & Sheiner, L. B. (1994). Comparison of the Akaike Information Criterion, the Schwarz criterion and the F test as guides to model selection. *Journal of Pharmacokinetics and Biopharmaceutics, 22*(5), 431–445. https://doi.org/org/10.1007/BF02353864.

Meyer, H. (1899). Zur theorie der alkoholnarkose. *Naunyn-Schmiedeberg's Archives of Pharmacology, 42*(2), 109–118.

Pascual, M., & Levin, S. A. (1999). From individuals to population densities: Searching for the intermediate scale of nontrivial determinism. *Ecology, 80*(7), 2225–2236. https://doi.org/10.1890/0012-9658(1999)080[2225: FITPDS]2.0.CO;2.

Randić, M. (2001). The connectivity index 25 years after. *Journal of Molecular Graphics and Modelling, 20*(1), 19–35. https://doi.org/10.1016/S1093-3263(01)00098-5.

Todeschini, R., & Consonni, V. (2009). *Molecular descriptors for chemoinformatics* (Vol. 41) (2 volume set). Weinheim: Wiley-VCH.

Sentences and Systems

Aldo Frigerio

1 Introduction

A key topic in philosophy of language concerns the conditions of possibility of a natural language—that is, the features a natural language not only possesses but must possess if it is to be considered a natural language. Since Frege, scholars have agreed that the semantics of a natural language is necessarily compositional. The principle of compositionality states that the meaning of a sentence is a function of the meanings of the parts of that sentence. Here, the term "function" highlights the existence of an algorithm by means of which the meaning of the whole can be derived, based on the meanings of the parts and the rules of composition. It seems clear that the principle of compositionality is, at least to some extent, a valid principle. By way of example, consider the text you are now reading. In all probability, you have never previously encountered some of these combinations of words. Nevertheless, you understand these unfamiliar sentences perfectly. If every new sentence were a new meaning unit and not dependent on its parts—that is, if the principle of compositionality were not in force—it would be difficult to explain why such sentences present no problem of interpretation, as the meaning of each new sentence would require a specific learning as any new word does.

However, this is not the case, and the principle of compositionality explains why we can readily understand new sentences: given the meanings of the parts, we are able to construct the meaning of the whole. This is what you are doing while reading this text. Because you know the meanings of the

A. Frigerio (✉)
Department of Philosophy, Catholic University of the Sacred Heart, Milano, Italy
e-mail: aldo.frigerio@unicatt.it

© Springer Nature Switzerland AG 2019
G. Minati et al. (eds.), *Systemics of Incompleteness and Quasi-Systems*,
Contemporary Systems Thinking,
https://doi.org/10.1007/978-3-030-15277-2_4

constituent words, you can understand the meaning of these sentences, even though they are new to you. This explains why dictionaries contain lists of words rather than lists of sentences; if sentences were the smallest units of meaning and did not depend on the meanings of their parts, learning a language would involve learning the meanings of sentences rather than a lexicon and grammar. One might conclude, then, that the principle of compositionality implies an atomistic view of meaning—that in order to know the meaning of more complex linguistic units, it suffices to know the meanings of the simplest units. The aim of this essay is to show that this interpretation of the principle of compositionality is incorrect.

Here, I contend that the atomistic view of meaning fails for at least two reasons.

1. Even if the principle of compositionality were valid without restriction, it would not follow that a sentence's meaning is the sum of the meanings that constitute it. Sentences have a syntactic structure that differs from their linear order, and that structure affects semantic interpretation. It follows that a sentence's meaning is not reducible to the sum of the meanings of its constituent words.
2. In any case, natural languages are not entirely compositional. The principle of compositionality is restricted by the fact that the correct understanding of sentences often depends on understanding the linguistic context.

2 Sentences Are Structured Entities

The first reason why sentences are not sums of words is that they are hierarchically structured entities.

2.1 Syntactic Structure

It has been demonstrated for a long time by very convincing arguments that the sentences used by speakers of a language have both a linear order, in which words follow each other,[1] and a further level of organization that may differ from this. This second level of organization is not linear but hierarchical, as the morphemes and words that form a sentence combine into increasingly larger, nested constituents. For instance, in the sentence "Two brothers of Paul will arrive soon", we can first distinguish two large constituents: the noun phrase (NP) "two brothers of Paul" and the verb phrase (VP) "will

---

[1] For written texts, "follow" is intended in a spatial sense; for oral texts, it is intended in a temporal sense.

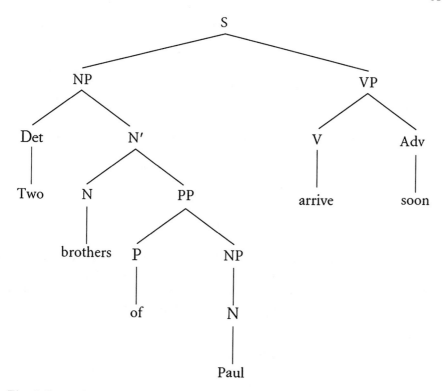

Fig. 1 Syntactic structure of the sentence "Two brothers of Paul will arrive soon"

arrive soon". These phrases are formed in turn by smaller constituents—for instance, the NP "two brothers of Paul" can be segmented into a determiner ("two") and another constituent ("brothers of Paul"). This latter constituent can be further segmented into a noun ("brothers") and a prepositional phrase ("of Paul"), and so on. This structure is usually represented by means of a tree diagram, as in Fig. 1, where S stands for sentence, N for noun, V for verb, P for preposition, PP for prepositional phrase, and Adv for adverb.

Such representations presuppose that languages are constituted by a finite number of discrete basic elements (phonemes, morphemes, words). This implies that their division into increasingly smaller constituents must end at a certain point—that is, there are elements that do not contain smaller elements. Additionally, representations of this kind presuppose that the rules that generate sentences are recursive—that is, that elements of a given kind can occur within elements of the same kind (For instance, the NP "Paul" occurs within the NP "two brothers of Paul"). Recursivity is a property of rules in which the rule can be applied to the result of applying the rule. This property explains the productivity of natural languages. Although they have a finite number of basic elements and a finite number of combinatorial rules, a potentially infinite number of sentences can be generated. The production of this infinite set from a finite initial set can be explained only if the rules that generate larger constituents from smaller ones are recursive.

Rules that enable the generation of sentences presuppose that words are assigned to different categories (noun, verb, preposition, adverb etc.), and that only some combinations of words from certain categories are allowed. For instance, one may combine an article with a noun but not with a verb (unless the verb is substantivized); while "the dog" is a grammatical phrase, "*the goes" is not. Permitted combinations of words from basic categories create larger constituents, which can be combined into still larger constituents, and so on—always on the basis of rules that permit only certain combinations. For instance, the rule governing how a noun phrase and a verb phrase form a sentence can be written as follows:

$$S \rightarrow NP + VP$$

The existence of such structures and rules shows that a sentence cannot be conceived simply as the sum of its constituent words. Beyond their superficial order, sentences have a syntactic structure that links the words into a whole. That structure is governed by rules that specify which relations among the parts are permitted and which are not. This shows that sentences are systems. To account for how language works, it is not sufficient to list its constituent elements (phonemes, morphemes, words) and their meanings; as well as knowing the elements, we must also know how those elements can be structured into wholes.

2.2 Isomorphism Between Syntax and Semantics

Chomsky (1957) characterizes the syntactic structure of sentences as independent of the semantics associated with this structure. However, it is possible to interpret Chomskyan theory in a different way, in which there is instead a more or less perfect correspondence between syntactic and semantic structure. On this view, two words are syntactically connected because of their semantic connection—that is, the connection between their meanings. Syntactic structure would then describe the order in which these meanings must be composed. This isomorphism between syntactic and semantic structure implies at least two prerequisites. First, semantic categories must correspond to syntactic categories—that is, the same *kind* of meaning must correspond to every string belonging to a particular syntactic category. On the other hand, certain semantic operations must correspond to syntactic operations. For example, a given composition of the meanings of V and NP must correspond to the syntactic operation V + NP. Taking account of this semantic-syntactic structure of natural languages, Partee et al. (1990) rephrased the principle of compositionality as follows: the meaning of a complex expression is a function of its constituents and the grammatical rules used to combine them.

3 Limits of the Principle of Compositionality

While formal and logical languages are usually entirely compositional, natural languages are not. In this section, I address some limitations of the principle of compositionality in natural languages, analyzing some phenomena in which the whole conversely determines the meanings of the parts.

3.1 Idioms

Certain natural language expressions, which grammarians refer to as *idioms*, do not abide by the principle of compositionality, and their meaning must be specifically learned. Such expressions may be groups of words or entire sentences—for example, "It's raining cats and dogs", "kicked the bucket" or "red herring". Their meaning is not compositional—that is, the meanings of the parts (and the rules of composition) do not suffice to explain the meaning of the whole.

3.2 Ambiguity and Polysemy

While idioms are of some relevance, languages are not for the most part idiomatic. However, there are other more pervasive phenomena that limit the principle of compositionality. These phenomena include ambiguity and polysemy. Many words have more than one meaning, and the precise sense in which such words are used is determined by the context, as in the following examples.

a. You parked across the street? That's *fine*.
b. If you park there, you'll get a *fine*.

The English word "fine" has more than one meaning, and when used in a particular occasion, it expresses only one of these meanings. In most cases, someone who hears or reads a sentence containing an ambiguous word can readily discern the intended meaning from the linguistic context. In this case, it is the context (the whole) that determines the meaning of the word (the part) rather than *vice versa*. Given that ambiguity and polysemy are widespread phenomena, this represents one important limitation to the principle of compositionality.

3.3 Anaphoric Pronouns

Another pervasive phenomenon concerns the referents of anaphoric pronouns, which are again determined by the immediate context, as in "Ann said to

Paul that *he* had to join *her* immediately". Anaphoric pronouns such as "he" or "her" have no reference outside the context in which they are used. They acquire a referent only from the linguistic context (the words that precede and follow). The proper names "Ann" and "Paul" appear in the linguistic context, serving as referents for the personal pronouns. Thus, the anaphoric pronoun is an example of the whole determining the meaning of a part. Indeed, it is the anaphoric pronoun's presence in one sentence rather than another that provides a certain referent for the pronoun itself. If we change the context, the referent is also changed. In this case, the assignment of the meaning proceeds *top-down* (rather than *bottom-up*).

3.4 Semantic Indeterminacy

Although the phenomenon of semantic indeterminacy bears some resemblance to ambiguity and polysemy, these concepts must be carefully distinguished. As described above, an ambiguous or polysemous word has more than one meaning, and context determines the intended meaning on a given occasion. In the case of semantic indeterminacy, however, there is only one indeterminate meaning, which is determined by the context of use. The following examples (cf. Searle 1980) serve to illustrate this point.

> Ann *cuts* the lawn.
> John *cuts* the cake.

We know that the operation of cutting a lawn is very different from cutting a cake; while the blades of grass are severed using a sickle or a lawnmower, a cake is cut into slices by a knife. It would be surprising if, to cut the lawn, Ann took a knife and performed very long incisions or took a scissors and cut the blades of grass one by one vertically. Similarly, it would be surprising if, to cut the cake, John used a lawnmower. Clearly, the meaning of the verb "to cut" is specified by the context. In general, it means "to divide something by means of a sharp tool". However, the ways in which the object is divided, the kind of tool used and how it is used are determined by the context and specifically by the object that is cut. For every object, our encyclopedic knowledge suggests the tool to be used to cut it and the ways in which it must be cut, lending the sentence a more determinate meaning. In such cases, the context specifies a meaning that the word would not have in another context.

Semantic indeterminacy is a widespread phenomenon, as for instance in these predications of color noted by Recanati (2004):

> *red* car
> *red* grapefruit
> *red* book.

A car can be judged to be red when most parts of its body are red, even though other parts such as wheels, mechanical parts, underside and interior may not be red. For a grapefruit to be described as red, it must be red internally, although its peel may be another color. A book with a red cover may be described as red, although its pages may be another color. In short, the parts of an object that must be red to predicate its redness depend on the object itself and on our encyclopedic knowledge.

3.5 Exophoric Pronouns

As a final example, consider the nature of exophoric pronouns, which refer to objects in the extra-linguistic context. Suppose that Ann and John are dining at a nice restaurant on the sea. The night is beautiful, and the temperature is perfect. They have just been served and have begun to eat. John says:

It's tasty, isn't it?

Clearly, John is referring to the food that has just been served. Suppose, however, that, he had uttered a different sentence:

It's beautiful, isn't it?

It seems likely that John is referring to the restaurant where they are eating or, more generally, to the experience they are sharing. As the situation in which the two sentences are uttered is the same, the difference in the referent of the pronoun "it" must be determined by the predicate of the two sentences. As the predicate "tasty" is usually applied to food, it is plausible to believe that John wishes to refer to the more salient food in that moment: the dish they are tasting. On the other hand, the predicate "beautiful" is usually applied to things that are delightful to look at or, more generally, to pleasant experiences, but not to the flavor of food. The beauty of the restaurant and the setting make reference to the place or the circumstance more probable. Therefore, the predicate can help in determining the referent of the exophoric pronouns.

While it is apparent that an *anaphoric* pronoun acquires its referent from the context, this is far less obvious in the case of *exophoric* pronouns, which usually refer to salient objects in the utterance context. In some cases, however, there are several candidate referents, and the extra-linguistic context will not suffice. In these cases, the predicate of the sentence can assist identification. Because many predicates are applicable only to certain categories of objects, the predicate is likely to provide information about the category to which the referent belongs while excluding other possible candidates. Moreover, as participants in conversation are led to presuppose that their interlocutors speak truly, they are also assumed to predicate true things of the objects to which they refer.[2] If something predicated of a candidate referent

---

[2] Grice (1989) emphasizes that participants presuppose that their interlocutors say true and pertinent things.

of an exophoric pronoun is clearly false (although of the right category), it will probably be discarded, and an alternative candidate will be considered. In John's sentence "It's beautiful, isn't it?", the fact that the restaurant and the panorama are pleasing renders these plausible referents of the pronoun "it".

I conclude that the predicate can play an important role in selecting the referent of an exophoric pronoun because candidates that are from the wrong category or falsely predicated are commonly ruled out. This is another case in which the linguistic context is crucial for the determination of the meaning of a word. Again, the whole determines the meaning of the part rather than the other way around.

4 Conclusion

This paper defends two opposite theses. First, the centrality of the principle of compositionality has been asserted, as without the bottom-up processes of compositionality, we could not utter new sentences and hope to be understood. On the other hand, radical versions of the principle of compositionality cannot be accepted, and some aspects of natural language can be understood only if the context determines the meanings of the elements in a typically top-down process. At this juncture, a question naturally arises: are these two theses contradictory? While in-depth analysis of this problem is beyond the scope of this essay, we can suggest that there is at least one way of interpreting these theses as non-contradictory. To do so, we must understand the relation between words and context as a virtuous circle. It is because a word, literally and conventionally, has a certain meaning that a certain context is selected—that is, that a certain encyclopedic knowledge is mobilized and the recipient's attention is directed to a certain portion of the world. On the other hand, it is the context so activated that permits the meaning of a word to be determined. The mobilized encyclopedic knowledge and the fact that the meaning of a word must be composed with the meaning of another word triggers the process of determination of the two words' meanings. If it is true that, in a sentence, the whole depends on its parts and the parts on the whole, then the sentence is a system that we cannot dissect into its separate parts without losing something essential.

References

Chomsky, N. (1957). *Syntactic structures*. The Hague, Paris: Mouton.
Grice, P. (1989). *Studies in the ways of words*. Cambridge, MA: Harvard University Press.

Partee, B. H., Meulen, A. G. ter, & Wall, R. E. (1990). *Mathematical methods in linguistics*. Dordrecht: Kluwer Academic Publishers.

Recanati, F. (2004). *Literal meaning*. Cambridge: Cambridge University Press.

Searle, J. (1980). The background of meaning. In J. Searle, F. Kiefer & M. Bierwisch (Eds.), *Speech acts theory and pragmatics* (pp. 221–232). Dordrecht: Reidel.

Does Systemics Still Need Theories? Theory-Less Knowledge?

Gianfranco Minati

1 Introduction

Understanding is a difficult matter (Maturana and Varela 1992; Von Foerster 2003). It should answer, for instance, questions such as why? and how? and then find solutions. A brief "definition" would be the ability to provide effective explicit symbolic cognitive representations usable, for instance, to anticipate, avoid, design, modify, and regulate phenomena. This often coincides with the availability of theories and models constituting explicit symbolic approaches as opposed to non-explicit, non-symbolic approaches such as networks, statistical, and sub-symbolic approaches. However, the concept of theory may have a more generalised meaning such as a corpus, a coherent group of tested or verifiable general propositions allowing extrapolations and predictions, and even falsification. Well-known examples are Darwin's Theory of Evolution, the Theory of Relativity, Quantum Field Theory, and Big Bang theory currently the prevailing cosmological explanatory theory for the birth of the universe.

The contrast considered here is between symbolic representations possessing formal properties, e.g., equations, versus representations acquiring properties only when performed with input data, i.e., after processing. For instance, a neural network acquires properties when operating rather than when only formally considered, as against statistics.

Understanding is even considered as the *evolutionary mission of Homo Sapiens Sapiens* (Henshilwood and Marean 2003; McBrearty and Brooks 2000).

G. Minati (✉)
AIRS / Italian Systems Society, Milano, Italy
e-mail: gianfranco.minati@AIRS.it

© Springer Nature Switzerland AG 2019
G. Minati et al. (eds.), *Systemics of Incompleteness and Quasi-Systems*,
Contemporary Systems Thinking,
https://doi.org/10.1007/978-3-030-15277-2_5

Understanding is the way of interacting with, using, and playing with Nature, as a conceptual analogy to eating. In some religious views, e.g., Judaism, the ability to understand, to know, follows ancestral events, such as eating fruit from the tree of knowledge.

One should attempt to clarify the difference with other related concepts such as unintelligible, or incomprehensible expressing with different nuances the, possibly temporary, ineffectiveness of attempts to understand.

One may also consider the contrast between ideal and non-ideal models.

Ideal models are based on strong hypotheses, and not dependent upon initial conditions or boundaries. Ideal approaches assume that it is possible to zip the essential characteristics of phenomena into a set of ideal equations, examples including the well-known Maxwell and thermodynamic equations.

The non-ideal class of models consists of data-driven approaches, i.e., clustering retrospectively by finding emergent correspondences without looking for their respect of pre-established ones. This is the case, for instance, for cellular automata, genetic algorithms, meta-structures (i.e., considering cluster and intra-cluster properties, variance), neural networks, power laws, scale-freeness, and topology (Bunde and Havlin 2012; Estrada 2016; Minati and Licata 2012, 2013, 2015; Ruelle 2008; Schroeder 2009). Non-ideal models can also be based on considering lucky choices which are then studied through computer simulations (Minati and Licata 2015, pp. 50–51).

An example of non-classical understanding, without the availability of theories, is given by approaches introduced with connectionism, first introduced in 1921 by the psychologist Edward Lee Thorndike (1874–1949) to designate a way of representing human and animal behaviour as based on connections between stimuli and responses (Thorndike 1921). This is currently used in the study of brain connectivity. Moreover, the concept is used, for instance, for so-called sub-symbolic computation of which neural networks is a typical example. The process is digitally, symbolically performed by a computer, a classical Turing machine. However, the symbolic computations acquire, in turn, subsequent non-symbolic, i.e., sub-symbolic, properties. For instance, the ability to learn which is typical of neural networks. This allows for the possibility of computational emergence and emergent computation (Licata and Minati 2016).

Other examples are statistical correspondences, network properties, and evidence-based medicine.

2 End of Theory?

This is the title of an article published by Anderson (2008).

The problem relates to a cognitive research strategy to produce knowledge. There are more and more cases of knowledge produced without the search for, or availability of, theories, by using concordances and correspondences in a data deluge often termed Big Data (Calude and Longo 2016), using data-driven approaches with very large databases.

The reason why we discuss this issue here is that General Systems(s) Theory is more and more generalised in Systemics intended as a corpus of a multitude of approaches all based on the concept of system.

The concept of incompleteness (Minati 2016a) and quasiness for Systemics may be dealt with theoretically (Minati and Pessa 2018). However, systemic incompleteness and quasiness may be considered as aspects of non-theoretical, non-symbolic, non-explicit representations of processes typical of Systemics. That is, systemic properties without referring to the theoretical concept of system based on interrelations and interactions among elements constructivistically intended as constitutive. The typical example is the usage of the concept of emergence, with its multiple modelling using different approaches but without a theory of emergence. Can we continue to use partial multiple models, deal with and use emergence without a robust theory?

Are we using an epistemological, phenomenological, empiricist-like approach, looking for high probabilities, and falsification-independent approaches, relegating the search for theories to philosophical studies?

Probably the availability of large quantities of data, computers for simulations, multiple representations, and scale-free approaches allowing for effective models and approaches will have to coexist with the lack of theories. However the problem is to consider whether:

- Models and approaches presuppose, or are converging towards, suitable robust, effective theory(ies);
- Such theories do not necessarily exist;
- The search for such theories is scientifically unnecessary and perhaps even irrelevant.

This is opening the way to a completely data-driven science, where the focus is on the map, where the map is even richer with data and more effective than the territory itself, and experiments are performed on models (Sagiroglu and Sinanc 2013).

3 Multi-Dimensional Knowledge?

Paraphrasing Von Foerster (2003), there is no information, nor anomalies in the environment. If a given phenomenon looks strange, this means that the theoretical framework used to interpret this phenomenon is inappropriate. This cognitive process of reformulation of the model is labelled abduction, and its aim is to "normalize" anomalies.

Charles Sanders Peirce (1839–1914), defines his concept of abduction in the following way:

> Abduction is the process of forming an explanatory hypothesis. It is the only logical operation which introduces any new idea (Peirce 1998).

An issue may relate to different non-equivalent or even incompatible understandings, rather than to possibly converging, with subsequent levels of refining.

The point is that we live in an age where understanding seems to be no longer fully required, focusing on usage of data.

However, the rational mind and science are supposedly devoted to this task, following the inherent nature of Homo Sapiens Sapiens. This is related to the so-called intelligence of matter (Bazaluk 2016), the comprehensibility of the universe (Maxwell 1998), and, probably, to consciousness (Blackmore 2013). However, cognitive representations may be very different and of a different nature in different disciplines. For instance, understanding in biology, chemistry, economics, mathematics, medicine, physics, psychology, quantum science, and sociology may be different culturally, phenomenologically, empirically or theoretically.

This contribution focuses on the legitimacy of the (interdisciplinary?) interchange between these dimensions and on the legitimacy of the specific or cross-usability of approaches lacking or having only poor explicit symbolic understandings. Is the source of such legitimacy cognitive or phenomenological?

What is the nature of such legitimacy? Homologation, effectiveness, and ratification for a unique best approach or for taking part in multiple approaches?

Are we to decide whether the cognitive strategy for knowledge coincides or not with the search for a suitable theory?

4 Complexity: Knowledge to Solve and Knowledge to Manage

There is an important difference between dealing with problems considered as not yet understood, i.e., no theories are available at the moment, but solvable in principle, versus problems which are non-understandable in principle, e.g., considered as having no theoretical representations and, consequently, no available solutions, thus being only manageable, i.e., to orient, start, or vary the phenomena (Minati 2016b).

> The dynamics of complex systems is known only a posteriori and the idea to zip the essential characteristics of change into a set of ideal equations, typically a Lagrangian or Hamiltonian formulation based on general symmetry or conservation principles, is unsuitable. As, for instance, power laws and scale-freeness are clues of complexity,

i.e., the occurrence of processes of emergence and self-organisation, properties of the behaviour of systems selecting from among equivalent possibilities, for instance, respecting the degrees of freedom, may profile complex behaviours. (Minati and Licata 2015, pp. 54–55).

When dealing with an amount of data, typically very large as in the case of Big Data there are various strategies possible:

1. Search for ideal representations, effective to completely represent, simulate, and forecast the process generating such data. In this case real data are considered coincident with solutions of suitable explicit, symbolic formal models, i.e., suitable equations often intended as laws, such as Maxwell's equations, equations of mechanics, thermodynamics or relativity.
2. Apply connectionist approaches, e.g., use networks and neural networks.
3. Simultaneously use multiple approaches, correspondences and correlations, evidence based.

Points 2 and 3 relate to explicit, i.e., symbolic, intractability of large amounts of data, non-zippable into a small number of degrees of freedom (variables), equations, and non-computable (solvable) over a small and finite time period. On the other hand, such intractability may be considered to be related to levels of complexity.

While point 1 focuses on the possibility of finding solutions to formal representations (maps), points 2 and 3 focus on the possibility of using multiple approaches allowing multiple, possibly equivalent, ways to manage (Watzlawick et al. 1974).

A typical example of the second and third cases is given by attempts to influence the properties of emergent collective behaviours. Properties of emergent collective behaviours cannot be explicitly decided, as a problem to be solved, but rather be induced by using suitable approaches such as acting on environmental conditions or introducing perturbations Minati and Pessa (2018).

5 Acting as Blind, but Sensitive to a Large Amount of Detail?

Can the lack of theorisations be considered tout-court as the decay, renunciation of the cognitive strategy to explicitly, i.e., symbolically, understand? Is non-symbolic knowledge a complete lack of knowledge or, rather, a different kind of knowledge more appropriate for the properties of complexity?

There are thus many examples of cases where we deal with a lack of explicit, symbolic understanding which, however, does not affect empirical cross-usages showing the effectiveness of connectionism, evidence-based medicine, homologous variables, incompleteness, lack of a theory of emergence versus its properties, network representations, and uncertainty principles. *More realistically we should consider future contexts mixing theoretically symbolic and non-symbolic approaches allowing soft-theorisations and incomplete-theorisations.*

We may also question whether our environment is cognitively symbolically understandable for evolutionary reasons (Maturana and Varela 1992).

However, as for biodiversity and genetic reserves, some abilities do not correspond linearly to the necessity of dealing with the current world environment and we are probably delegating some abilities to cognitively design evolutionary changes and steps.

6 Conclusion: The Source of Logical Consistency

We conclude this non-terminable issue by mentioning a mysterious property of reality which may be expected to again play a fundamental role in our cognitive adventure.

We refer to the mysterious logical consistence which makes non-demonstrable what instead requires breakthroughs, new incompatible approaches. A typical example is given by Euclid's famous fifth postulate. The content of the postulate seems obvious, but several attempts to demonstrate it failed.

The Jesuit Gerolamo Saccheri (1667–1733), Professor of Mathematics at the University of Pavia, involuntarily opened the way for non-Euclidean geometries when trying to prove the validity of the fifth postulate per absurdum. Considering as absurdum the Euclidean geometry with the negation of the fifth postulate, he developed a series of theorems he expected to be contradictory. In reality, his attempt did not lead to contradictions. What is the source (the reason?) for this consistence?

Such consistence is less surprising in physics when dealing with experiments where, however, scientific experiments are viewed as questions to reality which is supposed to respond by making them happen (the case of ether). However, as there are no answers without questions, on the other hand, events may turn into answers if we abductively invent the proper question of which they can be the answer.

We are facing unsolved problems in Mathematics such as the Millennium Problems listed by the Clay Mathematics Institute (URL). However, we have conceptual contexts for which unanswered questions may be understood as inappropriate, representing unsuitable expectations, rather than unsolved problems. This may be the case, for instance, when looking for regularities in the distribution of prime numbers, while, in reality, such hypothetical regularities would be a disaster for applications such as stochastic generation in simulations and for cryptography.

This mysterious consistence is trying to tell us something, i.e., that we should look for something else?

We may consider this logical consistency as an idealistic property, of a Pythagorean nature? We may rather consider this consistence as having the same nature such as in physics, the differences occurring only when considering cognitive processes of the production of knowledge idealistically, autonomous from Nature, assumed to be of a different nature from Nature. The cognitive strategy is that we produce and do not discover knowledge as one of our evolutionary roles.

- Is the question *"why?"* suitable only for well-defined problems and related to simplified, schematic versions of intractable, complex contexts?
- Does complexity require a new kind of knowledge as a coherent corpus of concordances and correspondences?
- Should we stop to look for theories for emergence or, rather, *use* better the multiple approaches available and the enormous quantities of data collected and generated by maps (models) through simulations?

References

Anderson, C. (2008). The end of theory: The data deluge makes the scientific method obsolete. *Wired Magazine, 16*(7), 16–07.

Bazaluk, O. (2016). *The theory of evolution.* Newcastle: Cambridge Scholars Publishing.

Blackmore, S. (2013). *Consciousness: An introduction.* New York: Routledge.

Bunde, A., & Havlin, S. (2012). *Fractals and disordered systems.* New York: Springer.

Calude, C. S., & Longo, G. (2016). The deluge of spurious correlations in big data. *Foundations of Science,* 1–18. https://doi.org/10.1007/s10699-016-9489-4.

Estrada, E. (2016). *The structure of complex networks: Theory and applications.* Oxford: Oxford University Press.

Henshilwood, C. S., & Marean, C. W. (2003). The origin of modern human behavior: Critique of the models and their test implications. *Current Anthropology, 44,* 627–651.

Licata, I., & Minati, G. (2016). Emergence, computation and the freedom degree loss information principle in complex systems. *Foundations of Science, 21*(3), 1–19. http://link.springer.com/article/10.1007/s10699-016-9503-x.

Maturana, H. R., & Varela, F. J. (1992). *The tree of knowledge: The biological roots of human understanding.* Boston: Shamhala Publications.

Maxwell, N. (1998). *The comprehensibility of the Universe: A new conception of science.* New York: Oxford University Press.

McBrearty, S., & Brooks, A. (2000). The revolution that wasn't: A new interpretation of the origin of modern humans. *Journal of Human Evolution, 39*, 453–563.

Minati, G. (2016a). Knowledge to manage the knowledge society: The concept of theoretical incompleteness. *Systems, 4*(3), 26. http://www.mdpi.com/2079-8954/4/3/26/pdf.

Minati, G. (2016b). General system(s) theory 2.0: A brief outline. In G. Minati, M. R. Abram & E. Pessa (Eds.), *Towards a post-bertalanffy systemics* (pp. 211–219). Cham: Springer.

Minati, G., & Licata, I. (2012). Meta-structural properties in collective behaviours. *International Journal of General Systems, 41*, 289–311.

Minati, G., & Licata, I. (2013). Emergence as mesoscopic coherence. *Systems, 1*(4), 50–65. http://www.mdpi.com/2079-8954/1/4/50.

Minati, G., & Licata, I. (2015). Meta-structures as multidynamics systems approach. Some introductory outlines. *Journal on Systemics, Cybernetics and Informatics, 13*(4), 35–38. http://www.iiisci.org/journal/sci/issue.asp?is=ISS1504.

Minati, G., & Pessa, E. (2018). *From collective beings to quasi-systems.* New York: Springer.

Peirce, C. S. (1998). Harvard lectures on pragmatism. In N. Houser, J. R. Eller, A. C. Lewis, A. De Tienne, C. L. Clark & D. B. Davis (Eds.), *The essential Peirce: Selected philosophical writings, 1893–1913* (Chapters 10–16, pp. 133–241). Bloomington: Indiana University Press.

Ruelle, D. (2008). *Chaotic evolution and attractors.* Cambridge: Cambridge University Press.

Sagiroglu, S., & Sinanc, D. (2013). Big data: A review. In: *2013 International Conference on Collaboration Technologies and Systems (CTS)* (pp. 42–47). IEEE Xplore Digital Library. Available at: https://www.researchgate.net/publication/261456895_Big_data_A_review.

Schroeder, M. (2009). *Fractals, chaos, power laws: Minutes from an infinite paradise.* New York: Dover.

Thorndike, E. K. (1921). *The teacher's world book.* New York: Teachers College.

Von Foerster, H. (2003). *Understanding understanding: Essays on cybernetics and cognition.* New York: Springer.

Watzlawick, P., Weakland, J. H., & Fisch, R. (1974). *Change-Principles of Problem Formation and Problem Resolution.* New York: Norton.

Web Resources

http://www.claymath.org/millennium-problems

Part III
The Concept of Incompleteness and Quasiness in Science and Philosophy

On the Complexity of Baroque Music and Implications on Robotics and Creativity

Andrea Roli

1 Introduction

The complexity of music is a quite elusive property, yet somehow acknowledged among human beings. It is often put in tight relation to *criticality* (Kauffman 1993; Roli et al. 2018)—i.e. the edge of chaos—as a commonly appreciated piece of music is neither totally regular and predictable nor completely erratic and chaotic, but it is rather a mixture of regularity and surprise (Montuori 2003). Indeed, listeners typically feel satisfied with the possibility of predicting the music they are listening to, but they are also captured by some amount of surprise.

We understand that quantifying art might turn out to be a meaningless process unless research hypotheses, procedures and limits of possible conclusions drawn are clearly stated. The attempts to measure the complexity of music, or to compute some quantity of a specific feature of it, abound both with the aim of classifying or comparing music and composing, also automatically. We believe that some considerations on the complexity of music from the perspective of information theory and dynamical systems may be of help to elucidate the limits of measuring music and providing suggestions for the development of more effective—and principled—approaches for attempting such endeavour, as well as for suggesting new ways for understanding the cognitive processes concerning music and for improving artificial creativity systems. In this contribution we mainly focus on baroque music, which we believe is a paradigmatic example. However, our considerations can be

A. Roli (✉)
Department of Computer Science and Engineering, *Alma Mater Studiorum* Università di Bologna, Cesena, Italy
e-mail: andrea.roli@unibo.it

© Springer Nature Switzerland AG 2019
G. Minati et al. (eds.), *Systemics of Incompleteness and Quasi-Systems*,
Contemporary Systems Thinking,
https://doi.org/10.1007/978-3-030-15277-2_6

extended to other music genres[1] and also artificial complex systems, such as robotics.

2 Baroque Music as an Emerging Process

One of the main peculiarities of baroque music is that the actual piece of music is a *process* (Minati 2002) that emerges from the dynamical interaction between two sources of information: the musical score and the performance practice. The music written in the score is just a skeleton of what will be played and the performance practice will guide the performer—and composer at the same time—so as to bring to life the essential information contained in the written notation (Haynes 2007). Performance practice can be seen as a coherent and composite set of constraints and rules that provides the environment in which the performers execute written music. Emerging phenomena in baroque music have already been nicely discussed elsewhere (Pietrocini 2009, 2012), for example in the case of *basso continuo*. Here we would like to illustrate this aspect by means of an example concerning *diminutions*, i.e. ornaments and variations of a musical line (Veilhan 1979). Let us consider the *incipit* ("Grave") of Sonata no. 1 from Op. 5 by Arcangelo Corelli, as it appears in the first edition (see Fig. 1). We can observe two lines: the upper one for the leading voice—the violin—and the lower one for basso continuo (e.g. harpsichord). The melody written in the upper line is just a sketchy indication of the music that has to be played, which has to be executed by adding several kinds of variations. The actual execution is then the result of performers' interpretation, according to the performance practice of that time and of course their musical tastes and skills. A subsequent edition of Op. 5 contains also the diminutions according to Corelli's interpretation, as illustrated in Fig. 2 where we can appreciate the comparison between the original line and the one with diminutions. This is one of the rare documents in which direct examples of performance practice are provided. For most baroque music, these rules and guidelines have to be inferred from indirect sources.

Besides musicological considerations, here we would like to focus on a crucial phenomenon: the emergence of actual music from a set of instructions (the score) and a frame of constraints in which to apply the instructions (the performance practice). This property is not limited to baroque music. Indeed, we envisage a striking analogy between the execution of baroque music and the behaviour of a robot: the actual robot behaviour emerges from the interaction between the robot control program and the environment (including the physical properties of the robot itself). The same control program may give rise to completely differing behaviours depending on the environment or some

---

[1] Whatever this expression may mean.

Fig. 1 Incipit of Sonata no. 1 from Op. 5 by Arcangelo Corelli, as it appears in the first edition (Gasparo Pietra Santa, Rome, 1700)

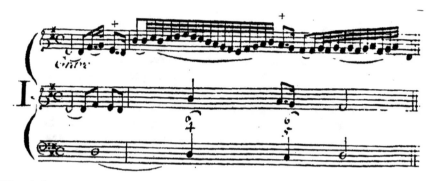

Fig. 2 Incipit of Sonata no. 1 from Op. 5 by Arcangelo Corelli, as it appears in the third edition (Estienne Roger, Amsterdam, ca.1723) with diminutions approved by Corelli himself

physical characteristics of the robot (Pfeifer and Bongard 2007; Pfeifer and Scheier 1999). Therefore, the robot's behaviour cannot be reduced to internal control alone: it is the result of the interaction between the robot and the world (i.e. among neural control system, robot's body and environment). This analogy suggests us that baroque music can be seen as such a kind of emerging phenomenon, as it is the result of the dynamical interaction between a simplified, incomplete *code* (music score) and the *environment* (performance practice). Moreover, note that the control program of a robot is inherently and necessarily incomplete, because it is based on an abstraction of the real world. As a consequence, as for robotic systems, inferences made only on the basis of the musical score (the code) might not be corrected. Examples of this situation are abundant in early music and may involve basso continuo, diminutions, temperaments, instruments ensembles, etc.[2] It is interesting to observe that similar considerations are also beautifully discussed in a famous

---

[2] An exception should maybe be raised for J. S. Bach's music, which is extremely detailed, including diminutions and precise indications for basso continuo. Indeed, Bach's music somehow concludes the times of early music. However, the levels of Bach's music are so many and intertwined, involving also affects and rhetoric (Haynes and Burgess 2016), that we can anyway assert that our considerations hold also in this case.

and inspiring book by Noble (2006), where biological cell's behaviour is described as a phenomenon emerging from the interaction between the genetic code and its environment.

3 Complexity of Baroque Music

Several ways to estimate the complexity of music have been proposed, mainly based on the computation of entropies or other measures from information theory, either directly from an encoding of the score or of a model (e.g. a neural network) trained to perform a target music (Abdallah and Plumbley 2009; Cox 2010; Liou et al. 2015; Manzara et al. 1992). We remark that the aim of this paper is to emphasise a property concerning the complexity of (baroque) music that has been overlooked so far, rather than to provide a survey of approaches for measuring complexity of music. These approaches are certainly valuable and help to investigate some specific aspects of a piece of music and may support the automatic generation of music. We claim that all the methods that take into account the musical sole score might miss some fundamental properties, which come from the dynamical character of music.

From an information-theoretical perspective, in processes such as baroque music the information is conveyed by the intertwined combination of musical text and performance practice: should we be able to retrieve just one of the two, we weren't able to get the whole message. Therefore, if one wants to compute some information measure on a piece of music and, in particular, some complexity measure, the sole music score is likely to provide a rather partial and limited picture. Here we would like to propose a complementary approach, which explicitly takes into account the interplay between performance practice and written music. First of all, a representative sample of executions of a piece of music should be analysed; to this purpose, automatic transcriptions from audio data may be of great help. But also features extracted directly from audio files should be considered. This information has to be complemented by a set of human-defined features (such as knowledge on basso continuo and diminutions practice). To this aim, techniques combining both model construction (e.g. Markov chain) and knowledge representation can be used. Hence, complexity may be estimated by both the constructed models (Crutchfield 1994; Grassberger 1986; Shalizi 2006) and the properties of the constraints composing the performance practice hypothesised. In fact, the set of constraints of performance practice[3] are currently inferred by several sources (such as treatises) and are therefore hypotheses for an environment, rather than true facts. *En pas-*

---

[3] Constraints may be prescriptive but also playing the role of preferences, acting as a bias towards specific choices.

sant, we observe that the reconstruction of a plausible set of constraints for reproducing a target emerging phenomenon is a fundamental issue in systemics and is assuming particular relevance also in robotics and niche construction.

We believe that, although not easy to be implemented, this approach may be of help in cognitive studies focusing on music (Meyer 1957; Rohrmeier and Koelsch 2012).

4 Creativity from Incompleteness

In the previous sections we have discussed what we believe is a crucial property of early music and also some artificial systems, such as robotics. Having put the spotlight onto this emerging phenomenon makes it quite natural to identify a new approach towards creativity, which takes its strength from the very inherent incompleteness of music description.

To the best of our knowledge, current approaches for music creation are generally based on machine learning or statistical techniques, which tune some model parameters through a training phase based on musical examples (Dubnov et al. 2003; Miranda 2002; Montuori 2003). The essence of the process is to extract features that characterise a given kind of music and to combine those features with ingenuity and also a bit of randomness—just to give some flavour of surprise.

A complementary approach consists in turning the incompleteness of both musical score and performance practice into a creative force. We may provide the description of a backbone of a behaviour (e.g. a musical score) and constraints may be generated, for example by means of an evolutionary technique. In addition, a set of elementary constraints may be defined in advance, along with rules for combining them, and an automatic design technique can be used to find a suitable context of performance practice in which the piece of music will be played. Clearly, the question as to what does exactly "suitable" mean is not trivial whatsoever and it has to do with expectation, memory, surprise, aesthetics, cognitive processes and many other aspects of music (Agres et al. 2018; Meyer 1957; Pearce and Wiggins 2006; Wolf 1976).

We conclude this section by observing that this idea is of course directly applicable also to the robotic field. Instead of defining a robot behaviour in terms of precise prescriptive instructions in a programming language, it may learn the constraints to apply to its basic, incomplete behaviour. Contrarily to automatic generation of music, in this case finding a suitable set of constraints is somehow less difficult, as it is possible to define either a task-dependent objective function or a task-agnostic merit factor based on some general quality we require the robot to have and that can be measured by means of information-theoretical measures (e.g. the mutual information between the wheels of the robot). We would like to stress that, like in the case of music, robots may anyway behave differently even if respecting the same set of constraints because of the incompleteness of constraints, possibly showing unexpected behaviours.

5 Coda

In the last resort, it is the job of the musician to create his own interpretation. There is not just one possible interpretation, but many. Freedom of choice, however, must always remain within the limits of the style and practice of the time. (Veilhan 1979)

The aim of this work is to call the attention to a specific property of baroque music—and also other systems—which we think has not been yet considered in its full potential. We believe that this perspective may help researchers and musicians on the one hand to address the issue of estimating the complexity of music in a grounded context and on the other hand to devise new methods for artificial creativity. In our view, this latter objective is particularly relevant when artificial creativity is used as a mean for exploring natural creativity processes and understand why humans are so attracted and influenced by this intangible yet so concrete phenomenon known as music.

The reader should understand that the purpose of this paper was not to provide experimental results, which, we hope, may come in the future. First of all, we plan to define a method that helps providing some information on the complexity of a piece of music, not with the goal of reducing music to numbers, but to provide quantitative support to qualitative conjectures. In parallel, we plan to explore the way of artificial creativity by incompleteness, both in music and robotics.

Acknowledgements This work would not have been possible without insightful discussions about music and its engaging complexity the author had with Francesca Camagni, Michele Pasotti and Gabriele Raspanti. Thanks also to Massimo Franceschet, whose works show how science and art can be beautifully combined.

References

Abdallah, S., & Plumbley, M. (2009). Information dynamics: Patterns of expectation and surprise in the perception of music. *Connection Science, 21*(2–3), 89–117.

Agres, K., Abdallah, S., & Pearce, M. (2018). Information-theoretic properties of auditory sequences dynamically influence expectation and memory. *Cognitive Science, 42*(1), 43–76.

Cox, G. (2010). On the relationship between entropy and meaning in music: An exploration with recurrent neural networks. *Proceedings of the Annual Meeting of the Cognitive Science Society, 32*, 429–434.

Crutchfield, J. (1994). The calculi of emergence: Computation, dynamics, and induction. *Physica D, 75*, 11–54.

Dubnov, S., Assayag, G., Lartillot, O., & Bejerano, G. (2003). Using machine-learning methods for musical style modeling. *IEEE Computer, 36* (10), 73–80.

Grassberger, P. (1986). Toward a quantitative theory of self-generated complexity. *International Journal of Theoretical Physics, 25*(9), 907–938.

Haynes, B. (2007). *The end of early music: A period performer's history of music for the twenty-first century.* Oxford: Oxford University Press.

Haynes, B., & Burgess, G. (2016). *The pathetick musician.* New York, NY: Oxford University Press.

Kauffman, S. A. (1993). *The origins of order: Self-organization and selection in evolution.* New York/Oxford: Oxford University Press.

Liou, C. Y., Simak, A., & Cheng, W. C. (2015). Complexity analysis of music. *Complexity, 21*(S1), 263–268.

Manzara, L., Witten, I., & James, M. (1992). On the entropy of music: An experiment with Bach chorale melodies. *Leonardo Music Journal, 2*(1), 81–88.

Meyer, L. (1957). Meaning in music and information theory. *The Journal of Aesthetics and Art Criticism, 15*(4), 412–424.

Minati, G. (2002). Music and systems architecture. In: *Proceedings of the 5th European Systems Science Congress*, October 16–19, Hersonissos, Iraklion, Crete. http://www.afscet.asso.fr/resSystemica/Crete02/Minati.pdf.

Miranda, E. (2002). *Composing music with computers.* Burlington: Elsevier.

Montuori, A. (2003). The complexity of improvisation and the improvisation of complexity: Social science, art and creativity. *Human Relations, 56*(2), 237–255.

Noble, D. (2006). *The music of life.* Oxford: Oxford University Press.

Pearce, M., & Wiggins, G. (2006). Expectation in melody: The influence of context and learning. *Music Perception: An Interdisciplinary Journal, 23*(5), 377–405.

Pfeifer, R., & Bongard, J. (2007). *How the body shapes the way we think.* Cambridge, MA: The MIT Press.

Pfeifer, R., & Scheier, C. (1999). *Understanding intelligence.* Cambridge, MA: The MIT Press.

Pietrocini, E. (2009). Music: Creativity and structure transitions. In G. Minati, M. R. Abram & E. Pessa (Eds.), *Processes of emergence of systems and systemic properties. Towards a general theory of emergence* (pp. 723–744). Singapore: World Scientific.

Pietrocini, E. (2012). Music: Emergence and metastructural properties in the practice of the thorough bass. In G. Minati, M. R. Abram & E. Pessa (Eds.), *Methods, models, simulations and approaches. Towards a general theory of change* (pp. 633–646). Singapore: World Scientific.

Rohrmeier, M., & Koelsch, S. (2012). Predictive information processing in music cognition. A critical review. *International Journal of Psychophysiology, 83*, 164–175.

Roli, A., Villani, M., Filisetti, A., & Serra, R. (2018). Dynamical criticality: Overview and open questions. *Journal of Systems Science and Complexity, 13*, 647–663. A preliminary version of the paper is available as arXiv:1512.05259v2.

Shalizi, C. R. (2006). Methods and techniques of complex systems science: An overview. ArXiv:nlin/0307015.

Veilhan, J. (1979). *The rules of musical interpretation in the baroque era.* Paris: Leduc.

Wolf, T. (1976). A cognitive model of musical sight-reading. *Journal of Psycholinguistic Research, 5*(2), 143–171.

Music: Creativity and New Technologies. A Systemic Approach Towards Multimedia Project and Sound Design

Emanuela Pietrocini and Maurizio Lopa

1 Mousikè Téchne

> Deities do not mingle with men: mediation occurs through demons. Those who have competence thereof are demonic; those who are experts in technai and craft are merely mechanics ($\beta\acute{\alpha}\nu\alpha\upsilon\sigma\sigma\varsigma$).[1]

> Production is any activity carried out by technai: all craftsmen, to be rigorous, deserve to be called poets. Commonly, though, what is called poetry is only a small part of such activities, the one pertaining mousiké and metre.[2]

Despite the invention of recording techniques and the relatively recent advent of digital are perceived as the deepest revolutions in the course of music history, in fact technology has constantly accompanied and influenced musical practice, moulding its creative and commercial dynamics. Technological advancements which had been developed for processes of different nature, have found their own practical application in the music field. For example, the expertise acquired in the field of mining and of the processing of metals have made it possible to use metal strings and to build wind instruments of modern design; movable metal type printing, inaugurated by Gutenberg with his 1455 Bible is replicated in the music field by Ottaviano Petrucci first and then by Pierre Attaignant, thus making it possible to produce, in the years around 1520, low cost printed music accessible to an increasingly wider audience.

E. Pietrocini (✉) · M. Lopa
Schola Palatina, Early Music Department, La Vertuosa Compagnia de' Musici di Roma, ISIA of Pescara, Multimedia Design, Pescara, Italy

---

[1] Plato, *Symposium*, ed. by G. Colli, (1979), 203a.

[2] Plato, cit., 205c.

© Springer Nature Switzerland AG 2019

G. Minati et al. (eds.), *Systemics of Incompleteness and Quasi-Systems*, Contemporary Systems Thinking, https://doi.org/10.1007/978-3-030-15277-2_7

117

For centuries, Music has continued its journey expressing itself with new forms, new styles, new composition techniques and, for a long period of time, technology has remained—along with printers, instrument makers, costume designers, stage technicians—in a more or less passive way, at the service of music.

Yet something new starts moving with the advent of modern science. In 1600 William Gilbert, continuing the work of Italian physicist Girolamo Cardano, publishes the treatise *"De Magnete, Magneticisque Corporibus"* in which we find the term "electricus" for the first time. From that moment on everything happens at an increasingly fast pace: in 1660 Otto von Guericke invents what can be considered the first electric generator. In 1729 Stephen Gray studies the conductivity of bodies; the terms "conductor" and "isolating" are introduced by Jean Théophile Desaguiliers in 1740. In 1746 the first electric condenser sees the light.

In 1830 Michael Faraday builds the first electromagnetic generator of electric current; simultaneously Samuel Finley Breese Morse, utilizing the passage of electricity in a conducting thread goes so far as to invent the telegraph; from 1833 it is a continuous flourishing of researches, experiments and enhancements which will lead to the creation of the telephone: the voice can now be transmitted thanks to the electric current. Technology, stimulated by scientific progress, has reached communication; the βάναυσος (banausos), Plato's craftsman, has become a demonic man.

During the decade following the presentation of the first telephone devices created by Antonio Meucci and Alexander Graham Bell, we find Louis Aimé Augustin Le Prince, Thomas Alva Edison and the Lumière brothers, connected to the invention and development of the first systems of recording and projection of images in motion. The cylinder phonograph, a system of recording and reproduction of sound that uses wax cylinders was patented by Edison in 1878 (Fig. 1).

2 Cinema

It was evident from the very beginning that the natural completion to films was a musical commentary; initially it was conferred to the improvisational expertise of a pianist who, eyes on the screen, would underline the actions and expressions with music. But technology was already able to record and reproduce—despite its qualitative limitations—both sounds and images. Soon we moved to the reproduction, during the films, of musical pieces previously recorded on wax cylinders. Only one big obstacle was left: the synchronization between the projected image and the sound composed of music, dialogues and noises.

Fig. 1 Edison wax cylinder record with cardboard box (left)

The end of the era of silent films coincides with the release of the film "*The Jazz Singer*" produced by Warner Bros. in 1927. For the very first time an actor would speak from the screen and the audience could hear his voice synchronized with the images. The synchronization technique used was *sound-on-disk* and the equipment used, the *Vitaphone* (Miceli 1982), allowed the mechanical coupling of a wax cylinder gramophone and the film projector (Fig. 2).

It was a technique that, although innovative, presented quite a number of complications: being a system that implied the simultaneous use of two different devices, one for the images and the other for the sound, the precision of the synchronization was partly designated to the expertise of the projection technician. Moreover, the wax cylinders of gramophones would deteriorate very rapidly, thus making it necessary to replace them only after a few dozen reproductions (Liebman 2003).

Because of the high costs of production and of the initial lower quality of the audio, a new technology was trying to make its way with difficulty: the so-called *sound-on-film* which, taking inspiration from the studies of Alexander Graham Bell around 1880 on the possibility of transforming the sound impulses into luminous impulses, allowed to print the audio track directly on the side of the celluloid containing the images. Of great importance in this sector are the researches by Ernst W. Rühmer, who, in 1901 presents in Germany a method by which one can fix the sound information on the sensitive emulsion

Fig. 2 The Vitaphone projection system

of the celluloid thanks to a ray of light coming from an arch lamp connected, through a modulator, with a microphone. On this line, researches continued during the following decade not only in Germany but also in Britain and the United States, where in 1925 Fox purchased the exclusive for the *Tri-ergon* system (Miceli 1982, p. 171).

As a matter of fact, the introduction of sound brought to surface a series of problems that technicians, directors and actors found themselves to solve once again with the help of technology. The first systems of recording would not allow to intervene on sound material with editing operations. Microphones had to be on the scene and record the actors' dialogues with a technique similar to the modern *direct sound*. But while in our days this can rely on directional microphones and in some cases on radio microphones to catch the actors' dialogues with a total precision, free from interferences, the first microphones would record also the noise of the not so silent movie cameras which therefore had to be placed in suitable soundproof areas, to the detriment of mobility and freedom of take. The actors themselves were forced into unnatural movements and positions in order to have their voices recorded correctly. Only around 1930 there appeared the first movie cameras fitted with *blimp*, a soundproof shell which dramatically decreased the noise; in 1931 the first directional microphones produced by Western Electric and RCA were finally introduced.

3 The Film Boom

The effects on the audience of the new union of sound and image were, to say the least, explosive. The year following the launch of the film "*The jazz singer*", announced by the press as the first Vitaphone creation, Warner Bros. profits soared from two million to about fourteen million dollars (Gomery 1985); in order to compete for such a promising market, the first film distribution companies were created; United Artists, MGM, Columbia, Paramount, formally establishing the birth of the industry of entertainment and of mass communication.

If on the one hand the introduction of sound had given wings to the film industry, on the other the merciless market laws imposed dramatic changes. The musicians who improvised musical commentaries in the era of silent films and that made each show slightly different from the other, were destined to disappear (according to calculations, in the very first years of the advent of sound films, over 20,000 musicians were left without a job). Dozens of film stars saw their careers destroyed because of their unpleasant voices or of their poor diction (Oderman 2000, p. 188).

In 1929 the American Federation of Musicians openly declared its opposition to the widespread use of recorded music to the detriment of live performances with commercial. This is how, on a page of the Pittsburgh Press, the image of a can with the label reading "Canned Music/Big Noise Brand/Guaranteed to Produce No Intellectual or Emotional Reaction Whatever" appeared (Fig. 3). And the following text:

> This is the case of Art vs. Mechanical Music in theatres. The defendant stands accused in front of the American people of attempted corruption of musical appreciation and discouragement of musical education. Theatres in many cities are offering synchronized mechanical music as a substitute for Real Music. If the theatre-going public accepts this vitiation of its entertainment program a deplorable decline in the

Fig. 3 Canned music on trial

Art of Music is inevitable. Musical authorities know that the soul of the Art is lost
in mechanization. It cannot be otherwise because the quality of music is dependent
on the mood of the artist, upon the human contact, without which the essence of
intellectual stimulation and emotional rapture is lost [...].[3]

On the other hand, the new technologies imposed the creation of more and
more specialised professionals which were increasingly far from the world of
those improvised technicians or visionary inventors and dreamers which had
characterised the pioneering era of film-making.

---

[3] Advertisement published by American Federation of Musicians, Pittsburgh Press, 1929.
Digital Collection Ad*Access. Duke University Libraries. (Retrieved December 9, 2009).

4 Media and Music

We gestate in Sound, and are born into Sight
Cinema gestated in Sight, and was born into Sound.[4]

First performed live, then reproduced with wax cylinders or records, music initially plays a rather complementary role in the newborn world of multimedia. It has its own language, its own syntax and can develop, drive and create emotions in complete freedom: it was film-making which needed music, not vice versa. When music offered its service to film-making, thus becoming part of a multimedia product, it simply had to "dress" the actions and the film situations underlining and amplifying their mood according to a simple rule: happy scene/happy music, sad scene/sad music with already known elements which, in a sense, were ready for use.

The introduction of the symphonic poem by Richard Strauss "*Also sprach Zarathustra*", composed in 1896, is wonderfully appropriate to emotionally reinforce one of the most famous scenes of the film "*2001: A Space Odyssey*" produced 72 years later, in 1968, by Stanley Kubrick, thus creating a fantastic puzzle which uses, in our days, music from the previous century to give strength to a scene that takes place in the near future. From a systemic point of view, this highlights how the interaction established among self-sufficient systems, structured and validated with others which are still being structured and of little defined nature, still mainly unstable and with varying properties indicate a strong dynamicity, capable to provide the observers with extremely new patterns which allow to detect emergency at high levels of complexity and to acknowledge coherence among shared properties which, other than characterizing the birth of a new entity (Multimedia), are transferred permanently to each one of the interacting systems.

5 Multimedia Sound Design

Film-making as a representation of reality and film-making as a new means that makes it possible to give shape to the most daring imagination; according to Siegfried Kracauer, two peculiar tendencies of the experimented language of photography—the realistic and the formative—can be highlighted immediately and precociously. The prototypes are, on one hand, the Lumière brothers, strictly realistic; on the other, Georges Méliès, surreal and creative inventor of the first special effects (Kracauer 1960, p. 30). This is how Félix Mesguich, considered as one of the first professional cameramen, describes the work of the Lumière brothers, for whom he worked frequently:

---

[4] Walter Murch, from the foreword to Chion (1990, p. VII).

> As I see it, the Lumiere Brothers had established the true domain of the cinema in the right manner. The novel, the theater, suffice for the study of the human heart. The cinema is the dynamism of life, of nature and its manifestations, of the crowd and its eddies. All that asserts itself through movement depends on it. Its lens opens on the world.[5]

If it is true that music is suitable for an emotional characterisation of situations and characters, a realistic representation of the world cannot exclude noises: although the term "Sound Design" was still not in use in those times, in all theatre, radio and film productions a number of specialists in the creation of sounds and noises—mostly vocal or mechanical—were already being employed.

The description of the sound effects created for the performance—in 1926 at the Garrick Theatre in London—of the comedy *"The Ghost Train"* written in 1923 by Arnold Ridley is particularly enlightening: a group of travellers who get stuck overnight in a small railway station in the countryside, lives moments of terror because of the passage of a ghost train.

> The train effect in *"The Ghost Train"* at the Garrik Theatre in 1926 was produced with the following apparatus: a garden roller pushed over slats of wood nailed to the stage at regular intervals, to represent the train passing over the rail joints; three cylinders of compressed air, one to blow the whistle, the second for an uninterrupted steam effect, and the third used with a tin, covering and uncovering the jet rhytmically, for the exhaust; a large tank; a large thunder-sheet; a thick, oval thunder-sheet and mallet; a whistle, for distant whistle effect; a side-drum and a small padded mallet; a bass drum; some heavy chains; and sand-paper for distant puffing effect. Seven men were employed to work it all.[6]

Transferred to the world of film-making, the whole of the activities related to the production and recording of sounds and noises is called "Foley", from the name of sound-effects artist Jack Donovan Foley, who started working at the Universal Studios in 1914. Nowadays, although the foley's activity still keeps many of the handicraft aspects that characterised its beginnings, the modern sound designer can rely on technical means which have enormously widened its possibility: to the "organic" sounds of pioneers are added the "synthetic" ones made possible by a new generation of synthesizers available and accessible to any small recording studios, not to mention the possibility of electronically modifying the natural sounds.

The ease with which today one can separate the sound from its source and re-associate it to a new image, thanks to the availability of increasingly performing computers and dedicated software, is accompanied by an ever growing attention to the synchretic aspect of sound and makes it possible that the one and the other actually come into a dynamic relationship. Walter Murch writes:

---

[5] Hauser (1951).

[6] Napier (1936, pp. 39–40).

The most successful sounds seem not only to alter what the audience sees but to go further and trigger a kind of *conceptual resonance* between image and sound: the sound makes us see the image differently, and this new images makes us hear the sound differently, which in turns makes us see something else in the image, which makes us hear different things in the sound, and so on.[7]

It is what Murch defines *synchresis*:

the spontaneous and irresistible mental fusion, completely free of any logic, that happens between a sound and a visual when these occur at exactly the same time.[8]

6 Multimedia and Complexity

From a linguistic point of view, the term multimedia can be defined as one of the most common Latinisms of English mediation related to information and communication technologies. Used both as a noun and as an adjective, it refers to the use of multiple means to represent and convey information, contents, expressive elements.

In fact, the term multimedia indicates more a process rather than an object or a quality, since the fields and the areas of application of what we define as multimedia products are numerous. First of all, as already highlighted in the previous paragraphs, multimedia can be referred to any artistic/expressive form that embraces a number of different representative modalities[9]: theatre, cinema, opera, television, radio, live performances are born as ante *litteram* multimedia products.

Moreover, as previously underlined, the evolution of forms and languages, the new expressive and communication requests connected to social and economic changes, have determined a stunning development of dedicated technologies, so much that today, the meaning—in the field of information technology—of the term multimedia refers to a particular interactive electronic title, with the possibility of non sequential navigation which avails itself of different modalities of communication: music, text, images, animations, video clips, etc.[10]

Multimedia is also in the industrial and economic-business field, where we have companies operating in more fields of communication and at different levels, from production to distribution, to network management, to editions.

How can we therefore generalise the definition of multimedia product? The question is relevant to the goal of this essay because if we think about the phenomenological aspect in the multiplicity and variety of its manifestations we will have to limit ourselves to the mere classification, in acknowledging the

---

[7] Walter Murch, from the foreword to Chion (1990, p. XXII).

[8] Walter Murch, from the foreword to Chion (1990, pp. XVIII/XIX).

[9] Trotta (2002, p. 182).

[10] Trotta (2002, p. 182).

pervasive-ness of the multimedia object in all the current forms of communication, including the educational technologies and scientific information. What we believe can contribute to draw a unitary vision—as we already hinted at—is considering the multimedia product as an entity generated by a dynamic process of relations and networks of relations among systems at different levels of complexity.

In fact, as highlighted in the previous chapters, the establishment of connections and interactions has followed, in time, an apparently uneven path, not informed by relationships of causality: the creation of new patterns, instruments, forms and languages seems to be the result of local requests, due to technical, communication or market-related needs, to the industrial development, to social changes. What is actual is that the discovery and the use of those same technologies occurred in different parts of the world, through different or similar paths, simultaneously or after a few years, without having any common line of research and very often in the absence of any shared scientific data and resources. If observed globally, the process actually describes a convergent pace, with a very strong acceleration towards generalized patterns. This is evident in the emerging collective behaviours, which are established as a consequence of the interaction without any explicit project of the interacting agents (Minati and Pessa 2006). Multimedia is nowadays an indicative element of collectivity; it identifies the system of communication of components, their action on the local environment and the collective intelligence.

In this sense, as of today multimedia communication can be represented as a topological model of network composed of basic elements (knots) connected by means of transmission and information; it is actually the topological structure which determines the possible compatible behavioural scenarios (Pessa 2002, 2013). In the same way, it is possible to suppose the pertinence to a pattern of description according to which multimedia is represented as an emergent meta-system.

7 Interactivity and Creativity

The logical openness which characterises the topological pattern allows the observer to detect emergency in the continuous structural change of multimedia. The search for new tools, forms and contents moves parallel to technological innovation, at times using it, other times inspiring its development. This is how in the past few decades the interaction between user and multimedia product has become the centre of research, thus turning the same concepts of use and communication upside down. This is how interactive multimedia systems were born, which, in a wide variety of application domains and since they are based on a substratum of technical, scientific and humanistic knowledge, use the combined action of multiple media in order to create an information technology application designed for the user and around the user in which the interaction with the same user becomes the main aspect and modality through which the goals of the application are reached (Camurri and Volpe 2004, p. 2).

The intention in this context is not that of penetrating the details of this eminent development, but to draw the attention to the characteristics which, in our view, and at this evolutionary stage, denote the meta-stability of the system, its continuous becoming. The interactive component brings the function of the observer to a peculiar plane: the survey of emergencies, therefore the condition of existence and acknowledgment of these according to a pattern (Minati 2010, 2013) occurs simultaneously with their own generation and may start a process of irreversible change of the environmental conditions, of the same object, of its goals.

One example among many is represented by the activity of the *Makers*: real non professional "digital craftsmen", they design and self-build digital devices, interfaces, applications, electronic tools and robotic creations according to non conventional techniques and patterns, re-using materials, re-modulating functions and destinations, operating in a cooperative way and sharing resources. The *Makers* are a true cultural movement proposing a new model of development of technological innovation and of economy which focuses on the creative *vis*, on the production of low cost technologies, recycling, sustainability. It is not by accident, we believe, that the Maker phenomenon started spreading, growing and developing from 2008 to 2009, in conditions of deep political and economic crisis.

It is also true, however, that the multimedia product can be considered interactive already from the advent of videogame: the Second World Was had just come to an end and the first device, designed by Thomas T. Goldsmith Jr. and Estle Ray Mann in 1947, simulated the launch of a missile towards a target, thus taking inspiration from military radar screens. This is quite a peculiarity: the step determining a real change in state is originated, in a period of crisis, from entertainment, from a game, from the need of a professor, the physician William Higinbotham, who in 1958 tried to motivate and entertain his pupils by adopting a system that, through the intelligent use of the oscilloscope, made them participate in an interactive way to his lessons, by simulating the physical laws present in a tennis match: this is how "Tennis for Two" was born (Nibali 2014).

8 Quasity and Multimedia Design

In terms of creativity, the definition of intuitive emergency (Crutchfield 1994), although unrefined and incomplete, still keeps its special charm and poetic imprint; in a way it evokes with greater effectiveness that particular condition in which, facing the unpredicted results of a process of change in which coherence can be identified, what emerges is wonder, surprise and astonishment (Licata 2013). Like of any work of art, the intuition of beauty is the acknowledgement of coherence (Pietrocini 2013) accompanied to a state of undetermined and anxious perception, nearly of disquiet. In a sense, it is the perceptive call to uncertainty, the intrinsic awareness of the transient nature of what is observed, the expectancy of inevitable events which may lead to the collapse of the actual form in order to embrace and set up new possible stabilities. Multimedia seems actually to amount to a Quasi-System (Minati 2013). Designing multimedia, under this aspect, is first of all an intuitive process; it is the ability to understand and manage the incomplete as a resource.

The qualities of the multimedia designer, as they are illustrated in professional profiles, delineate an expertise based on "multi-faceted talent": communication and people skills in order to understand and interpret the needs of clients, as well as to effectively coordinate the team of specialists (graphic designers, technicians, programmers, developers etc.); economic and marketing competences, knowledge of the main operating systems and of the fundamental software and hardware elements; project-related abilities including the creation of prototypes, simulations and evaluations, even a good knowledge of the English language (business English). In the end, but not always, what is sought is "to have a creative tendency". Probably, this is one of the reasons why contemporary production abounds with stereotypes: commercials, film-making, websites, videogames, music and an incredible quantity of *app* are so undifferentiated that they no longer kindle any attention, thus losing all effectiveness.

In the presumption to interpret correctly the needs of the audience, often one overrates—even excessively—the statistical certainties and conventions. "Louder, better": such exclamation causes roars of laughter among the students of a course in Sound Design, but it identifies what is universally applied to induce the audience to draw all their attention on sound or music: a war of decibel in film theaters, at concerts and gigs, in public premises where the "music noise" in the background has reached intolerable levels and prevents any form of verbal communication, whilst remaining mainly ignored in most cases.

Obviously all this has nothing to do with creativity; we are talking simply about the unaware and ordinary use of tools which have extraordinary potentials.

9 Conclusions

The evolutionary path of multimedia, as illustrated in particular in the first chapter of this essay, describes a continuous oscillation between technical and expressive needs, between handicraft products and works of art, between industry and cultural fashion. Likely, what is evident is the permeability of contents, information and strategies which have mutually influenced and characterised opposite positions and fields in a transverse way.

The opposition between positions that are openly pragmatic and aspirations of some aesthetic dignity which come to surface in the development process, describes from a systemic point of view a particular aspect of the process of generation of coherences between multiple and overlapping systems: what may happen is in fact the coexistence of processes and countable outcomes, mainly of finalistic and probabilistic nature, as well as non-countable ones, connected to uncertainty. From this point of view, the incoherence detected locally would represent the start of new emergency processes and, because of these, the tendency to establish new coherences.

The multimedia Creative is a designer of systemic coherence, is capable of managing the incomplete as a resource and to represent the *Un-Finished* in a finished product. He/she is capable of building an image or a sound with the suggestion of another image or sound, or to evoke an image using a sound, as Bruno Munari affirms,

> in order not to reveal itself at once, letting doubt emerge in the mind of the audience, so that each member of the audience may form their own personal image.[11]

In aiming at this, one must give up the certainty of a universally positive outcome.

It may be said that not necessarily must the multimedia Creative produce works of art; on the contrary, in the majority of cases he/she will have to contend with the needs of productivity and of the market, still guaranteeing maximum effectiveness and efficiency.

We are persuaded that one position does not exclude the other: there are plenty of examples of ingenious intuitions, gathered by accident and without a precise intent but developed with artistic intelligence, which have led to enormous developments and likewise profits.

The multimedia audience is now more than ever an active part of the work; it is an interacting agent of the product, it identifies the functions of the systemic observer and as such possesses cognitive patterns to acknowledge coherence. It makes its own choices, and in time we wish that it will design paths that are increasingly less aligned to predictions.

References

Chion, M. (1990). *Audio-vision, sound on screen.* New York: Columbia University Press.

Crutchfield, J.P. (1994) *The calculi of emergence: Computation, dynamics and induction, Physica D, Santa Fe Institute* (pp. 11–54).

---

[11] Munari (1997, p. 71).

Gomery, D. (1985). The coming of sound: Technological change in the American film industry. In A. Utterson (Ed.), (2005), *Technology and culture – The film reader* (pp. 53–67). Oxford/New York: Routledge/Taylor & Francis.

Hauser, A. (1951). *The social history of art* (Vol. II). London: Routledge & Kegan Paul.

Kracauer, S. (1960). *Theory of film: The redemption of physical reality* (p. 30). New York: Oxford University Press.

Licata, I. (2013). Incertezza. Un approccio sistemico. In L. Urbani Ulivi (Ed.), *Strutture di Mondo, il Pensiero Sistemico come Specchio di una Realtà Complessa* (Vol. 2, pp. 35–71). Bologna: Il Mulino.

Liebman, R. (2003). *Vitaphone films: A catalogue of the features and shorts.* Jefferson: McFarland & Company.

Miceli, S. (1982). *La musica nel film, arte e artigianato.* Firenze: La Nuova Italia.

Minati, G. (2010). Sistemi: Origini, ricerca e prospettive. In L. Urbani Ulivi (Ed.), *Strutture di Mondo, il Pensiero Sistemico come Specchio di una Realtà Complessa* (Vol. 1, pp. 15–46). Bologna: Il Mulino.

Minati, G. (2013). Note di sintesi: novità, contributi, prospettive di ricerca dell'approccio sistemico. In L. Urbani Ulivi (Ed.), *Strutture di Mondo, il Pensiero Sistemico come Specchio di una Realtà Complessa* (Vol. 2, pp. 315–336). Bologna: Il Mulino.

Minati, G., & Pessa, E. (2006). *Collective beings.* New York: Springer.

Munari, B. (1997). *Arte come mestiere.* Roma, Bari: Laterza.

Napier, F. (1936). *Noises off: A handbook of sound effects.* London: Frederick Muller Ltd.

Oderman, S. (2000). *Lillian Gish: A life on stage and screen.* Jefferson, NC: McFarland & Company.

Pessa, E. (2002). What is emergence? In G. Minati & E. Pessa (Eds.), *Emergence in complex cognitive, social and biological systems* (pp. 379–382). New York: Kluwer Academic/Plenum Publishers.

Pessa, E. (2013). Emergenza, metastrutture e sistemi gerarchici: Verso una nuova teoria generale dei sistemi. In L. Urbani Ulivi (Ed.), *Strutture di Mondo, il Pensiero Sistemico come Specchio di una Realtà Complessa* (Vol. 2, pp. 73–88). Bologna: Il Mulino.

Pietrocini, E. (2013). Musica dell'Emergenza. Prospettive di ricerca sistemica in musica e in musicologia. In L. Urbani Ulivi (Ed.), *Strutture di Mondo, il Pensiero Sistemico come Specchio di una Realtà Complessa* (Vol. 2, pp. 289–313). Bologna: Il Mulino.

Plato. *Symposium*, Edited by G. Colli (1979). Milano: Adelphi.

Trotta, M. (2002). *La pubblicità.* Napoli: Ellissi.

Web Resources

Camurri, A., & Volpe, G. (2004). *Sistemi multimediali interattivi* (p. 2). ftp:// www.infomus.org/pub/PPM/Publications/AMI-CamurriVolpe.pdf

Nibali, S. (2014). *Storia ed Evoluzione dei Videogiochi Dai Coin-Op ai Social-Games* (p. 7). https://www.dmi.unict.it/~faro/appunti/ informatica/Videogames.pdf

Architecture and Systemics: Performance Revisited

Carlotta Fontana

1 The Industrialization Rule: All Men Have the Same Function

When industrialization took command (Giedion 1948) performance slowly became a word for architects, too. The book by Sigfried Giedion, a landmark in twentieth century architectural literature, had an even more significant sub-title: "a contribution to anonymous history". Giedion examines the effects of mechanization in everyday life, tracing the outline of a social history of technology which, at the time, represented a critical breakthrough, not always understood and even resented in a number of academic circles. Claiming that in the University

> chairs of anonymous history ought to be created,

and blaming the destruction of documents about the early stages of industrialization

> (Later periods will not understand these act of destruction, this murder of history),

Giedion wanted to show how inventions, mass-production and the work of ordinary people in the industrial era

> are continually shaping and re-shaping the patterns of life

in an unprecedented way, at every possible level. He also wanted to open a research field to find an answer to the question:

> what does mechanization mean to man?

C. Fontana (✉)
Dipartimento di Architettura e Studi Urbani (DAStU), Politecnico di Milano, Milano, Italy
e-mail: carlotta.fontana@polimi.it

© Springer Nature Switzerland AG 2019
G. Minati et al. (eds.), *Systemics of Incompleteness and Quasi-Systems*,
Contemporary Systems Thinking,
https://doi.org/10.1007/978-3-030-15277-2_8

in order to investigate the dangers of a way of life in which control by humans over products becomes increasingly more difficult, people become increasingly dependent upon production and, in general,

> man is overpowered by means.

Marking a significant distance from his much-praised previous work (Giedion 1941), *Mechanization takes command*, published shortly after the apocalypse of WW2, recalls an analogy between mass-production and mass-destruction, and the horrors of organizational efficiency applied to extermination.

> Future generations will perhaps designate this period as one of mechanized barbarism, the most repulsive barbarism of all.

Thus, while claiming a well balanced attitude toward the historical condition of "mechanization", Giedion questions the unabashed optimism about the idea of progress: after WW2,

> men have become frightened by progress, changed from a hope to a menace [...] before our eyes our cities have swollen into amorphous agglomerations. Their traffic has become chaotic, and so has production.

Giedion aims at defining mechanization's place in history, in society and in culture, while rejecting the mechanistic conception of the world. Such conception, he argues, has been shattered in every sphere—from physics to biology, psychology and art. A systemic, *holistic* way of conceptualizing

> domains having to do with the human organism

is far more promising; the book ends with a wonderfully forward-looking list of "new balances" required: between individual and community, the world as a whole and local issues, the spheres of knowledge, and

> between the human organism [...], its organic environment and its artificial surroundings.[1]

In fact, for centuries before industrialization, Architecture had represented a high-rank applied art, quite often supported by highly refined formal prescriptions, always supported in the construction phase by robust technical knowledge improved by experience along centuries. The practice of Architecture was destined to major buildings, promoted by public interest and wealthy clients, representing multiple social, civic, religious values. The main part of the built environment grew up in layers over time—without architects nor engineers[2] - taking its shape according to the geo-climatic peculiarities

---

[1] All previous quotations from Giedion (1948).

[2] As Christopher Alexander wrote:

> Architects are responsible for no more than perhaps 5 percent of all the buildings in the world. Most buildings [...] which give the world its form [...] come from the work of thousands of housewives, the officials in the building department, local bankers, carpenters, public works departments, gardeners, painters, city councils, families [...]. (Alexander 1966).

of the place and to the local resources, activities, customs, technical skills. At all scales, the construction process was slow, the means and materials mostly local, the knowledge and techniques improved over time by trial and error and handed down by tradition, through apprenticeship.

Industrialization, speeding up all human activities, undermined the foundation of architectural culture and knowledge. At the end of nineteenth century, massive urbanization forced the architectural culture to face unusual problems, under different aspects. Of course, an important and much-debated question was how to express the new aesthetic and symbolic values of the industrial age. But the big issue, the truly unusual problem, was how to design large quantities of low-cost housing of acceptable quality, new services and urban equipment to meet the collective needs of a changed society. It also entailed taking into account the rising to the forefront of large masses that, in few decades, had moved from the country fields to the city factories and workshops. Architectural culture, after World War 1, was committed to defining the minimum housing requirements to accommodate these new clients, both numerous and unknown. The studies by Alexander Klein in Berlin, those by Greta Schütte-Lihotsky in Frankfurt in the '20s, to name just a few, tried to integrate Taylorist-inspired ideas into the design process, with the objective of giving everyone an efficient, comfortable and pleasant home despite the financial constraints. These designers analysed the usual activities that take place in the house, measuring time and ergonomic relationships between movements, paths and equipment, committed to the idea of improving the efficiency, the health and the well-being of their unknown and anonymous "clients". This meant applying the industrial conception of functional analysis and organization to the production and reproduction of labour power which customarily take place in the environment where a family lives.[3]

In order to satisfy the housing needs of this new mass-entity, it was not possible to investigate the needs of a specific client. It became essential to trace—or to imagine—the significant elements common to countless, faceless individuals whose customs and ways would be increasingly leveled out by life in the industrial city. These people were identified as "users", expected to find satisfaction by living in well-equipped functional spaces. The study of repeatable typological solutions, suitable for buildings constructed by means of fast techniques and new materials available through industrial production, implied the "construction" of an average user, whose uniform behaviour and

---

[3] Studies on the *Existenzminimum*, as it was termed in German, were carried out in the 1920s both in capitalist Europe and in the newly born Soviet Union, with different degrees of insight about the women's role.

aspirations represented the foundation of the industrial and rationalist idea of *standard*.

> All men have the same organism and the same function. All men have the same needs,

claimed Le Corbusier in 1923,[4] while, as early as in 1932, Hitchcock and Johnson criticizing the idealism of the European functionalists, remarked that they aimed to satisfy the needs that one should have, rather than actual ones:

> Functionalism is absolute as an idea rather than a reality [...] The Siedlungen implies preparation not for a given family but for a typical family. This statistical monster [...] has no personal existence and cannot defend himself against the sociological theories of the architects [...] Europeans build for some proletarian superman of the future.[5]

2 Performance: Congruence Between Premises and Conclusions

In the 1950s "rational design" developed in the US and UK, drawing on the experience gained in industrial sectors to reduce errors, uncertainty, risks, costs, and time. During the '40s and in war production, a number of techniques of analysis and control for various processes—planning, industrial design and production—had been developed. Reduction of error entails the capability to integrate and manage the relationships and information flows between different actors in a complex process. Decision-making techniques were deployed along the lines of Operational Research: O. R. represents a method of mathematical analysis to identify and break down one specific general problem in sub-problems, in order to define a sequence of decisions capable of achieving *performance improvement* in both the process and its final product. Thus defined, the decision sequence can be summarized in a mathematical model that allows evaluating different solutions by modifying certain variables.[6]

The rational methodology applied to the program/project/production flow, refers to information theories and cybernetics, the

> science of control and communication, in animals and machines.[7]

Cybernetics, recalling the assertions of contemporary science on the impossibility of studying complex systems by reducing them to their simplest components, searches for methods capable to analyze and control systems of *extreme*

---

[4] Le Corbusier (1923).

[5] Hitchcock and Johnson (1932).

[6] Broadbent (1973, pp. 182–183).

[7] Wiener (1948).

intrinsic complexity[8] and focuses its analysis on the relationships between the elements of a system and their role.

During the 1960s, the ideas of *input*, *output* and *feedback* became familiar to rational architectural design, with different regional variations between European countries and the US.[9]

Asimow (1962) outlined a method describing industrial design in terms of information process, whose steps are

> the gathering, handling and creative organizing of information relevant to the problem situation; it prescribes the derivation of decisions which are optimized, *communicated and tested, or otherwise evaluated; it has an iterative character* [...].

The rational design process was generally structured in phases modeled on a decision sequence with feedbacks, often represented by flow diagrams.[10]

The Hochschule fur Gestaltung, established in Ulm in 1949,[11] provided a further important contribution. The Ulm School promoted a system-based, formalized approach to architectural design, merging the Bauhaus tradition of a meaningful industrial-age artistic production with the optimization aims of Operational Research. Along this line, the relationship between humans and human-designed-and-built environment at different scales entailed scientific analysis and a design approach where the main variables link the users' needs—which are closely connected to their environment—and the functional requirements of their activities.[12]

The approach called *metadesign*[13] represented a formalization of the design process which could generate models of design behaviour apt to deal with uncertain and changing situations. It was conceived as an

> ordered set of operations to achieve congruence between premises and conclusions, through systematic processing tools, and to knowingly define the limits of design alternatives compatible with the problem.[14]

The procedure takes into account both the analytical phase and the synthetic, conceptual one, providing

---

[8] Ashby (1956), Italian translation, pp. 12–13.

[9] The large theoretical production of the Anglo-American area since the 60s is represented by studies that have been largely translated. In Italy: Asimow (1962), Jones (1963, 1970), Gregory (1966) and Archer (1965).

[10] Broadbent (1973, p. 257).

[11] Tomàs Maldonado, professor at the Ulm school from 1954 to 1967 directed it from 1956 to 1960, established the disciplinary and academic field of *Environmental Design*, within the frame of a wider "design philosophy" based on analytical methodologies. He had a fundamental influence on design theories in Italy: he was professor of Environmental Design at the University of Bologna (1976–1984) and at the Politecnico di Milano (1985–1994) where he greatly contributed to establish the school of Industrial Design.

[12] In Italy, this approach to design in architecture gave life to the academic discipline "Tecnologia dell'Architettura" (Architectural Technology), established in 1969.

[13] Andreis Van Onck brought forward the idea while at Ulm in 1963.

[14] Boaga and Giuffrè (1975).

the organization of a system of spatial requirements descending from human activities, both specific and in their mutual relationship, which by concretizing and quantifying these requirements in relation to any specific context, brings forward a *field of design variations* (dimensional, typological, etc.) from which solutions can be derived that correspond to the general objectives of the customer and user.[15]

Generally speaking, the rational design approach proceeds from the preliminary analysis of the user's needs according to the system of activities to be provided, to the definition of a program containing specific requirements, to the construction of a model representing an environmental spatial system which properly meets the requirements coming from the organization of the established activities. This environmental subsystem of spaces, and the technological subsystem physically containing it, represent a complex building organism: a dynamic system that, in performing its functions, continually processes matter, energy and information that flow in and out of its physical boundaries.

At this stage, rational design theories agreed that the designer's goals should be expressed in terms of performance,[16] which had to be specified in a set of criteria. The conjoined terms of *need-requirement-performance* were at the core of this idea.

3 The Natural Face of Function: Something New Under the Sun

The models of "rational design drawn from industry" were criticized as early as in the 1960s (Broadbent 1973). It was noted, and widely debated, that a decision process—consisting of a single sequence—is very different from a design process, which is

the way of structuring the order in which a vast number of decisions may be made[17]

and requires the capacity of going round the cycle several times. From another point of view, the need of a more formalized method to help design accomplish the new tasks posed by mass-building production, represented an updated version of the old debate—architecture contended between the realms of Art and Science. J.C. Jones wrote:

The method is primarily a means of resolving a conflict that exists between logical analysis and creative thought. The difficulty is that imagination does not work well unless it is free to alternate between all aspects of the problem, in any order and at any time, whereas logical analysis breaks down if there is the least departure from

---

[15] Magnaghi (1973).

[16] Broadbent (1973, p. 293).

[17] Broadbent (1973, p. 256).

a systematic step-by-step sequence [...] so systematic design is primarily a means of keeping logic and imagination separated by external rather than internal means.[18]

Jones's assumptions were widely shared in a time when the idea that logic and imagination, as well as reason and feelings, represent worlds apart within the human mind, was commonly accepted.

By applying this distinction coherently, most rational design theories did not take into account the issue of form as priority. The layout contrived by the meta-design process could do as a sort of "generative cue" for the building plan. As for the building's appearance, the commonplace idea was that it should represent its purpose, complying the slogan "Form Follows Function". The origin of this principle, as it is now widely recognised,[19] can be traced back to a long article written in 1896 by Louis Sullivan, the great American architect, with the title: *The Tall Office Building Artistically Considered.*[20] Sullivan writes:

> The architects of this land and generation are now brought face to face something new under the sun, namely, that evolution and integration of social conditions, that special grouping of them, that results in a demand for the erection of tall office buildings. It is not my purpose to discuss the social conditions; I accept them as the fact and I say at once that the design of the tall office building must be recognized and confronted at the outset as a problem to be solved, a vital problem, pressing for a true solution. [...] It has come in answer to a call, for in it a new grouping of social conditions has found a habitation and a name.

From this rational approach Sullivan proceeds to explain the architectural nature of the problem:

> How shall we impart to this sterile pile, this crude, harsh, brutal agglomeration, this stark, staring exclamation of eternal strife, the graciousness of those highest forms of sensibility and culture that rest on the lower and fierce passions? How shall we proclaim, from the dizzy height of this strange, weird, modern housetop, the peaceful evangel of sentiment, of beauty: the cult of a higher life?

Sullivan highlights the architect's own task, that is, finding the answer to a question that is both aesthetic and ethic. The problem is unprecedented, as tall building are; therefore, architects cannot resort to traditional rules, to the established "working tools" of the current profession. Instead, one should follow one's "natural instinct" and, after establishing the functional and technological structure of the tall building, one shall understand which parts of the building will need a special aesthetic connotation, within a harmonious overall composition, according to their own purpose and to their relationship with the city. Sullivan advocated

> the erection of buildings finely shaped and charming in their sobriety,

---

[18] Quoted in Broadbent (1973, p. 257).

[19] Cf. Wikipedia, 2018, entry "Form follows function".

[20] Sullivan (1896).

against any academic ornamentation, but his article has not the polemic tone and the dry wit of Adolf Loos's invective.[21] He rather includes decoration in the formal issue, which represents a higher order of enrichment, entailing a moral character and edifying tasks. In fact, formal accomplishment allows the designer to advance the stage of the economic-functional program

> beyond the imagined sinister building of the speculator-engineer-builder combination.

Once the material aspects of the construction are resolved in the design draft, the architect must reason "with equal depth, innocence and audacity" on the aspects concerning the spiritual nature, and therefore the feelings and emotions, that this kind of building should express and arouse. To accomplish this goal, architects should get rid of academic teaching. not wasting their time with

> theories and foreign forms declined with an American accent.

They should rather observe nature and consider the wonderful variety of natural forms:

> All things in nature have a shape, that is to say, a form, an outward resemblance, that tells us what they are, that distinguishes them from ourselves and from each other. Unfailingly in nature these shapes express the inner life, the native quality, of the animal, tree, bird, fish, that they present to us; they are so characteristic, so recognizable, that we say, simply, it is natural. [...] Unceasingly the essence of things is taking shape in the matter of things, and this unspeakable process we call birth and growth. Awhile the spirit and the matter fade away together, and it is this that we call decadence, death. [...] Whether it be the sweeping eagle in his flight, or the open appleblossom, the toiling work-horse, the blithe swan, the branching oak, the winding stream at its base, the drifting clouds, over all the coursing sun, *form ever follows function, and this is the law.* [...] It is the pervading law of all things organic and inorganic, of all things physical and metaphysical, of all things human and all things superhuman, of all true manifestations of the head, of the heart, of the soul, that the life is recognizable in its expression, that *form ever follows function.* This is the law.

The 3 Fs cliché "Form Follows Function" totally betrays Sullivan's ethical and poetic stance, but all the same it became very popular in the mainstream culture of post-War professional architects and, above all, in the practice of speculative building developments all over the world. Thus, it represented an easy target for the "anti-modern" reaction that burst out in the late 60s. In his *Form Follows Fiasco* (Blake 1977) author and architect Peter Blake proclaimed:

> Most of the time the form is nothing but a probable hypothesis of the function. Most of the times in good (or more likely in bad) the form follows the current rates of the bank loan. Most of the times in modern architecture, the form is anti-functional. Most of the time these three assertions can be true.

---

[21] Loos (1908).

4 Form Follows Nature and Time: Over Many, Many Years

> The ultimate object of design is form. [...] If the world were totally regular an homogeneous, there would be no forces, and no forms. Everything would be amorphous. But an irregular world tries to compensate for its own irregularities by fitting itself to them, and thereby takes on form [...] we speak of these irregularities as the functional origins of the forms.[22]

Christopher Alexander published the work that gave him international fame (Alexander 1964) in the golden years of rational design research. Form was its central subject, defining the design problem itself:

> [...] every design problem begins with an effort to achieve fitness between two entities: the form in question and its context. The form is the solution to the problem; the context defines the problem. In other words, when we speak of design, the real object of discussion is not the form alone, but the ensemble comprising the form and its context.

Alexander calls *good fit* the property of such an *ensemble*, which represents the design goal:

> If the ensemble is a truckdriver plus a traffic sign, the graphic design of the sign must fit the demands made on it by the driver's eye. An object like a kettle has to fit the context of its use and the technical context of its production cycle.

Alexander's idea of function is far more complex than the representation given by the diagrams of system engineering, which shape functional programs in the design process. As Sullivan did at the end of the nineteenth century, he refers to nature itself and to the thought of D'Arcy Thompson,[23] who

> has even called form the *diagram of forces* for the irregularities.

Alexander argues

> In a problem of design we want to satisfy the mutual demands which the two [elements of the ensemble] make on one another.

How can we find the good fit for the ensemble of a human settlement plus its physical and social context? It is a completely new design problem: in common practice engineering—i.e. a stated arrangement of iron filings in a certain position of a given magnetic field—we can judge the fit between form and context by either testing the form directly against the context, or by describing the characteristics of both terms mathematically and calculating the fit or lack of fit. In our case, we are unable to give an adequate description of our context, which is too complex, neither can we wait and see if our formal solution represents the correct one:

---

[22] Alexander (1964).

[23] D'Arcy Thompson (1917).

> Yet we certainly need a way of evaluating the fit of a form which does not rely on the
> experiment of actually trying the form out in the real world context. Trial-and-error
> design is an admirable method. But it is just real world trial and error which we are
> trying to replace by a symbolic method, *because real trial and error is too expensive
> and too slow* [...].[24]

Alexander was fully engaged in rational design research. He looked for the
formal rules of aggregation that could be abstracted by analysing "real life"
human settlements and trying to translate their complex relationships into
formal terms, using graphs and set theory, in order to discover their underlying order. His early efforts proved unsatisfactory and he rejected some of
this approach not much later (Alexander 1966). Nevertheless, and in spite of
this failure in defining a design method, his work brought into full light some
very good points and questions:

> Understanding the field of the context and inventing a form to fit are really two
> aspects of the same process. It is because the context is obscure that we cannot give
> a direct, fully coherent criterion for the fit we are trying to achieve [...] How is it,
> cognitively, that we experience the sensation of fit?

The consideration implies that we will never be able to make an exhaustive
and finite list of positive requirements, which in real life represent a potentially infinite set. To approach the question, Alexander suggest a simple way
of picking a finite set of requirements, by thinking of them in terms of *misfits*.
He claims that it is easier to understand how and where a situation is not
satisfactory:

> This is because it is through misfit that the problem originally brings itself to our
> attention. We take just those relations between form and context which obtrude
> most strongly, which demand attention most clearly, which seem most likely to go
> wrong. *We cannot do better than this.*[25]

This represents a sort of "fuzzy approach" towards the properties of good
design: not a rigid list of requirement/performance prescriptions, which could
never be totally exhaustive, but rather a path of good advice against events
"most likely to go wrong".

In the second half of the twentieth century, many studies investigated the
entities that give the built environment its form.[26] Even a simplified outline
would exceed the scope of these notes; I will restrict to briefly mentioning
two instances linked by one theme, to be further developed elsewhere: the
first instance refers to Alexander's reflections after his profound re-thinking
of the approach brought forward in the *Notes on the Synthesis of Form*; the
second one refers to the studies by Saverio Muratori, who in the 1950s, inves-

---

[24] Alexander (1964).

[25] Alexander (1966).

[26] See among the others: in the UK (Martin and Steadman 1971; Martin and March 1972;
Steadman 1975; Hillier and Hanson 1984). In the USA (Lynch 1960, 1981). In Italy the
question of *Forma Urbis* is at the core of many important studies, see Rossi (1966) and
Aymonino (1977).

tigated building typology and urban morphology according to their geographical, historical and functional peculiarities. In both cases, Time appears as the great Master builder to which human settlements owe their most durable, best configuration.

In *A City is not a Tree*, Alexander recognizes that the vast majority of people, and a good number of architects as well, prefer old buildings and old cities to new ones. He calls new cities, deliberately planned and designed, *artificial cities*, while *natural cities* are

> those cities which have arisen more or less spontaneously over many, many years.[27]

He demonstrates that the formal organisation of "natural cities" is a *semi-lattice*, the structure of living things where different activities can overlap and interact while belonging to different subsets, in opposition to the structure of "artificial cities", which can be represented like a *tree diagram*, where every subset separately stems like the branches of a tree. Alexander does not elaborate the issue of time; he just notes that also planned cities may become "natural" over time—like Liverpool and New York. In fact, a number of Roman towns had their origin as military camps, which is a typical tree organisation, and nevertheless, "over many many years" they acquired the more subtle and more complex structure of a semi-lattice. Alexander does not openly indicate Time as one of the entities—or forces—which give the built environment its form. Nevertheless, when he writes that any living reality, any real system whose existence actually makes the city live, must be provided a physical receptacle, he somewhat implies that Time, flowing "over many many years" provides exactly the opportunity of physical receptacles for systems that were not anticipated in the original plan.

Everything changes over time: Time, as a shaping force, destroys material things and overturns social structures—it breaks the boundaries which prevent overlapping. Alexander seems to admit that there's no possibility of planning a semi-lattice structure

> because designers, limited as they must be by the capacity of the mind to form intuitively accessible structure, cannot achieve the complexity of a semi-lattice in a single mental act [...] for the human mind, the tree is the easiest vehicle for complex thoughts.

Nevertheless,

> the city is not, cannot and must not be a tree. A city is a receptacle for life [...] If we make cities which are trees, they will cut our life within to pieces.[28]

For the designer, the puzzle remains unsolved.

On the other hand, the issue of time is prominent in the work of Saverio Muratori, who investigated the logic of morphogenesis in human settlements, with the main goal to provide a design tool for new developments in

---

[27] Alexander (1966).

[28] Alexander (1966).

old cities (Muratori 1959, 1967). His *operante storia urbana* ("operational urban history") reconstructs the organic link between human groups and their human-made environment combining material history and geography along with field work. He investigates typology and morphology in their material layering over time, within the mould of local geography. Typology and morphology gradually consolidate, they condition later transformations and are in turn continuously transformed—they embody and express motifs which are formal, functional, cultural, symbolic; their persistence is an indication of both adaptability and generative power.

5 About Affordance: Something Roughly Fit

At present, performance-based design seems to refer mostly to mechanical HVAC and structural systems, energy saving and, generally speaking, to design areas where quantitative assessment and risk control are involved. Here, functional quality and the "perfect fit" are properly required, and we must not forget that mechanical systems and services represent one layer of the building, with a more limited lifespan.[29]

The idea of performance-based design is inadequate when dealing with the human built environment, which is continuously evolving and results from the never-ending activities of generations over time. There is no way of controlling such flow, of eliminating uncertainty and flaws in its way.

Yet, the relationship between the "users"—people, communities, human groups and single beings—and the actions that shape their eco-techno systems, still require some kind of conceptualization which can help action appraisal and some form of "operational tool" for decision-making in the realm of common good. An interesting line of thought to approach *the ensemble of a human settlement plus its physical and social context* seems to be the elaboration of the idea of *affordance*, a term coined by environmental psychologist J.J. Gibson (1979) who derived it from the verb *to afford*:

> The affordances of the environment are what it offers the animal. What it provides or furnishes, either for good or ill. [...] It implies the complementarity of the animal and the environment.[30]

Affordance depends upon the physical properties of an environmental component which are measured relative to the properties of an animal in order to define if the animal can use it in its ecological niche: for example, a large solid surface which is conveniently rigid and flat offers to a terrestrial walking creature of appropriate weight and size the possibility to walk/run/rest on

---

[29] The idea that buildings are composed of layers (namely Shell, Services, Scenery and Sets) with different lifespans, calling for different design approaches and maintenance strategies, was introduced by British architect Frank Duffy of DEGW in the 1970s.

[30] Gibson (1979).

it. Affordance characterizes the relationship between observer/user and its environment in terms of *opportunities* and involves cognitive, cultural and social issues which are increasingly complex according to the species.[31]

The built environment in its development is subject to the shaping forces of human activities over time, with all the constraints and possibilities that Time and Nature put in its way. Over time, it becomes a goldmine of ever-changing *affordances*. The collective organizations of the human animal—communities—should be able to identify them to promote the species' survival.

References

Alexander, C. (1964). *Notes on the synthesis of form*. Cambridge: Harvard University Press.

Alexander, C. (1966). A city is not a tree. *Design, 206* (February), 46–55.

Archer, L. B. (1965). *Systematic method for designers*. London: H. M. Stationery Office.

Ashby, W. R. (1956). *An introduction to cybernetics*. London: Chapman and Hall. (Italian trans.: (1971). *Introduzione alla Cibernetica*. Torino: Einaudi).

Asimow, M. (1962). *Introduction to design*. Englewood Cliffs: Prentice-Hall.

Aymonino, C. (1977). *Lo studio dei fenomeni urbani*. Roma: Officina Edizioni.

Blake, P. (1977). *Form follows fiasco: Why modern architecture hasn't worked*. Boston: Little Brown.

Boaga, G., & Giuffrè, R. (1975). *Metodo e progetto*. Roma: Officina Edizioni.

Broadbent, G. (1973). *Design in architecture*. London: Wiley.

D'Arcy Thompson, W. (1917). *On growth and form*. Cambridge: University Press.

Gibson, J. J. (1979). *The ecological approach to visual perception*. Boston: Houghton Mifflin.

Giedion, S. (1941). *Space, time and architecture*. Cambridge: Harvard University Press.

Giedion, S. (1948). *Mechanization takes command*. New York: Oxford University Press.

Gregory, S. A. (Ed.) (1966). *The design method*. New York: Springer.

Hillier, B., & Hanson, J. (1984). *The social logic of space*. Cambridge/New York. Cambridge University Press.

Hitchcock, H. R., & Johnson, P. (1932). *The international style*. New York: W. W. Norton.

---

[31] In the design realm, the idea has been popularized by Norman (1988), and simplified to define the degree of interaction between a designed object and its user.

Jones, J. C. (1963). *A method of systematic design.* In J. C. Jones & D. J. Thornley (Eds.), *Conference on design methods, London, September 1962* (pp. 53–73). London: Pergamon Press.

Jones, J. C. (1970). *Design methods.* New York: Wiley.

Le Corbusier (1923). *Vers une Architecture.* Paris: Éditions Crès.

Loos, A. (1908). *Ornament and crime.*

Lynch, K. (1960). *The image of the city.* Cambridge: The MIT Press.

Lynch, K. (1981). *Good city form.* Cambridge: The MIT Press.

Magnaghi, A. (1973). *L'organizzazione del metaprogetto.* Milano: Angeli.

Martin, L., & March, L. (Eds.) (1972). *Urban space and structures.* Cambridge: Cambridge University Press.

Martin, L., & Steadman, P. (1971). *The geometry of environment.* London: RIBA Publications.

Muratori, S. (1959). *Studi per una operante storia urbana di Venezia.* Roma: Istituto Poligrafico dello Stato.

Muratori, S. (1967). *Civiltà e territorio.* Roma: Centro Studi di Storia Urbanistica.

Norman, D. A. (1988). *The psychology of everyday things.* New York: Basic Books.

Rossi, A. (1966). *L'architettura della città.* Padova, Marsilio.

Steadman, P. (1975). *Energy, environment and building.* New York: Cambridge University Press.

Sullivan, L. (March 1896). The tall office building artistically considered. *Lippincott's Monthly Magazine,* Philadelphia.

Wiener, N. (1948). *Cybernetics.* Cambridge: The MIT Press.

Systemic Ontology and Heidegger's Ontology: A Discussion on Systems and "Logos"

Elena Bartolini

Since its first appearance on the scientific scenery, systems thought brought within its considerations an implicit ontology, essentially based on relations, structure and unity. In his 1968 work, Von Bertalanffy defines systems as:

[...] sets of elements standing in interaction.[1]

Nevertheless, while describing which are the aims of a general system theory and proposing some possible progresses, he does not focus on the implications of such ontological assumptions. Most recently, the Research Group on System Thought at the Catholic University in Milan has provided many suggestions on a metaphysical reflection as well as on some ontological explanations about systems.[2] However, few are the inquiries specifically addressed to the importance of those structured relations that are ontologically constitutive of systems.[3] Moreover neither the definition of relation, especially structured relations, is object of clarification even though its pivotal role in such thought. The principal aim of the present paper is to discuss these points, comparing a systemic ontology with some Greek terms assumed through Heidegger's interpretation. In such a proposal, incompleteness will be motivated as a constitutive element of this particular ontological perspective.

E. Bartolini (✉)
University of Milano – Bicocca, Milano, Italy
e-mail: e.bartolini@campus.unimib.it

[1] Von Bertalanffy (1969, p. 58).

[2] See the three volumes edited by Urbani Ulivi (2010, 2013, 2015) especially the contributions from Urbani Ulivi, Giuliani, Minati, Vitiello and Del Giudice. For some considerations on a systemic anthropology from a philosophical point of view see also Bartolini (2015).

[3] For example, the volume edited by Hooker (2011) represents a notable effort in showing the "revolutionary" contribution of systems thought, but it focuses only on Sciences and on Philosophy of Science, without considering other possible implications in Humanities.

© Springer Nature Switzerland AG 2019

G. Minati et al. (eds.), *Systemics of Incompleteness and Quasi-Systems*,
Contemporary Systems Thinking,
https://doi.org/10.1007/978-3-030-15277-2_9

According to its etymology, ontology is preliminarily a study of what is, it focuses on what indicated while affirming that there is something, that something is. Even though the employment of this term is quite recent, usually it is through ontology that some philosophical concepts from Ancient philosophy are investigated for further developments. However, Heidegger warns about the risks of an ontology led in a traditional way. In *Introduction to Metaphysics*,[4] he claims:

> The term "ontology" [...] designates the development of the traditional doctrine of beings into a philosophical discipline and a branch of the philosophical system. But the traditional doctrine is the academic analysis and ordering of what for Plato and Aristotle, and again for Kant, was a question, though to be sure a question that was no longer originary. [...] In this case "ontology" means the effort to put Being into words, and to do so by passing through the question of how it stands with Being [not just with beings as such].

Even thought he highlights which could be the negative side of a traditional ontology, he confirms the possibility of a new one really attentive to Being and not only to beings, how happened instead in the history of western metaphysics—according to his perspective. In the same context, Heidegger follows:

> We ask the question — How does it stand with Being? What is the meaning of Being? — not in order to compose an ontology in the traditional style [...]. The point is to restore the historical Dasein of human beings — and this also always means our ownmost future Dasein, in the whole of the history that is allotted to us — back to the power of Being that is to be opened up originally.[5]

Not only a different ontology is desiderable, but it should also consider the "historical *Dasein*", which is not assumed in its past but rather in its present as pivotal intersection between what was and the possibilities of what could be. It is crucial to notice that here Heidegger presents a connection among *Dasein*, Being and its originary openness.

Considering the challenge to think a new ontology, how should it be? Shall we intend it as wholly knowable and determinable? in other words, the new ontology needed to describe reality in its shades shall be considered a complete one or not? The term "complete" derives from the Latin *completus*, past participle of *complere*, meaning "to fill up", then employed to indicate "fulfill, finish a task". Something complete is something accomplished, thus what tends to a satisfaction, to a balanced state. When something is complete is more understandable, since its constitutive elements are stable, entirely defined. Being confined, what is complete is somehow close. Is systems ontology a complete ontology?

In *Metaphysics* Z 17, Aristotle, who in this book discusses *ousia* (οὐσία) intended as the main connotation of being,[6] states

---

[4] Fried and Polt (2014, pp. 43–44).

[5] Fried and Polt (2014, p. 46).

[6] Sachs (1999, p. 117, 1028 a).

> But then there is what is composed of something in such a way that the whole is one, in the manner not of a heap but of a syllable — and the syllable is not the letters, nor are B plus A the same as the syllable BA [...]; therefore there is something that is the syllable, not only the letters, the vowel and the consonant, but also something else.[7]

This claim has been usually adopted in comparison with the statement of Anderson, Noble Prize in 1977, who attested:

> More is different.[8]

Both these quotes aim to highlight the impossibility to find into the components those linear consequences of what is. Hence, what is deducible, combined with the previous considerations about ontology, is that a systemic ontology cannot be only attentive to the single being, to any kind of item present in front of us, but rather should consider the multiple levels through which reality presents itself,[9] namely the hierarchical structure of it.[10] Thus, given the central role of structural relations it seems possible to affirm that systems are based on a certain kind of relational ontology.

Is there in the history of philosophy a concept that describes structures, relations and their dynamics? Greek thought names it *logos* (λόγος). In this sense, Bateson speaks about the "patterns which connects"[11] and, following his reasoning, Baracchi suggests a sort of continuity from the concept of *system* (σύστημα), to that of *logos*.[12] The first one indicates something "connected with itself and cohesive"[13]: systems result to be characterized by an excess due to the existence of internal and mutual relations among its parts.[14] Thus, it guarantees to the system

> a dynamics, through the presence of states, and provides account for the possible complexity of its behavior.[15]

Whereas the second term is usually translated as discourse, sentence or reason, but its origin refers to the verb *leghein* (λέγειν) meaning

---

[7] Sachs (1999, p. 152, 1041 b).

[8] Anderson (1972, p. 393).

[9] For an essential description of this specific topic see for example Laszlo (1972, pp. 165–180).

[10] The adjective "hierarchical" is not adopted here with a connotation of value in which higher level is ontologically superior to the lower one or vice versa: it only recognizes the presence of a relational structure.

[11] Bateson (1979, p. 8).

[12] Baracchi (2013, pp. 204–219, see especially pp. 206–212).

[13] Baracchi (2013, p. 206).

[14] Mari (2011, p. 586).

[15] Mari (2011, p. 586, my translation).

a connection that protects and preserves: linking, gathering, articulating so as to hold the differing together while saving it as such, as differing.[16]

Consequently, Baracchi concludes, *logos*

bespeaks relation, correlation, a fitting together from which arise configurations of meaning, a union that literally makes sense, bring sense forth and lets it be illuminated.[17]

Since these considerations, she underlines a similarity between organisms, i.e. living systems, with discourse and reason:

This generative arrangement, which is the bearer and locus of sense, equally defines linguistic articulation, the work of rationality, and the *organized structures* (whether internal or external, whether visible or invisible) of life.[18]

Hence, structured relationships, ordered and organized, are essential to life itself. Elsewhere, Baracchi concisely states:

Bestowing order means giving life.[19]

The world surrounding us presents itself in a way that

the hanging together of the world is a matter of communing and communications: the world conveys itself to itself, speaks to itself, as it were, pervaded by the ripples of information at once (in)forming and transforming it.[20]

It seems that the specific way in which what is reaches our senses, our same way to be too, is characterized by relations, namely structures, and movement, namely dynamic changing. The two instances are both a connotation of the Greek term *logos* which demonstrates to be the ordered appearance of being, the in-formed way.[21] More clearly,

[...] it is precisely in this pulsating and rippling motility that the world emerges, as the body of the all: one and choral, the fabric of unitary yet vibrant becoming — above all, alive.[22]

Minati, trying to explain the appearance of independent organisms from the physical point of view, clarifies that such event is possible thanks to what he calls the rupture of the symmetry,

[...] considering all auto-organized phenomena as a consequence of the quantum phenomenon of symmetry breaking.[23]

---

[16] Baracchi (2013, p. 211).

[17] Baracchi (2013, p. 211).

[18] Baracchi (2013, p. 211, italics mine).

[19] Baracchi (2016, p. 24, my translation).

[20] Baracchi (2013, p. 206).

[21] For a more detailed examination of this issue, see Urbani Ulivi (2014) and Bartolini (2014).

[22] Baracchi (2013, p. 206).

[23] Minati (2010, p. 36, my translation).

The symmetry discussed here is the one related to those equations describing the dynamic of the system[24]: when the symmetry is spontaneously broken what is observable is that the state of the system presents a certain kind of order.[25] The same concept of information is here pivotal, because, as recalled by Vitiello,

> to the order is associated a higher degree of information [...] which is not present in the case of a symmetrical configuration.[26]

In systems thought, information is not conceived as a simple message, but rather the element that literally in-forms, it is to say that allows the emergence of a new structure. If, thanks to a new information, the symmetry is broken, the position of every element is not exchangeable: its same place is significative for the order of the system, it is crucial to distinguish it from the others.[27] But, if such happening is possible, thus signifies that there should be the *possibility* for unpredicted configurations. It is to say, there should be the possibility of actualizing new potentialities. *Physis* ($\varphi\acute{\upsilon}\sigma\iota\varsigma$) is the Greek word indicating

> the event of standing forth, arising from the concealed and thus enabling the concealed to take its stand for the first time.[28]

According to the translation proposed by Heidegger, this Greek term refers to the force through which beings become observable, taken from the concealing. In the *Basic concepts of Aristotelian Philosophy*, Heidegger exhorts to consider a *natural entity* ($\phi\acute{\upsilon}\sigma\epsilon\iota~\acute{o}\nu$) as

> a being that is what it is from out of itself on the basis of its genuine possibilities.[29]

The break of the symmetry, through the introduction of information, has as consequences order and life, i.e. *logos*. Only starting from an incomplete ontology, in which there is no given order yet, such emerging is possible. Only where there is power, it is to say no closure or completeness, boundaries can be traced, order can be established, life could be. But, at the same time, once instituted this order persists. *Physis* and *logos*: (re)newing and maintenance. *Logos* is the here and now display of *physis*. *Physis* is that unpredicted source for beings' appearances. Both are sides of what is, aspects of Being[30].

---

[24] Vitiello (Cf. 2010, pp. 111–113).

[25] (Ibidem, my translation).

[26] (Ibidem, my translation).

[27] (Ibidem).

[28] Fried and Polt (2014, p. 16).

[29] Metcalf and Tanzer (2009, pp. 32–33).

[30] See Fried and Polt (2014, p. 15):

Phusis is Being itself, by virtue of which beings first become and remain observable,

and Fried and Polt (2014, p. 145):

Effectively, since its etymology, one of the ways in which *logos* can be intended is meaning "to make manifest", "to allow to appear".[31] In this sense

> it can only be understood if its essential relation to *physis* is borne in mind.[32]

More clearly:

> the power which emerges from concealment must be gathered together, one.[33]

In this sense, logos-discourse is founded on *Logos*-Being: making manifest something into our human verbal or corporeal expression is due to the previous existence of such dynamical displaying of *physis* into *logos*, conceived as the ordered but unpredictable interrelation surrounding us. Hence, the language, not only in the specific connotation of speech but also in its wider meaning,[34] is possible because man, defined as *zoon logon echon* (ζοον λογου εχων), is the one capable to observe the worldly interrelations and is able to interfere with them, changing the constraints or creating new ones. Therefore, as attested by Baracchi,

> In its most basic sense, well exceeding the exercise of the human calculative capacities, rationality is relationality, the meaningful bonding that discloses aspects inaccessible through the examination on unrelated components.[35]

Logos is not only reason, if with it what is intended is a definite predictable knowledge. It is rather the mediation, the *ratio* as *proportio*, of the multiple and various appearances of beings.

From the assumptions of such an ontology, I choose two terms to indicate the main consequences for the interpretation of the human in this sense, inspired by the work of Baracchi[36] and Urbani Ulivi[37]: architecture and unic-

---

> *Logos* is constant gathering, the gatheredness of beings that stands in itself, that is, Being.

[31] Fay (1977, p. 95).

[32] (Ibidem); here it is clear the reference to Heraclitus fragment 50.

[33] For what concerns the relation between *logos* and *ousia*, see also Heidegger's words in Metcalf and Tanzer (2009, p. 15):

> The *logos* as *horismos* (ὁρισμός) addresses beings in their *ousia*, in their being there.

[34] Brogan and Warnek (1995, p. 103):

> [...] this is the structure we call *"language"*, speaking; but not understood as vocalizing, rather in the sense of a speaking that says something, means something [...]. *Logos* is discourse, the gathering laying open, unifying *making something known* [*Kundmachen*]; and indeed above all in the broad sense which also includes pleading, making a request, praying, questioning, wishing, commanding and like.

[35] Baracchi (2013, p. 211).

[36] Baracchi (2008).

[37] Urbani Ulivi (2010).

ity. If systems ontology is not closed in a sort of completeness, it means that through his own agency and freedom, namely his *ethos*, the human builds his own being, becoming a unique individual.

References

Anderson, P. W. (1972). More is different. Broken symmetry and the nature of hierarchical structure of science. *Science, 4047*(177), 393–396.

Baracchi, C. (2008). *Aristotle's ethics as first philosophy.* New York: Cambridge University Press.

Baracchi, C. (2013). The syntax of Life: Gregory Bateson and the "Platonic View". *Research in Phenomenology, 43*, 204–219.

Baracchi, C. (2016). *Amicizia.* Milano: Ugo Mursia Editore.

Bartolini, E. (2014). Lavori sistemici. Confronti in un privatissimum. *Rivista di Filosofia Neo-Scolastica, 106*(3), 687–695.

Bartolini, E. (2015). *Per un'antropologia sistemica. Studi sul "De Anima" di Aristotele,* Senago: Albo Versorio.

Bateson, G. (1979). *Mind and nature. A necessary unity.* New York: E. P. Dutton.

Brogan, W., & Warnek, P. (1995). *Martin Heidegger. Aristotle's metaphysics Θ 1–3. On the essence and actuality of force.* Bloomington and Indianapolis: Indiana University Press.

Fay, T. A. (1977). *Heidegger: The critique of logic.* The Hague: Springer.

Fried, G., & Polt, R. (Trans.) (2014). *Martin Heidegger. Introduction to metaphysics* (2nd ed.). New Haven/London: Yale University Press.

Hooker, C. (Ed.). (2011). *Philosophy of complex systems.* Oxford/Amsterdam/Waltham: Elsevier.

Laszlo, E. (1972). *Introduction to systems philosophy. Toward a new paradigm of contemporary thought.* New York: Harper & Row.

Mari, L. (2011). Qualche riflessione sulla retroazione. *Rivista di Filosofia Neo-Scolastica, 103*(4), 571–586.

Metcalf, R. D., & Tanzer, M. B. (Trans.) (2009). *Martin Heidegger. Basic concepts of Aristotelian Philosophy.* Bloomington and Indianapolis: Indiana University Press.

Minati, G. (2010). Sistemi: Origini e Prospettive. In L. Urbani Ulivi (Ed.), *Strutture di Mondo. Il Pensiero Sistemico come Specchio di una Realtà Complessa* (pp. 15–46). Bologna: Il Mulino.

Sachs, J. (Trans.) (1999). *Aristotle. Metaphysics* (2nd ed.). Santa Fe: Green Lion Press.

Urbani Ulivi, L. (2014). Approfondimenti sistemici. Seminari e privatissimum. *Rivista di Filosofia Neo-Scolastica, 106*(3), 453–465.

Urbani Ulivi, L. (Ed.). (2010). *Strutture di Mondo. Il Pensiero Sistemico come Specchio di una Realtà Complessa.* Bologna: Il Mulino.

Urbani Ulivi, L. (Ed.). (2013). *Strutture di Mondo. Il Pensiero Sistemico come Specchio di una Realtà Complessa.* (*Volume Secondo*). Bologna: Il Mulino.

Urbani Ulivi, L. (Ed.). (2015). *Strutture di Mondo. Il Pensiero Sistemico come Specchio di una Realtà Complessa.* (*Volume Terzo*). Bologna: Il Mulino.

Vitiello, G. (2010). Dissipazione e Coerenza nella Dinamica Cerebrale. In L. Urbani Ulivi (Ed.), *Strutture di Mondo. Il Pensiero Sistemico come Specchio di una Realtà Complessa* (pp. 111–113). Bologna: Il Mulino.

Von Bertalanffy, L. (1969). *General system theory. Foundations, development, applications.* New York: George Braziller.

The Idea of Incompleteness in the Internal Realism of Hilary Putnam

Antonio Lizzadri

1 Introduction

The evolution of the systemic thought from the "first" to the "second" Systemics requires a philosophical reformulation of the related ontological and epistemological issues. If traditional requests of the first Systemics—think about emergency or complexity—had found an adequate theoretical overview of the phenomenon inside an ontological and epistemological anti-reductionist[1] and holistic[2] framework, the detailed study and the problematization of the same requests—which brought the second Systemics to explore systems not suitable for complete modeling, due to their high level of complexity—can be sustained by a further theoretical effort.

The ambitious project of the second Systemics to define the structural dynamic of becoming reopens the great questions over the consistency of reality and the possibility of knowledge: if becoming is structural, what are the ontological borders beyond which the becoming would deny its own dynamic, and, ultimately the same idea of reality? Furthermore, from an epistemological point of view, how should we interpret the incompleteness of our models in order not to invalidate the possibility of knowledge itself?

Among the recent proposals in the contemporary debate about realism, the "internal realism" of Hilary Putnam seems to offer an adequate theoretical model to provide a solution to the problems over the consistency of reality and the possibility of knowledge. In fact, Putnam's internal realism

A. Lizzadri (✉)
Catholic University of the Sacred Heart, Milano, Italy
e-mail: antonio.lizzadri@unicatt.it

---

[1] Giordani (2010).

[2] Corvi (2010).

© Springer Nature Switzerland AG 2019
G. Minati et al. (eds.), *Systemics of Incompleteness and Quasi-Systems*,
Contemporary Systems Thinking,
https://doi.org/10.1007/978-3-030-15277-2_10

affirms a plastic conception of reality by virtue of interdependency between the prospective of the observer and the prospective of the participant; it respects the freedom of becoming without degenerating into an ontological anarchy, and it allows for the possibility of knowledge without degenerating into a radical skepticism.

This paper will delve into such matters by adopting Putnam's semantic approach, which can be defined as "quasi-systemic" since not only it recognizes meaning as an emerging property and as irreducible to the natural and social "semantic indicators" that constitute it, but also because it does not forecast to determine a priori a hierarchy of the same.

I shall argue that Putnam does not interpret such hierarchical incompleteness as indeterminacy of the reference, but as complementarity between the theory of meaning and the theory of understanding, by reason of the already mentioned interdependency between the prospective of the observer and the one of the participant.

In other words, the definition of the meaning is always going to be part of a linguistic procedure in which the anthropological complexity of the speaker plays a crucial role. In this sense, it will turn out that the essence of such hierarchical incompleteness is freedom.

2 The Meaning of Meaning Between Extension and Stereotype

Putnam's proposal seems to be of particular interest in the contemporary discussions on semantics due to its attempt to overcome the unsolvable dichotomies between scientist hyper-realism and anti-realistic deconstructionism. Despite their apparent differences, these two views share a complementary reductionism: on the one hand, the neopositivistic conception demands to reduce the meaning of a name to a "pack" of necessary material conditions, sufficient to determine its extension; on the other the anti-realistic conception reduces it to the use of the name in a specific context.

The main flaw of traditional semantics' theory is an abstract vision that does not take into consideration that some kind of pre-theoretic notion of meaning is presupposed by and antecedent to any attempt of strict theoretic definition.

According to Putnam, then, it is worth explicating the pre-theoretic meaning of meaning, always and inevitably operating in the linguistic practice.

This is the meaning Putnam's recurring mental experiments, which lead us beyond the conditions of possibility of our language, explicating them.

The most significant mental experiment for our purpose is "Twin Earth". Putnam supposes that in a spot of the galaxy there is a planet called Twin

Earth very similar to the Earth: on Twin Earth even the spoken language is English. The only difference is that the liquid called "water" is not H_2O, but a different liquid with a very long and complex chemical formula that Putnam shortens with the initials XYZ. However, XYZ is undistinguishable from water in standard conditions of pressure and temperature, it has the same taste, it is equally quenching and it fills the oceans, the lakes and the seas of Twin Earth on which it rains XYZ and so on.[3]

If it came out that on Twin Earth the water is XYZ, the terrestrial starship would probably send a similar report: "On Twin Earth the word 'water' means XYZ".

It is from this, apparently ordinary, example that Putnam derives an important lesson: our use of the word "meaning" in such circumstance shows how the extension is not a sufficient condition to determine the meaning. In fact, although it certainly represents one of the necessary conditions, no one would say that "on Twin Earth the meaning of the word 'water' is XYZ", identifying *tout court* the meaning with the extension through the copula ("the meaning *is* XYZ").

Actually speaking, anyone would say this only if the statement "water is XYZ" would be known by every single adult person on Twin Earth, but this would state that the extension is never on its own determining for the definition of meaning, but its decisiveness is always co-determined by the linguistic context, that is by the linguistic community considered as whole.

In other words, the more the knowledge of the meaning of a word is shared by the linguistic community, the more the extension it is determining for the definition of meaning.

However, according to Putnam, this does not mean that the extention is constructed by the community, nor that every single speaker should adequately own the knowledge of the deep structure associated to a word: generally speaking, the knowledge of the superficial features functional to life, i.e., what Putnam calls a "stereotype", is sufficient to the community to know and to understand the *real* meaning of a word.

In this respect, consider Putnam's example of gold. Let us imagine our linguistic community as a factory in which someone has got the "task" of wearing wedding bands, others the "task" of selling wedding bands, others the "task" of recognizing if something is actually gold or not. It is not necessary nor practical that all the individuals wearing wedding bands or debating over "gold standards" should be committed to the gold trade and vice versa.

In conclusion, if a hidden structure exists, it usually determines what being a member of natural kind means, but this does not prevent the deep structure from becoming useless in many other circumstances and the superficial features from becoming the useful ones.

---

[3] Putnam (1975).

3 Meaning, Understanding, Translation

As we can see, the semantics of Putnam can be rightfully defined "quasi-systemic", not only because it opposes a reductionist idea of meaning, but also because it doesn't demand from us to crystallize the newly-found complexity, determining a priori a hierarchy between natural and social factors.

After all, in order to avoid the anarchical degeneration of such "freedom" of meaning, Putnam puts forth a theory that ties together the natural and the social components of meaning.

In Putnam's opinion, the representational and the "use" theories of meaning are both inadequate only if they are assumed unilaterally and opposed to each other. On one hand, if we'd like to explain how we do actually need our theories to successfully—but without making a miracle out of it—guide our behaviour, it's crucial to consider them as a sort of the world's "map" that somehow represent it; on the other, even the "use" theory is essential, because it is an important warning about the map not having an absolute meaning in itself, but since it is used with a human perspective.[4]

Therefore, Putnam's theory of meaning as use doesn't want to deny the existence of an objective reality, but it reminds us that such reality exists for us and through us. A theory of reference, then, will not be able to forget a theory of understanding which sets the assignment of meaning to the human dynamics and in the same real context in which it happens. In Putnam's opinion, such actualization of the theory of meaning effectively happens in *translation processes*.[5] In fact, the operation that is accomplished during "translation" is the construction of a *global theory* which identifies the meaning of the words considering also the speaker's behavior, his own beliefs, desires and intents.

Therefore, in Putnam's opinion, a theory of translation does not simply have a linguistic or psychological significance but, first of all, it has a global philosophical importance because, trying to rationalize and to understand the speaker's behaviour starting with its relationship with reality, it is a "substantial metaphysical theory" which defines from a general point of view the relationship between language and reality at the same time, or rather—from this dynamic metaphysical perspective—the relationship between speakers and context/environment.

4 Internal Realism and Incompleteness

In *Realism and Reason*[6] Putnam defined his theory of meaning, which is at the same time a theory of understanding and translation, as "internal realism".

---

[4] Putnam (1978).

[5] Putnam (1976).

[6] Putnam (1977).

I will conclude this paper by outlining the general features of such metaphysical theory and by relating it to the systemic category of incompleteness, which will reveal in this connection its deep explicatory power.

First, internal realism is incomplete because, being a global theory of the relationship between speakers and environment, where the psychological description of common sense is also expressed in first person and made of feelings and intents, it will never be used completely modeled by the functional organization of the cerebral states described in third person. In fact, some abilities are too complex to be described by a theory. For example, in the case of translation, there is no precise explicit theory that would allow me to describe the translation ability that I acquired: even if my brain had a complete and totally formulated in some hypothetic "brain language" analytical hypothesis, I would not have it, nor would the scientists.

Let us suppose we had such a theory: what would happen? Would it be possible to love someone if we could calculate something like: "If I say X the probability that he/she will react in the Y way is 15%"? Would it be possible to have friends or enemies? Would it be possible to think of us as persons?

Putnam does not know the answers but he is sure that their knowledge would modify our nature in an unpredictable way. Being partially obscure to ourselves—that is, not being able to understand each other like we understand hydrogen atoms—is a constitutive fact that, by leaving the relationship between language/thought and reality undetermined, inevitably determines it.

In fact, the impossibility of identifying a necessary and univocal connection between language/thought and reality using a naturalistic criterion, due to the irreducibility of the language/thought to the functional organization of the cerebral states, represents itself a condition of possibility of the relationship between language/thought and reality. More precisely, if language/thought and reality never completely correspond, they correspond asymptotically to infinity.

In conclusion, Putnam's internal realism is a theory of meaning and a theory of understanding, which is at one and the same time also a theory of translation. It assumes a genetic conception of truth understood as a correspondence in which the linguistic community understands reality more and more while understanding itself.

In this way, Putnam proposes a revision of Kant's image of knowledge as "performance". In Kant's opinion the author, the I, appears as well as a character like in a comedy from Pirandello: the author in the performance isn't the real one, but the "empirical I". The real author is the "transcendental I". Putnam modifies Kant's image in this way: first of all, the performance is never individual but always a social one; furthermore, the authors in the performances are the real ones. This would be irrational if such stories were fictional just like Kant's phenomenon. In fact, a fictional character cannot be also a real author. In Putnam's opinion, our stories are real ones, even if incomplete.[7]

---

[7] Putnam (1977).

References

Corvi, R. (2010). Dall'olismo epistemologico al pensiero sistemico: Un percorso possibile? In L. Urbani Ulivi (Ed.), *Strutture di Mondo. Il Pensiero Sistemico come Specchio di una Realtà Complessa* (pp. 175–195). Bologna: Il Mulino.

Giordani, A. (2010). L'ontologia della sostanza alla luce della teoria dei sistemi In L. Urbani Ulivi (Ed.), *Strutture di Mondo. Il Pensiero Sistemico come Specchio di una Realtà Complessa* (pp. 197–229). Bologna: Il Mulino.

Putnam, H. (1975). The meaning of "Meaning", In K. Gunderson (Ed.), *Language, mind and knowledge* (pp. 131–193). Minneapolis: University of Minnesota Press. Republished in H. Putnam (1975), *Mind, Language and Reality: Philosophical Papers*, (Volume II), Cambridge, New York: Cambridge University Press. (Italian translation by R. Cordeschi: Putnam H. (2004), Il significato di "significato". In Putnam H., *Mente, linguaggio e realtà*, (3rd edition) (pp. 239–297). Milano, Italy: Adelphi).

Putnam, H. (1976). Meaning and knowledge, (from John Locke Lectures at Oxford University). In H. Putnam (1978), *Meaning and the moral sciences* (pp. 7–80). London, Boston: Routledge & Kegan Paul. (Italian translation: Putnam H. (1982), Significato e conoscenza. In *Verità e Etica*. Milano, Italy: Il Saggiatore).

Putnam, H. (1977). Realism and reason. In *Proceedings of the American Philosophical Association* (Vol. 50, pp. 483–498). Republished in H. Putnam (1978), *Meaning and the moral sciences* (pp. 123–128). London, Boston: Routledge & Kegan Paul. (Italian translation: Puntnam H. (1982), Realismo e Ragione. In *Verità e Etica*. Milano, Italy: Il Saggiatore).

Putnam, H. (1978). Reference and understanding, in H. Putnam (1978), *Meaning and the moral sciences* (pp. 97–119). London, Boston: Routledge & Kegan Paul. (Italian translation: Putnam H. (1982), Riferimento e comprensione. In *Verità e Etica*. Milano, Italy: Il Saggiatore).

Part IV
Incompleteness and Quasiness in Post-Bertalanffy Systemics Complexity

Are Dynamically Undecidable Systems Ubiquitous?

Marco Giunti

1 Introduction

One of the main tenets of Wolfram's *A New Kind of Science* (Wolfram 2002) is the Principle of Computational Equivalence (PCE): "Almost all processes that are not obviously simple can be viewed as computations of equivalent sophistication" (pp. 716–717). PCE, together with the seemingly uncontroversial premise that the behavior of a computationally universal system is not obviously simple, entails that computationally universal systems should be ubiquitous, for they would almost coincide with the very many systems that display some form of complex behavior. A further consequence of this fact is that dynamically undecidable systems should be ubiquitous as well, for it is well known that the long term behavior of any computationally universal system is in general undecidable (consider for instance the halting problem, or related undecidability results).

In this paper, I propose an independent argument for the ubiquity of computational universality and, as a consequence, dynamical undecidability as well. My argument does not presuppose PCE and, in essence, goes as follows.

In the first place, I briefly recall a number of relevant facts. (i) Computationally universal systems are a special kind of computational systems, which in turn can be thought as a special kind of discrete deterministic dynamical systems; (ii) the property of computational universality is based on the relation of emulation between dynamical systems, which is a quite

M. Giunti (✉)
ALOPHIS - Applied LOgic Philosophy and HIstory of Science, Dipartimento di Pedagogia Psicologia Filosofia, Università di Cagliari, Cagliari, Italy
e-mail: giunti@unica.it

© Springer Nature Switzerland AG 2019
G. Minati et al. (eds.), *Systemics of Incompleteness and Quasi-Systems*, Contemporary Systems Thinking,
https://doi.org/10.1007/978-3-030-15277-2_11

weak structure preserving mapping from the state space of the emulated system to the state space of the emulating one; (iii) the relevant state space structure preserved by this mapping reduces to the transition graph of the emulated system and, whenever the emulated system has discrete time, such a graph in turn reduces to the graph of the transition of duration one (1-step transition graph); (iv) like any graph, the state transition graph of any dynamical system exhaustively decomposes into a set of internally connected and mutually disconnected components; (v) for any discrete time dynamical system, such components are of just five mutually disjoint and jointly exhaustive types; (vi) for any computationally universal system, the problem of deciding the type of the component to which an arbitrary state belongs is undecidable (undecidability of the state classification problem). This result is in fact a further manifestation of the dynamical undecidability of computationally universal systems.

I then argue that (v) ensures a strong structural similarity between the 1-step transition graphs of any two computational systems, so that (a) reproducing the dynamics of an arbitrary computational system does not seem to require a system with especially unusual or extraordinary features. In addition, it must be kept in mind that (b) the structure preserving mapping needed for computational universality is emulation, which is itself a quite weak mapping. I finally conclude that, given (a) and (b), computational universality might very well hold under very weak conditions, so that computationally universal systems, and dynamically undecidable ones as well, might be much more widespread than thought before.

2 Computationally Universal Systems

A computational system is *computationally universal* if it is able to emulate, or exactly reproduce, the behavior of a whole class of systems that are known to compute all partial recursive functions. For example, the class of all Turing machines (TMs) is one such a class, and Turing in his famous paper (Turing 1936) wrote down (secs. 6–7) the table of a machine that emulates all TMs. That machine is thus computationally universal. Smith (1971) proved that, for any given TM, one can construct a cellular automaton that emulates the TM. Therefore, the cellular automaton that emulates a universal TM is computationally universal as well.

Computationally universal systems are in fact a proper subclass of the *computational systems*. By this term, we mean all those systems that are studied or described by standard (or elementary) computation theory. Computational systems thus include, besides Turing machines and cellular automata, many other systems of different types, such as: finite-state machines, program machines, monogenic Post canonical systems, tag systems, etc. (see Minsky 1967). However, computational systems *do not* include super-Turing systems like oracle machines, or even analog computers, for all these systems are not in the domain of *standard* computation theory.

In turn, computational systems can be thought as a proper subclass of the deterministic dynamical systems with both discrete state space and discrete time. Thus, for our present goals, it is best to start with a concept of dynamical system as general as possible. Giunti and Mazzola (2012) maintain that the minimal structure to be imposed on time that ensures an interesting notion of dynamical system is just that of a monoid. Accordingly, they define a dynamical system on a monoid as follows.

Definition 1 (Dynamical System on a Monoid)
Let T and M be non-empty sets.
DS *is a dynamical system on a monoid* $L := DS_L :=$

1. $DS = (M, (g^t)_{t \in T})$ and $L = (T, +)$;

   - any $t \in T$ is called *a duration*, T *the time set*, and $L = (T, +)$ *the time model*;
   - any $x \in M$ is called *a state*, M *the state space*, and $DS = (M, (g^t)_{t \in T})$ *the dynamical system*;

2. a. $+ : T \times T \to T$;
   - $+$ is called *the operation of addition of durations* or, briefly, *duration addition*;

   b. for any $t \in T$, $g^t : M \to M$;
   - for any $t \in T$, g^t is called *the (state) transition of duration t* or, briefly, *the t-transition* or *the t-advance*;

3. a. i. the $+$ operation has the identity element;
   - the identity element of the $+$ operation is indicated by "0", where $0 \in T$;

   ii. the $+$ operation is associative;
   - thus, by 2a, 3(a)i, and 3(a)ii, the time model $L = (T, +)$ is a monoid;

   b. i. for any $x \in M$, $g^0(x) = x$;
   ii. for any $x \in M$, for any $v, t \in T$, $g^{v+t}(x) = g^v(g^t(x))$.

As mentioned above, computational systems can be thought as a proper subclass of the *deterministic dynamical* systems (see Fig. 1) with both *discrete state space* and *discrete time*. According to Definition 1, this means that an arbitrary computational system can be identified with a DS_L whose state space M is, at most, countably infinite, and whose time model $L = (\mathbb{Z}^{\geq 0}, +)$, where $\mathbb{Z}^{\geq 0}$ is the set of the non-negative integers and $+$ is the usual operation of addition on integers.

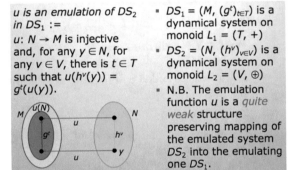

Fig. 1 Definition 1 conveys the general notion of a *deterministic dynamical system*

Within the figure (Fig. 1):

Determinism
(2.b) If x is the state at instant $i \in T$, $g^t(x)$ is the state of the system after a transition of duration $t \in T$.

condition (2.b)

Dynamics
(3.b.i) Whatever state x the system is in, the state transition of duration 0 does not modify that state.
(3.b.ii) Any transition of duration $v+t$ can be decomposed in two successive transitions, the first one of duration t, and the second one of duration v.

condition (3.b.i)

$g^0(x) = x$

condition (3.b.ii)

$g^{v+t}(x) = g^v(g^t(x))$

Fig. 2 Emulation between two dynamical systems on monoids

Within the figure (Fig. 2):

u is an emulation of DS_2 in DS_1 :=
u: N → M is injective and, for any $y \in N$, for any $v \in V$, there is $t \in T$ such that $u(h^v(y)) = g^t(u(y))$.

- $DS_1 = (M, (g^t)_{t \in T})$ is a dynamical system on monoid $L_1 = (T, +)$
- $DS_2 = (N, (h^v)_{v \in V})$ is a dynamical system on monoid $L_2 = (V, \oplus)$
- N.B. The emulation function u is a *quite weak* structure preserving mapping of the emulated system DS_2 into the emulating one DS_1.

3 Computational Universality and Emulation Between Dynamical Systems

We have said that computational universality is based on the fact that a computational system is able to *emulate*, or exactly reproduce, the dynamics of any system in a given class. In general, emulation obtains if a certain kind of structure preserving mapping holds between the emulated system and the emulating one. This kind of mapping can be precisely defined for any two dynamical systems on monoids as follows (see Fig. 2).

Definition 2 (Emulation for Dynamical Systems on Monoids)
Let $DS_1 = (M, (g^t)_{t \in T})$ and $DS_2 = (N, (h^v)_{v \in V})$ be a dynamical system on, respectively, monoid $L_1 = (T, +)$ and monoid $L_2 = (V, \oplus)$.
u is an emulation of DS_2 in DS_1 := u : N → M is injective and, for any $y \in N$, for any $v \in V$, there is $t \in T$ such that $u(h^v(y)) = g^t(u(y))$.

It is important to note that, according to Definition 2, the emulation function u is a *quite weak* structure preserving mapping of the emulated system DS_2 into the emulating one DS_1. For state transition durations are not usually preserved (i.e., in general, t may be different from v), even when the time models L_2 and L_1 are identical. In addition, the duration t of the state transition g^t that, through the mapping u, corresponds to h^v may depend on the state y to which h^v applies, and not just on its duration v.

4 Transition Graphs and the Five Types of Their Components in Discrete Time Dynamical Systems

The dynamical structure of the state space of an arbitrary DS_L is represented by its *transition graph*, as shown in Fig. 3. The transition graph of a DS_L in fact conveys all the information about its dynamics, so that the system itself can be identified with its transition graph. It has been shown (Giunti and Mazzola 2012) that the transition graph of any dynamical system on a monoid is a category.

Whenever a DS_L has *discrete* time model $L = (\mathbb{Z}^{\geq 0}, +)$, it is sufficient to consider its *1-step transition graph* (see Fig. 4), for the 0-transition is the identity function on the state space and any other state transition of duration $k \in \mathbb{Z}^{\geq 1}$ is obtained by iterating k times the state transition of duration 1.

4.1 The Five Possible Types of Each Connected Component of the 1-Step Graph of an Arbitrary Discrete Time DS_L

Like any graph, the state transition graph of an arbitrary DS_L exhaustively decomposes into a set of internally connected and mutually disconnected

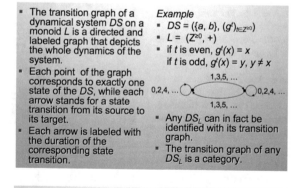

Fig. 3 The transition graph of a DS_L

- The transition graph of a dynamical system *DS* on a monoid *L* is a directed and labeled graph that depicts the whole dynamics of the system.
- Each point of the graph corresponds to exactly one state of the *DS*, while each arrow stands for a state transition from its source to its target.
- Each arrow is labeled with the duration of the corresponding state transition.

Example
- $DS = (\{a, b\}, (g^t)_{t \in \mathbb{Z}^{\geq 0}})$
- $L = (\mathbb{Z}^{\geq 0}, +)$
- if t is even, $g^t(x) = x$
  if t is odd, $g^t(x) = y, y \neq x$

- Any DS_L can in fact be identified with its transition graph.
- The transition graph of any DS_L is a category.

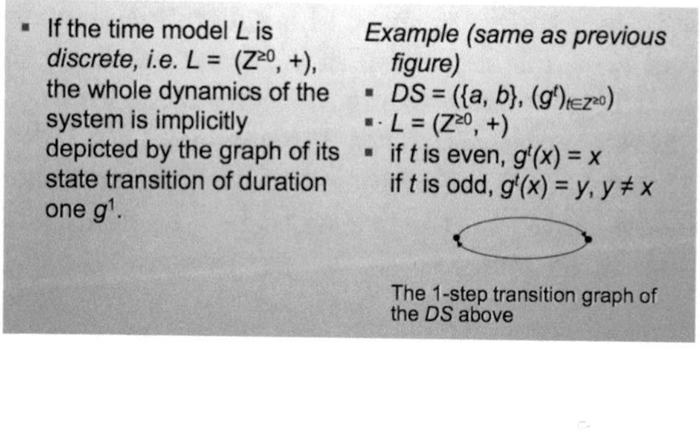

Fig. 4 The 1-step transition graph of a DS_L with discrete time model $L = (\mathbb{Z}^{\geq 0}, +)$

- If the time model *L* is discrete, i.e. $L = (\mathbb{Z}^{\geq 0}, +)$, the whole dynamics of the system is implicitly depicted by the graph of its state transition of duration one g^1.

Example (same as previous figure)
- $DS = (\{a, b\}, (g^t)_{t \in \mathbb{Z}^{\geq 0}})$
- $L = (\mathbb{Z}^{\geq 0}, +)$
- if t is even, $g^t(x) = x$
  if t is odd, $g^t(x) = y, y \neq x$

The 1-step transition graph of the *DS* above

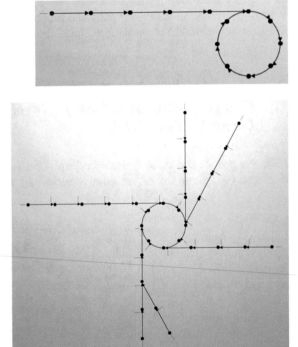

Fig. 5 Type 1: a directed cycle of n points ($n \in \mathbb{Z}^{\geq 1}$)

Fig. 6 Type 2: a directed line, infinite in two directions or just in the direction of its orientation

Fig. 7 Type 3: a directed cycle of n points ($n \in \mathbb{Z}^{\geq 1}$) to which exactly one directed line finite in the direction of its orientation, and possibly infinite in the opposite direction, is attached

Fig. 8 Type 4: a directed cycle of n points ($n \in \mathbb{Z}^{\geq 1}$) to which the roots of a finite number of possibly infinite trees (either with respect to the number of levels: infinite height; or to the number of points in a level: infinite thickness) are attached; either (a) at least two trees attach to the cycle, or (b) the unique tree attached to it has different branches (i.e., it is not just a directed line)

components, and it has been shown that each of these connected components is in fact the state space of a subsystem that contains both its whole future and its whole past (Giunti 2016; Giunti 2017, Theorem 1).

It is then not difficult to realize that, for any DS_L whose time model L is discrete, each connected component of its 1-step transition graph is always of one out of five possible types, which are shown in Figs. 5, 6, 7, 8, and 9 below. Note that the five types, besides being jointly exhaustive, are mutually disjoint as well.

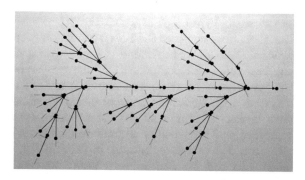

Fig. 9 Type 5: a directed line infinite in two directions or just in the direction of its orientation, to which the roots of a possibly infinite number of possibly infinite trees (in either height or thickness) are attached

4.2 Undecidability of the State Classification Problem for a Computationally Universal System

Given an arbitrary $DS_L = (M, (g^t)_{t \in \mathbb{Z}^{\geq 0}})$ with discrete time model $L = (\mathbb{Z}^{\geq 0}, +)$, the *state classification problem for DS_L* consists in deciding, for any state $x \in M$, the Type 1–5 of the connected component (of the 1-step graph) to which the state x belongs.

It is then almost immediate to realize that the state classification problem for any computationally universal system is undecidable. To see why, let us recall, in the first place, that the *halting* of a computational system DS_L can always be identified with its reaching a fixed point in state space or, in other words, entering a cycle made of exactly one state (the final or halting state).

In the second place, if the classification problem for a computational system DS_L is decidable, its halting problem is decidable as well. In fact, on the one hand, if the state x is classified as Type 2 or 5, the DS_L will not halt when started in state x. On the other hand, if the state x is classified as Type 1, 3 or 4, there is an obvious decision procedure to establish whether the DS_L is going to halt when started in state x. In fact, if the Type is 1, the system halts iff the state after the first step is x itself. If, instead, the Type is 3 or 4, after starting the system in state x, we must wait for a finite time until the system reaches a state twice, and the system halts iff it comes back to that state in just one step.

It thus follows that the classification problem for a computationally universal system DS_L is undecidable, for its halting problem is undecidable.

5 Summing up the Argument

If we now take a closer look to the five Types shown in Figs. 5, 6, 7, 8, and 9, we notice that Types 3 and 4 are in fact obtained by combining one copy of Type 1 with one or more copies of Type 2, while Type 5 is obtained by just combining many copies of Type 2. This means that Types 1 and 2 can be thought as the *basic* Types, out of which the other three *complex* Types are built. But then, if we consider the 1-step transition graphs of two arbitrary computational systems, we are bound to find a strong structural similarity between any two connected components of the two graphs. Hence, (a) reproducing the dynamics of an arbitrary computational system does not seem to require a system with especially unusual or extraordinary features. In addition, it must be kept in mind that (b) the structure preserving mapping needed for computational universality is emulation, which is itself a quite weak mapping (see Sect. 3). Therefore, because of (a) and (b), computational universality might very well hold under very weak conditions, so that computationally universal systems, and dynamically undecidable ones as well, might be much more widespread than usually thought.

References

Giunti, M. (2016). Decomposing dynamical systems. In G. Minati, M. Abram, & E. Pessa (Eds.), *Towards a post-bertalanffy systemics* (pp. 65–79). Cham: Springer.

Giunti, M. (2017). Decomposing dynamical systems (revised and enriched 10 Sept 2017). https://doi.org/10.13140/RG.2.2.21663.05283/1

Giunti, M., & Mazzola, C. (2012). Dynamical systems on monoids: Toward a general theory of deterministic systems and motion. In G. Minati, M. Abram, & E. Pessa (Eds.), *Methods, models, simulations and approaches towards a general theory of change* (pp. 173–185). Singapore: World Scientific.

Minsky, M. L. (1967). *Computation: Finite and infinite machines.* London: Prentice-Hall.

Smith, A. R. (1971). Simple computation-universal cellular spaces. *Journal of the Association for Computing Machinery, 18*(3), 339–353.

Turing, A. (1936). On computable numbers, with an application to the Entscheidungs problem. *Proceedings of the London Mathematical Society, Series 2, 42*, 230–265.

Wolfram, S. (2002). *A new kind of science.* Champaign: Wolfram Media.

A View of Criticality in the Ising Model Through the Relevance Index

Andrea Roli, Marco Villani, and Roberto Serra

1 Introduction

The *Relevance Index* (RI) had been originally introduced to identify key features of the organisation of complex dynamical systems, and it has proven able to provide useful results in various kinds of models, including e.g. those of gene regulatory networks and protein-protein interactions. The method can be applied directly to data and does not need to resort to models, possibly helping to uncover some non-trivial features of the underlying dynamical organisation. The RI is based upon Shannon entropies and can be used to identify groups of variables that change in a coordinated fashion, while they are less integrated with the rest of the system. These groups of integrated variables make it possible to provide an aggregate description of the system, at levels higher than that of the single variables and it can be applied also to networks, that are widespread in complex biological and social systems. In previous work, we have found that the RI can also be used to identify critical states in complex systems (Roli et al. 2017). We showed that the average RI, computed across random samples of cells of a given size in the Ising lattice, attains its maximum at the critical temperature.

A. Roli
Department of Engineering and Computer Science (DISI), Alma Mater Studiorum University of Bologna, Cesena, Italy
e-mail: andrea.roli@unibo.it

M. Villani (✉) · R. Serra
Department of Physics, Informatics and Mathematics, University of Modena e Reggio Emilia, Modena, Italy
European Centre for Living Technology, Ca Minich, Venezia, Italy
e-mail: marco.villani@unimore.it; rserra@unimore.it

© Springer Nature Switzerland AG 2019
G. Minati et al. (eds.), *Systemics of Incompleteness and Quasi-Systems*,
Contemporary Systems Thinking,
https://doi.org/10.1007/978-3-030-15277-2_12

In this contribution we present an in-depth analysis of the RI values across all subset sizes. Results show that a parameter defined as a function of subset size and RI is strictly correlated to the susceptibility of the system, which in turn assumes its maximum at the critical temperature. These results provide further evidence to the hypothesis that the RI is a powerful measure for capturing criticality and they also suggest that a possible explanation for this is that larger subsets are more correlated at criticality, as a consequence of all-range correlations typical of critical points in phase transitions.

After a brief introduction to the RI in Sect. 2 and a summary of previous results in Sect. 3, we will presents and discuss the new results on the Ising model in Sect. 4. In Sect. 5 we conclude the contribution and outline lines for future work.

2 The Relevance Index

The main concepts related to RI had been conceived in the work on biological neural networks by Tononi et al. (1998), who introduced several measures, among them the *Cluster Index*. The RI is an extension of this latter measure, that can be applied to dynamical systems (Filisetti et al. 2015; Roli et al. 2016; Villani et al. 2014, 2015). The purpose of the RI is to identify sets of variables that behave in a coordinated way in a dynamical system; the variables that belong to the set are *integrated* with the other variables of the set, much more than with the others. Since these subsets are possible candidates as higher-level entities, to be used to describe the system organisation, they will be called *relevant subsets* (omitting for brevity the specification that they are *candidates*). A quantitative measure, well suited for identifying them, is defined as follows—the presentation below follows the one given in Villani et al. (2014).

Let U be a system whose elements are discrete variables that change in time, and suppose that the time series of their values are known. The *Relevance Index* $r(S)$ of $S \subset U$ is defined as the ratio between the *integration* of S and the *mutual information* between S and the rest of the system:

$$r(S) = \frac{I(S)}{M(S; U \setminus S)} = \frac{\sum_{x \in S} H(x) - H(S)}{H(S) + H(S|U \setminus S)} \tag{1}$$

where $H(x)$ is the Shannon entropy of x and $H(S)$ is joint entropy of the set of variables in S.

When the RI is applied to identify relevant subsets, it is necessary to compare sets of different sizes. However, entropies scale with system size, so this requires considerable ingenuity. Following the original work of Tononi, a "RI method" has been developed for this purpose, where a statistical index is computed that allows meaningful comparisons of sets of different sizes:

$$T_c(S_k) = \frac{r(S_k) - \langle r_h \rangle}{\sigma(r_h)} \tag{2}$$

where $\langle r_h \rangle$ and $\sigma(r_h)$ are respectively the average and the standard deviation of the RI of a sample of subsets of size k extracted from a reference system U_h randomly generated according to the frequency of each single state in U. It is worth noting that the aim of the reference system is that of quantify the finite size effects affecting the information theoretical measures on a random instance of a system with finite size.

In principle, a list of candidate relevant sets can be obtained by computing the RI and the of every possible subset of variables in U and ranking the subsets by T_c values. The subsets occupying the first positions are most likely to play a relevant role in system dynamics. For large-size systems, exhaustive enumeration is computationally impractical as it requires to enumerate the power set of U. In this case, we resort to sampling or to heuristic algorithms.

3 RI and Criticality

Criticality usually refers to the existence of two qualitatively different behaviours that a system can show, depending upon the values of some parameters and it is then associated to parameter values that separate these qualitatively different behaviours. However, slightly different meanings of the word can be found in the literature, two major cases being (a) the one related to phase transitions and (b) dynamical criticality, sometimes called the "edge of chaos". In the former case, the different behaviours refer to equilibrium states that can be observed by varying the value of a macroscopic external parameter. In the latter case, the different behaviours are characterised by their dynamical properties: the attractors that describe the asymptotic behaviour of the system can be ordered states, like fixed points or limit cycles, or chaotic states. These two meanings are related but not identical.[1]

In a previous work, we have shown that the RI can be used to locate critical regions in complex systems (Roli et al. 2017). In our experiments we considered two different kinds of systems: the Ising model for phase transitions, and the Random Boolean Network model for dynamical criticality. For both the models, we computed the RI of randomly sampled groups of variables of varying size. Our main finding is that the RI is able to satisfactorily locate the critical points in both cases. An excerpt of the results on the Ising model is shown in Fig. 1.

---

[1] See Roli et al. (to appear) for a detailed review on the subject.

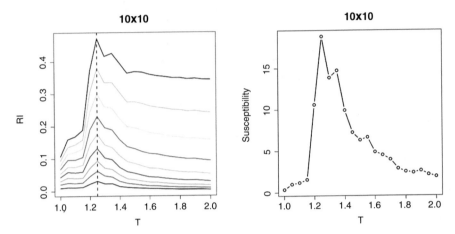

Fig. 1 Left: plot of RI for 10×10 Ising lattice. The median of the average RI values for groups of size 2 to 10 is plotted against T. The curves shift up with group size. The peaks of RI correspond to the susceptibility peak, which in turn corresponds to the critical value of the control parameter. **Right:** plot of median susceptibility values for 10×10 Ising lattice. The peak of susceptibility empirically identifies the critical T value

4 RI of the Ising Model

The Ising model is a notable example of a system that can undergo a phase transition as a function of a control parameter (Binney et al. 1992; Brush 1967; Stanley 1971). Let us consider a d-dimensional lattice of N atoms characterised by a *spin*, which can be either *up* (+1) or *down* (−1). The atoms exert short-range forces on each other and each atom tends to align its spin according to the values of its first neighbours. An external field may also be considered, which biases the orientation of the atoms. The energy of the system is defined as follows:

$$E = -\frac{1}{2} \sum_{\langle i,j \rangle} J \, s_i \, s_j + B \sum_i s_i \tag{3}$$

where s_i is the spin of atom i, $J > 0$ is a parameter accounting for the coupling between atoms, $\langle i,j \rangle$ denotes the set of all neighbouring pairs and B is a parameter playing the role of an external field. The system can be studied by means of usual statistical mechanics methods and it can be assessed whether it undergoes a phase transition; Onsager (1944) proved that the $d = 2$ model can undergo a phase transition under the hypothesis that $B = 0$.

In this work, we consider the two-dimensional model, with $B = 0$. We performed Monte Carlo simulations at constant temperature T. The Monte Carlo algorithm used is a classical Metropolis algorithm with Boltzmann distribution (see Algorithm 1).

Algorithm 1 Monte Carlo simulation of a 2d Ising model. Adapted from Solé (2011)

---

while maximum number of iterations not reached **do**

    Choose a random atom s_i

    Compute the energy change ΔE associated to the flip $s_i \leftarrow -s_i$

    Generate a random number r in $[0,1]$ with uniform distribution

    if $r < e^{-\frac{\Delta E}{k_B T}}$ **then**

        $s_i \leftarrow -s_i$

    end if

end while

The temperature is the *control parameter*, while the *order parameter* is the so-called *magnetisation*:

$$\mu = \frac{1}{N} \sum_i s_i \qquad (4)$$

For low values of T, the steady state of the system will be composed of atoms mostly frozen at the same spin and the time average of the magnetisation $\langle \mu(T) \rangle$ will be close either to 1 or -1; for high values of T the spins will randomly flip and it will be $\langle \mu(T) \rangle \approx 0$. For values close to the critical temperature T_c, a phase transition occurs: the system magnetisation undergoes a change in its possible steady state values.

In our experiments, we considered Ising lattices of L^2 spins, with $L \in \{10, 12\}$, arranged on a torus. We set $k_B = 2$, hence we expect a phase transition around the value $T_c \approx 1.13$. For each lattice size we run simulations with values of T spanning the range $[1, 2]$ at steps of 0.05. In finite size Ising models, the critical value of temperature is expected to deviate from the theoretical value. Therefore, the actual critical temperature value was estimated by computing the *susceptibility* (Christensen and Moloney 2005), defined as:

$$\chi = \frac{1}{TN} \left(\langle \mu^2 \rangle - \langle \mu \rangle^2 \right) \qquad (5)$$

where T is the temperature, N the number of atoms, μ the magnetisation of the system at a given time step and angular brackets denote the time average. The peak of χ may be used to identify the actual critical temperature value for finite instances. In Fig. 2 the susceptibility is plotted against the temperature value. As we can observe, the critical values are around $T = 1.25$ for both the lattice sizes considered, which is slightly higher than the theoretical one. This discrepancy is due to the finite size of the systems. This specific value will be taken as the critical one in the Ising models of our experiments.

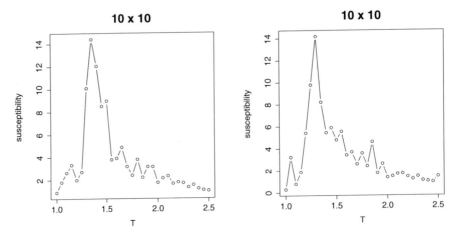

Fig. 2 Plots of susceptibility values vs. temperature

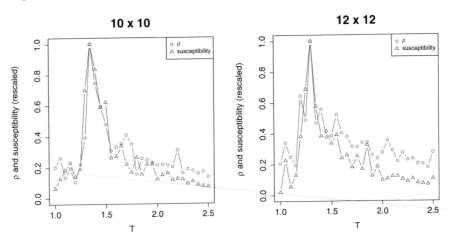

Fig. 3 Plots of susceptibility values and ρ as a function of T

Simulations were run until the transient was expired and we recorded 10^4 lattice configurations every L^2 steps. We computed RI and T_c for 10^3 randomly sampled subsets for each size between two and L^2 and kept the best 10^3, i.e. those with the highest values of T_c.

The most relevant result we observed is that larger clusters with high T_c seems to be correlated to high susceptibility values. To assess this informal observation we introduce an index $\rho(S) := \langle |S| \times T_c(S) \rangle$, where $|S|$ is the size of the subset, $T_c(S)$ its RI statistical significance and $\langle \cdot \rangle$ denotes the average. The correlation between ρ and χ is striking, as shown in Fig. 3.

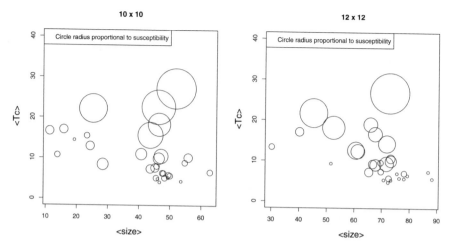

Fig. 4 Diagrams representing the susceptibility (proportional to the radius of the circle) at coordinates $(\langle size \rangle, \langle T_c \rangle)$

A further evidence for supporting this correlation is provided by considering separately the average size and average T_c of the best 1000 subsets. In Fig. 4 we show a diagram in which each circle has a centre at coordinates $(\langle size \rangle, \langle T_c \rangle)$ and radius proportional to χ. We observe that the largest circles are characterised by both high size and T_c.

5 Conclusion and Future Work

The results we observed by computing the RI on subsets of atoms of every size show that the most significant *dynamically relevant sets* in the Ising lattice tend to be larger and characterised by higher significance in correspondence to the critical temperature. Furthermore, an index defined as the product of size and T_c is shown to be highly correlated to the susceptibility, making it possible to locate the phase transition.

The relation between T_c and large subsets may be a consequence of the all-range correlations appearing at criticality in the Ising model. In future work we plan to further investigate this relation and assess to what extent the RI method can be used to detect early signals of criticality and dynamical change in complex systems.

References

Binney, J., Dowrick, N., Fisher, A., & Newman, M. (1992). *The theory of critical phenomena*. New York: Oxford University.

Brush, S. (1967). History of the Lenz-Ising model. *Reviews of Modern Physics, 39*(4), 883–893.

Christensen, K., & Moloney, R. (2005). *Complexity and criticality*. London: Imperial College.

Filisetti, A., Villani, M., Roli, A., Fiorucci, M., & Serra, R. (2015). Exploring the organisation of complex systems through the dynamical interactions among their relevant subsets. In P. Andrews, et al. (Eds.), *Proceedings of the European Conference on Artificial Life 2015 (ECAL 2015)* (pp. 286–293). Cambridge: The MIT.

Onsager, L. (1944). Crystal statistics. I. A two-dimensional model with an order-disorder transition. *Physical Review, 65*(3 and 4), 117–149.

Roli, A., Villani, M., Caprari, R., & Serra, R. (2017). Identifying critical states through the relevance index. *Entropy, 19*(2), 73.

Roli, A., Villani, M., Filisetti, A., & Serra, R. (2016). Beyond networks: Search for relevant subsets in complex systems. In G. Minati, M. Abram, & E. Pessa (Eds.), *Towards a post-bertalanffy systemics* (pp. 127–134). Cham: Springer.

Roli, A., Villani, M., Filisetti, A., & Serra, R. (to appear). Dynamical criticality: Overview and open questions. *Journal of Systems Science and Complexity* (A preliminary version of the paper is available as arXiv:1512.05259v2)

Solé, R. (2011). *Phase transitions*. Princeton: Princeton University.

Stanley, H. E. (1971). *Introduction to phase transitions and critical phenomena*. New York: Oxford University.

Tononi, G., McIntosh, A., Russel, D., & Edelman, G. (1998). Functional clustering: Identifying strongly interactive brain regions in neuroimaging data. *Neuroimage, 7*, 133–149.

Villani, M., Filisetti, A., Graudenzi, A., Damiani, C., Carletti, T., & Serra, R. (2014). Growth and division in a dynamic protocell model. *Life, 4*, 837–864.

Villani, M., Roli, A., Filisetti, A., Fiorucci, M., Poli, I., & Serra, R. (2015). The search for candidate relevant subsets of variables in complex systems. *Artificial Life, 21*(4), 412–431.

An Example of Quasi-System in the Generation and Transmission of Electrical Power

Umberto Di Caprio and Mario R. Abram

1 Introduction

The dynamical evolution of an electrical power system was the subject of many researches aiming at designing control systems and at planning the most effective operation procedures. In particular the involved processes present interesting elastic phenomena which may give origin to possible instabilities (Kimbak 1994).

These aspects became evident when great black-outs occurred in large electric networks. Many strategies were developed in order to prevent such events and the operating procedures were tuned taking into account the experience deriving from such dangerous events (Di Caprio and Saccomanno 1970).

The electric power system is an interesting example of a system composed by electric power generators interconnected to loads in elastic fashion by means of a large electric network. Many models were developed to examine the behavior of the entire system when particular events perturb it (Marconato 2008; Saccomanno 2003). When an interconnecting line is interrupted, the dynamic behavior evolves according to different possibilities which could bring to stable or unstable conditions. Indeed, due to instability, the system may evolve out of the normal operation range and then would lose the desired characteristics of the designed system.

In this paper we concentrate on a simplified model of a two machine electrical power system, with the aim to study its dynamic evolution, putting into evidence some critical conditions. Convenient simulations will be illustrated.

U. Di Caprio (✉)
ASDE (Associazione Dirigenti Enel), Roma, Italy

M. R. Abram
AIRS / Italian Systems Society, Milano, Italy

© Springer Nature Switzerland AG 2019
G. Minati et al. (eds.), *Systemics of Incompleteness and Quasi-Systems,*
Contemporary Systems Thinking,
https://doi.org/10.1007/978-3-030-15277-2_13

179

We concentrate upon a particular class of quasi-system, i.e. upon physical systems that are affected by instability problems due to which they operate outside standard conditions for which they were designed. In particular we consider electrical power systems which in anomalous operating conditions go outside the basin of attraction of their equilibrium points. An extreme case is represented by a power system undergoing black-out conditions, like it happened to occur in 2003 with regard to the Italian power system (Di Caprio 2006).

In Sect. 2 the equations used to model the system are recalled, while in Sect. 3 the simulation procedures are described and the results of simulation experiments are shown. In particular in Sect. 4 the possible evolution in "quasi system" is pointed out. Finally in Sect. 5 some conclusive remarks are illustrated.

2 Two Machines Electric Power System

As an example we can consider the case of a two machine electrical system. The simplified electrical power system is composed by two electric generators (synchronous machines) that supply two loads $\bar{Y}_{L1}$ and $\bar{Y}_{L2}$ and are interconnected by means of an electric line (Fig. 1). We consider the following definitions: (1) δ_1, δ_2: rotor angles of the machines; (2) P_{m1}, P_{m2}: mechanical power supplied to the two machines; (3) P_{e1}, P_{e2}: electrical power in output from the two machines; (4) M_1, M_2: equivalent masses of the machines (inertia constants); (5) P_{L1}, P_{L2}: electrical power absorbed by loads. Then the internal electromotive forces of the two machines in phasor form become:

$$\begin{cases} \bar{E}_1 = E_1 e^{j\delta_1} \\ \bar{E}_2 = E_2 e^{j\delta_2} \end{cases} \tag{1}$$

while the real part of the electric powers for the two machines are expressed by:

$$\begin{cases} P_{e1} = \mathrm{Re}\{\bar{E}_1 \cdot \bar{I}_1\} \\ P_{e2} = \mathrm{Re}\{\bar{E}_2 \cdot \bar{I}_2\} \end{cases} \tag{2}$$

The interconnecting electric line is modeled with a π quadrupole described by the admittances $\bar{Y}_{11}$, $\bar{Y}_{12}$ and $\bar{Y}_{22}$. According to a classic approach in electrical engineering, one defines the following circuit parameters:

$$\begin{cases} \bar{y}_{11} = \bar{Y}_{11} + \bar{Y}_{12} & \text{(self admittance of node 1)} \\ \bar{y}_{22} = \bar{Y}_{22} + \bar{Y}_{12} & \text{(self admittance of node 2)} \\ \bar{y}_{12} = -\bar{Y}_{12} & \text{(transadmittance)} \end{cases} \tag{3}$$

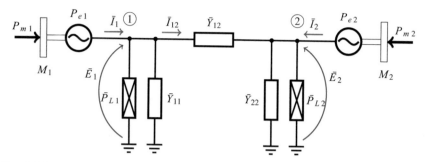

Fig. 1 Electric power system composed by two interconnected synchronous machines

starting from which one can deduct the currents:

$$\begin{cases} \bar{I}_1 = \bar{y}_{11}\bar{E}_1 + \bar{y}_{12}\bar{E}_2 \\ \bar{I}_2 = \bar{y}_{12}\bar{E}_1 + \bar{y}_{22}\bar{E}_2 \end{cases} \tag{4}$$

The dynamic evolution of the two generators is described my means of the following differential equations:

$$\begin{cases} \ddot{\delta}_1 = \dfrac{P_{m1} - P_{e1}}{M_1} \\ \ddot{\delta}_2 = \dfrac{P_{m2} - P_{e2}}{M_2} \end{cases} \tag{5}$$

In an electrical power system composed by more than one machine—and in particular, with that one composed by two machines, here analyzed—it is necessary to distinguish between relative motion and mean motion. The equation for the relative motion is:

$$\ddot{\delta}_1 - \ddot{\delta}_2 \triangleq \ddot{\delta}_{12} = \frac{P_{m1} - P_{e1}}{M_1} - \frac{P_{m2} - P_{e2}}{M_2} \tag{6}$$

It is always possible to assume one machine as the reference and to study the motion of the other with respect to the first (Di Caprio and Saccomanno 1970; Kundur 1994; Marconato 2008; Saccomanno 2003).

3 Simulations

The model of a two machines power system shown in the previous paragraph constitutes the nucleus of a simulation experiment. We use simulation with the goal to find the conditions under which critical behaviors come into evidence. The structural changes and instabilities implicates that the system goes out of standard operations; this means the real possibility that the power system goes into disruptive conditions.

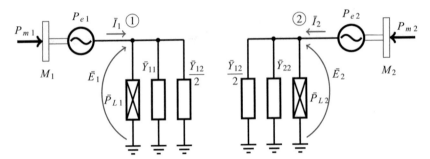

Fig. 2 Electric power system composed by two interconnected synchronous machines when in the medium of the power line occurred an interruption due to a short circuit

Table 1 Simulation example: (A) Parameters, (B) Dynamic initial conditions, (C) Perturbations

(A) Parameters	Node 1	Node 2
Electromotive forces	$E_1 \;= 1.0$	$E_2 \;= 1.0$
Mechanical powers	$P_{m1} = 1.0$	$P_{m2} = 1.0$
Inertia constants	$M_1 \;= 1.5$	$M_2 \;= 1.5$
Power loads	$P_{L1} \;= 1.2$	$P_{L2} \;= 1.1$
Line state	CLOSED	
Line length	100.0 [km]	
Line longitudinal admittance	$\bar{Y}_{12} = 0.0062 - 0.0230j$ [S]	
Line transversal admittance	$\bar{Y}_{11} + \bar{Y}_{22} = 2.82 \times 10^{-4} j$ [S]	

(B) Dynamic initial conditions	Node 1	Node 2
Machine 1	$\delta_1(0) = 0.0$ [rad]	$\omega_1(0) = 0.0$ [rad/s]
Machine 2	$\delta_2(0) = 0.0$ [rad]	$\omega_2(0) = 0.0$ [rad/s]

(C) Perturbations	Variable	From value	To value
$t = \;\;6.0$ [s]	Mechanical power P_{m1}	1.05	0.6
$t = 15.0$ [s]	Line state	CLOSED	OPEN
$t = 22.0$ [s]	Mechanical power P_{m1}	0.6	0.9

Above conditions are implemented by simulating a physical interruption with short circuit in the middle of the power line connecting the two machines and the loads (Fig. 2). In particular we can study the following phenomena: (1) Opening of the line (in the middle point); (2) Perturbations on mechanical power and/or loads; (3) Phase differences in initial conditions of the machines; (4) Random combinations of the above perturbations.

The simulation is performed using Matlab and Simulink software and shows the time evolution of the system under the two operating conditions described in Figs. 1 and 2. In Table 1 are reported, at the instant $t = 0$,

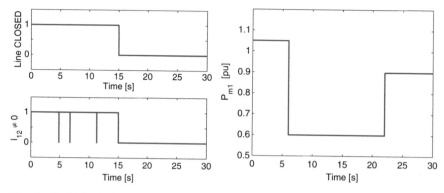

Fig. 3 (Left) Logical signals: line switching (1 for line CLOSED) and current $\bar{I}_{12}$ (1 for $\bar{I}_{12} \neq \bar{0}$). (Right) Perturbations on mechanic power P_{m1}

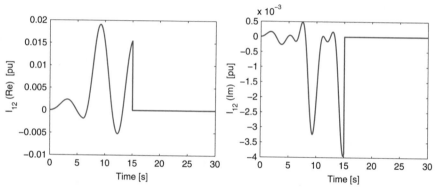

Fig. 4 Current $\bar{I}_{12}$ flowing between two machines: (Left) Real part; (Right) Imaginary part

(A) Parameters and variables, and (B) Dynamic Initial conditions, while (C) the sequence of perturbations (all the units are in Per Unit [pu]) executed during a simulation time $\Delta t = 30$ [s]. The evolution of some significative quantities describing the dynamics of the system is shown in Figs. 3, 4 and 5. In particular in Fig. 3 are shown the instants of the perturbations for the opening of the line (Fig. 3 (Left)) and the profile of the perturbations on the mechanical power P_{m1} (Fig. 3 (Right)). In Fig. 4 are reported the effects of the line opening on the current $\bar{I}_{12}$. Figure 5 shows how the perturbations on mechanical power and the structural changes, originated by the line opening, affect the trajectories in the state subspaces of the two machines.

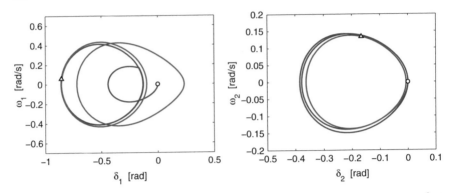

Fig. 5 Trajectories in state subspaces: (Left) space $\mathbf{D}_1^2$ of machine 1; (Right) space $\mathbf{D}_2^2$ of machine 2. (Open circle) $t = 0$, initial condition; (open triangle) $t = 15[\text{s}]$, line opening. Color of trajectories: (Blue) line connected; (Red) line open

4 Quasi Systems

If the description of a system is incomplete or inadequate, it is interesting to evaluate the idea of quasi-system. For a first examination it is useful to consider the state space and state subspaces in which the mathematical description of the fourth order electrical power system previously considered is established. With the help of the simplified visualizations in Fig. 6 we can fix the following description spaces:

1. *Global Description space.* The system operates and evolves on all the space $\mathbf{R}^4$, with $\mathbf{R}^4 = \mathbf{R}_1^2 \times \mathbf{R}_2^2$. Because the range of the state variables is unlimited, it is the natural environment in which we can study all the possible phenomena in the system evolution, as stability and instability conditions and structural changes (Fig. 6a, b).

2. *Design Conditions space.* The system operates and evolves inside the range of the design conditions. The range of state variables is limited to the values imposed by the desired operating conditions, then the environment in which we describe the evolution of the system is the space $\mathbf{D}^4 = \mathbf{D}_1^2 \times \mathbf{D}_2^2$, where $\mathbf{D}^4 \subset \mathbf{R}^4$, $\mathbf{D}_1^2 \subset \mathbf{R}_1^2$ and $\mathbf{D}_2^2 \subset \mathbf{R}_2^2$ (Fig. 6b).

   The spaces of design conditions contains the basins of attraction of the system, then it is evident that the stability and instability conditions of the system define the fundamental requirements and the constraints for design. As a consequence many phenomena, as instability, cannot be described into the design spaces.

The aforesaid distinctions show how the description of a system may be adequate or insufficient. Indeed this is one of the ideas supporting the need to introduce a characterization of a system description as a System (S) or a Quasi-system (Q).

Furthermore the relations between nonlinearity and instability constitute another interesting field of investigation (Anderson and Fouad 2002; Hahn 1967; Kimbak 1994; Kundur 1994; Marconato 2008; Saccomanno 2003). From the example we see how the nonlinearity can determine an evolution of the system outside the basin of attraction of the stable equilibrium points. When the two generators are connected the system evolves as a fourth order system for which the basin of attraction includes more than one equilibrium point. On the contrary, when the two generators are not connected, the global system may manifest a structural change and evolve toward two isolated second order systems. In this case the trajectories of the subsystems may evolve inside or outside the basin of attraction of the equilibrium point of each subspace.

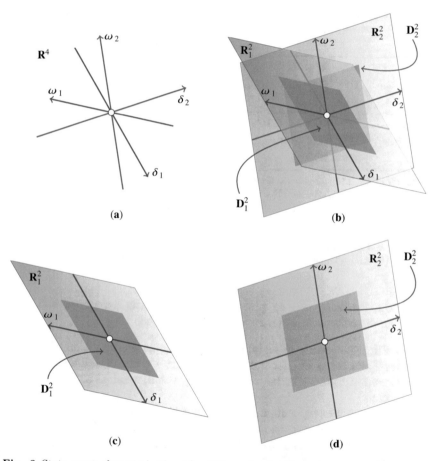

Fig. 6 State space characterization of system and subsystems for the electrical power system composed by two interconnected synchronous machines. (**a**) Global state space of the system; (**b**) State subspaces of the global state space. When a structural change separates the system into two independent subsystems, the subspaces become effective state spaces; (**c**) State space for machine 1; (**d**) State space for machine 2

Table 2 Descriptions of systems and relative state spaces with state variables

Type	System	State space	State variables
Global description	Global state space	$\mathbf{R}^4 = \mathbf{R}_1^2 \times \mathbf{R}_2^2$	$\delta_1, \omega_1, \delta_2, \omega_2$
	State subspace of machine 1	$\mathbf{R}_1^2 \subset \mathbf{R}^4$	δ_1, ω_1
	State subspace of machine 2	$\mathbf{R}_2^2 \subset \mathbf{R}^4$	δ_2, ω_2
Design conditions	State space	$\mathbf{D}^4 = \mathbf{D}_1^2 \times \mathbf{D}_2^2 \subset \mathbf{R}^4$	$\delta_1, \omega_1, \delta_2, \omega_2$
	State subspace of machine 1	$\mathbf{D}_1^2 \subset \mathbf{R}_1^2$	δ_1, ω_1
	State subspace of machine 2	$\mathbf{D}_2^2 \subset \mathbf{R}_2^2$	δ_2, ω_2

Table 3 Stability and Structure of a system influence the completeness of a description

Type	State space	Stability[a]		Structure[b]		Example[c]			
		(S)	(U)	(U)	(M)	(1)	(2)	(3)	(4)
Global description	$\mathbf{R}^4 = \mathbf{R}_1^2 \times \mathbf{R}_2^2$	S	S	S	S	S	S	S	S
	$\mathbf{R}_1^2 \subset \mathbf{R}^4$	Q	Q	Q	S	Q	Q	S	S
	$\mathbf{R}_2^2 \subset \mathbf{R}^4$	Q	Q	Q	S	Q	Q	S	S
Design conditions	$\mathbf{D}^4 = \mathbf{D}_1^2 \times \mathbf{D}_2^2 \subset \mathbf{R}^4$	S	Q	S	S	X	X	X	X
	$\mathbf{D}_1^2 \subset \mathbf{R}_1^2$	S	Q	Q	X	X	X	X	X
	$\mathbf{D}_2^2 \subset \mathbf{R}_2^2$	S	Q	Q	X	X	X	X	X

In different time intervals, the example can be characterized as Systems (S), Quasi-Systems (Q) and Systems and/or Quasi-Systems (X)
[a](S) Stable; (U) Unstable
[b](U) Unchanged; (M) modified
[c]Time Intervals: (1) 0.0–6.0 [s], Interconnected machines; (2) 6.0–15.0 [s], Interconnected machines and perturbation on P_{m1}; (3) 15.0–22.0 [s], separated machines after the line opening; (4) 22.0–30.0 [s], separated machines and perturbation on P_{m1}

Then, concentrating on the physical meaning of the dynamics, two main arguments become crucial (Kundur 1994; Marconato 2008):

1. *Stability.* The study of stability needs a complete description of the system. Then $\mathbf{R}^4$ is the natural space in which we can describe all the conditions for stability or instability during the dynamical evolution of the system.

   When instability phenomena are present, the system may evolve outside the basin of attraction of the stable equilibrium points (Hahn 1967; Kimbak 1994). In this case the evolution of the system goes outside the design space $\mathbf{D}^4$ which then becomes inadequate for the description of the system.

2. *Structure.* During the dynamic evolution, the structure of the system could become affected by a change. The presence of nonlinearities could

amplify the effects of these structural perturbations giving place to the raise of instabilities.

In the case of a structural change, the four-dimensional state space $\mathbf{R}^4$ of the global system reduces itself to two bidimensional and independent state subspaces $\mathbf{R}_1^2$ and $\mathbf{R}_2^2$. Then the subspaces become the new state spaces of the new independent subsystems. Also, the design conditions are defined into the new independent spaces $\mathbf{D}_1^2 \subset \mathbf{R}_1^2$ and $\mathbf{D}_2^2 \subset \mathbf{R}_2^2$ (Fig. 6c, d).

In Table 2 the different system descriptions are reported recalling the state spaces involved and the relative state variables (see also Fig. 6).

Instead Table 3 recalls how the stability and the structure of a system influence the completeness of its description. If we can give a complete description we speak of a system, if the description is not complete we speak of a quasi-system. The previous spaces are the natural environments from which we can characterize a description of systems and we are able to affirm *a priori* whether it is a system or quasi-systems. This is true for stability; instead for structure the characterization remains indefinite for Design Conditions because it depends from the stability of the system (it is marked as X).

Always in Table 3, for the simulated example the characterization of the description is reported for the four time intervals defined by the sequence of perturbations. For Global Description we can speak of systems (S) or quasi-systems (Q). In particular the structural changes due to the line opening, rise the subspaces $\mathbf{R}_1^2$ and $\mathbf{R}_2^2$ to state spaces for the complete description of two separate dynamical systems of the second order, then may be characterized as systems.

This fact is shown in Fig. 5, in which the state trajectories in intervals (1) and (2) are depicted in "blue" color (they are the projections on subspaces $\mathbf{R}_1^2$ and $\mathbf{R}_2^2$ of the state trajectory in space $\mathbf{R}^4$), while in intervals (3) and (4) are depicted in "red" color (they are two state trajectories into two different and independent state spaces $\mathbf{R}_1^2$ and $\mathbf{R}_2^2$). For these reasons the Global Description in time intervals (1) and (2) deals with systems (S) and quasi-systems (Q), while in time intervals (3) and (4) it considers only systems (S).

For the Design Conditions, the simulations show a stable evolution of the electrical power system in all the time intervals (1)–(4); then all the characterizations should be marked as systems (S). We marked (X) because the characterizations as systems or quasi-systems are possible depending of the stability of the system. In fact it is enough to change the system's parameters or the values and timing of perturbations to introduce instabilities. As a consequence the design conditions are no more respected and a characterization as quasi-systems (Q) should impose.

5 Conclusions

The simulation of an electrical power system composed by two interconnected synchronous generators constitutes an useful example to show how the adopted mathematical descriptions may be more or less adequate to represent the dynamical evolution of a system.

In particular the mathematical spaces in which the models of the dynamical systems are described constitute the natural environment in which we can study the dynamical evolution of a system under perturbations of structure leading to stability or instability conditions.

A description which enables to study the evolution of systems in all the operating conditions may be considered complete and we can use it to characterize a "system". When the description is unable to describe all the characteristics of a system's model, it is inadequate and it contains elements of uncertainty. In this case it cannot be assumed as a system; we call it a "quasi-system".

In our example we pointed out two main descriptions to show the concepts: the Global description and Design Conditions description.

With reference to engineering applications, the Design Conditions characterize a stable system evolving in normal operation in which the state and parameters assume only limited values. Considering Global Description all the operating conditions may be represented under all stability or instability states and the possible structural changes.

We saw how using the different mathematical spaces that host our descriptions, we can derive a qualitative characterization in terms of Systems (S) or in terms of Quasi-systems (S). In particular we saw how stability conditions may influence directly these characterizations.

References

Anderson, P. M., & Fouad, A. A. (2002). *Power systems control and stability* (2nd ed.). Piscataway: IEEE/Wiley.

Di Caprio, U. (2006). Typical emergencies in electric power systems. In G. Minati, E. Pessa, & M. Abram (Eds.), *Systemics of emergence. Research and development* (pp. 293–310). New York: Springer.

Di Caprio, U., & Saccomanno, F. (1970). Non-linear stability analysis of multimachine power systems. *Ricerche di Automatica, 1*(1), 2–29.

Hahn, W. (1967). *Stability of motion*. Berlin: Springer.

Kimbak, P. (1994). *Power systems stability*. New York: McGraw-Hill.

Kundur, P. (1994). *Power systems stability and control*. New York: McGraw-Hill.

Marconato, R. (2008). *Electric power systems (2nd Edition). Volume 3: Dynamic behaviour, stability and emergency controls.* Milano: CEI, Comitato Elettrotecnico Italiano.

Saccomanno, F. (2003). *Electric power systems. Analysis and control.* Piscataway: IEEE/Wiley.

Part V
Incompleteness and Quasiness in Social Systems

The Psychopathological Process as a System of Dysfunction and Systemic Compensation with Top-Down Modulation

Pier Luigi Marconi, Maria Petronilla Penna, and Eliano Pessa

1 Introduction

As it has been evidenced in the last years, the understanding of cognitive functions and dysfunctions requires an approach based on a suitable knowledge about the organization of brain large-scale neurocognitive networks (see Bressler and Menon 2010; Meehan and Bressler 2012; Bressler and Kelso 2016). In this regard it is to be remarked that neurocognitive networks cannot be described as equivalent to systems made by a given number (eventually constant) of elementary units (neurons) working in a cooperative way to produce a given macroscopic cognitive behaviour.

The first model proposed by Bressler and collaborators in 2010 (*Triple Network Model*: see Menon 2011; further experimental evidence of this model has

P. L. Marconi (✉)
Dipartimento di Psicologia Dinamica e Clinica, Università di Roma "La Sapienza", Roma, Italy

Italian Systems Society (AIRS), Milano, Italy
e-mail: pierluigi.marconi@uniroma1.it

M. P. Penna
Dipartimento di Pedagogia, Psicologia, Filosofia, Università degli Studi di Cagliari, Cagliari, Italy

Italian Systems Society (AIRS), Milano, Italy
e-mail: maria.pietronilla@unica.it

E. Pessa
Department of Brain and Behavioral Science, University of Pavia, Pavia, Italy
e-mail: eliano.pessa@unipv.it

© Springer Nature Switzerland AG 2019
G. Minati et al. (eds.), *Systemics of Incompleteness and Quasi-Systems*,
Contemporary Systems Thinking,
https://doi.org/10.1007/978-3-030-15277-2_14

been given by Wu et al. 2016), included three large-scale brain networks. This model describes both brain normal and pathological operation in terms of three core neurocognitive networks, that is the Default-Mode Network (DMN), the Salience Network (SN), and the Central Executive Network (CEN).

A later model was introduced by Raichle (2011), which added to the main three brain network of Bressler the Dorsal Attention Network (DAN) plus other input networks. However the Triple Network Model still remain valid, especially in studies about higher cognitive functions. These networks are mainly functional, being different from time to time and from a macroscopic process to another, being temporarily recruited according to the global needs of the brain.

In such a way we can observe a hierarchic systemic organization of the brain cortex, that from columns architecture, is integrated in dipoles which include large patch of the cortex. Such dipoles are functionally integrated in brain networks focused in global cognitive task as inner world processing, stimuli valence attribution, brain activity and behavioural response control and external stimuli processing.

The activities of these functional network are harmonized each other to reach the global goal of the mind/brain whole system to sustain the integration, the wellness and the persistence of individuals into the natural and social system.

The neurocognitive networks cannot be considered as networks in the usual sense, but rather temporary aggregates of smaller subsystems, sharing sometimes the local positions, other times the functionality, and, still other times, the interconnections. It would be more correct to define them as "quasi-networks" or "quasi-systems".

The internal harmonization of brain is warranted by the high level of degree of freedom of these "quasi-networks", which sustain the adaptation and the resilience capabilities of the whole human body-mind system.

When we have a failure of these capabilities, mainly linked to a reduction of such degrees of freedom, the disease state appears, with the linked risk to lose inner integration and resilience.

We can suppose that in disease the emergent properties, that we can find in healthy people, may have a top down influence to force a reaction of the whole brain to restore an adaptive behaviour by alternative way of functioning.

Observed from outside the brain dynamics of such a networks under "disease conditions" leads to the different sets of behaviours, signs and symptoms that we can include in the concept of clinical syndromes, whose ultimate goal is to survive, maintaining the best as possible adaptation to the environment. In agreement with a systemic approach to mental illness, it is possible to classify the components of these syndromes in five subsets, which can be related to the sequence of events of the pathogenesis of a disorder: system overload (or subjective discomfort), external integration dysfunction (social dysfunction), loss of some emerging functions (negative symptoms), reappearance of less adaptable primary functions (positive symptoms), activation of top-down reactions to induce compensatory activities of collateral systems (Marconi 2014).

At the base of these processes it could be the reduced integration of one or more of the subsystems, which could lead to the loss of flexibility of the general system as a whole (the mind), with the appearance of the five psychopathological components described above.

At present time we can study the brain dynamics related to the functional activation of brain networks, by means of computerized electrophysiology. The cortical electrical activity is related to the activity of stockades of pyramidal neurons which act as a whole as a detectable dipole. When inputs stimulates pyramidal cells a positive wave may be recorded on the scalp, while a negative one may be recorded when the cortex sends outputs by pyramidal cells.

Using neurophysiology we can study not only the time course of such an activity (input/output) but also how strong such an activity may be modulated by long range inputs. In fact the local activity is more linked to high frequency EEG oscillations, while the long range modulation are related to the low frequency EEG oscillation. We can study the coupling of these two activities, supposing that this high frequency/low frequency coupling may be an expression of network integration. Typically it is studied as phase amplitude coupling between the low frequency (2–12 Hz, Alfa, Delta, Theta Band) phase and the high frequency (40–100 Hz, Gamma Band) amplitude (*Cross Frequency Modulation*, CfM).

Daniel Siegel had supposed that the brain dynamics of maltreated people can be disrupted. Child maltreatment can be due either by parent inattention to child or by psychological or physical abuse. People with a personal history of child maltreatment clinically present emotional dysregulation, problems on social relationships, but also difficulties in error management and valence attribution.

We can suppose that some brain networks may have a lack of integration in relation to controls. However it can be also expected that different networks can be activated compensating the dysfunctional ones. These activities can be an expression of a top down influence which try to maintain the global functionality of the actual behaviour (response), even if it can be done in a less effective way.

To confirm this hypothesis we have observed data got from some neurophysiological studies performed with high-density EEG recording during neuropsychological tests.

Then we have constructed some artificial quasi-network modeling the observed system with a variable functional activation to observe which inner network dynamics can explain the observed data and the compensation effect.

2 Experimental Data

EEG event related data were recorded in 27 subjects (13 with child maltreatment history and 14 controls). Mean age in both group was similar (34.7 years vs 33.4 years in maltreated and controls respectively), and female were prevalent (3:1 in maltreated and 5:3 in controls). No difference was found at the cognitive performance, but maltreated had a different lower duration in studies, and less probable to have reached university degrees. The maltreated group was confirmed by psychometrics having more emotional dysregulation, more emotional distress, less functioning in everyday activities and in social relationships, even if they stated otherwise (i.e. a quality of life quite similar to control people). This confirms all the clinical components of the maltreated child syndrome: emotional distress and dysregulation, poor scholar functioning, relational dysfunction and real life attribute distortion.

The EEG was recorded during a neuropsychological task, in which subjects were asked to attribute an affective valence (appetitive or aversive) to infrequent affective stressing slide presented on the screen among a sequence of neutral slides. The task had three elements:

1. The judgment had to be done on infrequent slides (5% appetitive, 5% aversive).
2. The probe slide (on which the valence attribution had to be performed) was preceded by a cue slide which was alerting the subject about the probability of valence of the following slide (after a positive cue was possible just an appetitive or neutral slide only; similarly after a negative cue).
3. The response was pre activated as direction by the cue, but it needed also to be inhibited till the final valence attribution was completed.

So the cognitive task had to face the alert reaction (performed by the SN), the behaviour control (by CEN), the formal recognition of the content of the slide (by DMN) and then the final valence attribution (SN).

It is already known that the alert reaction and the relevance attribution tasks are linked to the event related EEG wave P_3 (alert P_{3a}, recorded in frontal areas, and relevance detection P_{3b}, in parieto-temporal areas), occurring between 250 and 450 ms after the stimulus, while the valence attribution is a more complex processing with positive/negative discrimination performed before the P_3 time windows and the valence attribution with secondary affective processing after the P_3 time windows. The *Last Positive Potential* (LPP) is the wave studied about such a late valence processing, where integration of physical characteristic of stimuli has to be associated to the affective valence given to it.

The high density EEG tracks were recorded at 250 Hz and processed with the MatLab package EEGLab. The Cross Frequency Modulation was computed using a EEGLab subroutine ad hoc modified. The rough EEG signals were treated to compute the independent sources (Independent Component

Analysis, ICA) and were excluded all the sources which computed dipole was with a residual variance more than 15% and the estimated location was outside the cortex, since the EEG signal is assumed to be originated mainly by the cortex pyramidal cells. As previously described, the brain dipoles can be considered as the functional subcomponents of the bran networks: they are characterized by a strong phase coherence of the electrical oscillations and by a similar brain location. We can assume that this phase coherence can be due to the external input with an intrinsic coherence in its properties, and/or to the effect of the internal network integration. We assume that this internal network integration may be detected by computing the CfM. This is done by applying on dipole signals the Hilbert transformation, and then computing correlation between high frequency amplitude and low frequency phase. Since our goal is to study the different brain dynamics of dysfunctional people in respect to controls, we have considered just only the dipole with significant different CfM between controls and maltreated people.

Observing the task behavioral performance of subjects, we can see that the presence of a cue help them to give an attribute valence to the probe. However it seems that aversive valence attribution is less demanding then appetitive valence attribution, since response time is significant lower after a priming cue (815 ± 108 ms, vs 691 ± 118 ms and 648 ± 118 ms after neutral, positive and negative prime respectively) (Fig. 1 (Right)). However we can see a difference between the two groups of people in respect to the accuracy of responses: maltreated people were less accurate after neutral cues and the most accurate after negative cues, while the controls were the most accurate after neutral cues and less accurate after positive cues (Fig. 1 (Left)).

This finding supports the hypothesis that aversive stimuli are processed faster and with more accuracy then appetitive ones as well as the presence of a cue facilitate valence attribution. However people with child maltreatment history trend to give answer even when neutral cues were shown, as the lack of a cue was lowering behavioral control. On the other hand control people

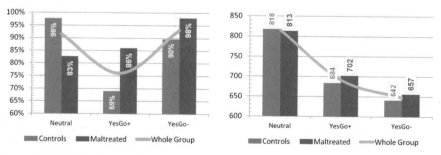

Fig. 1 (Left) Response accuracy: in controls condition effect $p = 0.005$; in whole group condition effect $p = 0.031$. (Right) Response time: in maltreated condition effect $p = 0.047$; in whole group condition effect $p = 0.038$

seem to be more selective and response controlled then maltreated people, especially in giving an appetitive valence.

These data support the hypothesis that the two groups have different ways to process inputs and affective valence.

Studying the brain dynamics behind such a behavior, nine dipoles had significant different CfM between two groups. The Maltreated group activated more dipoles included in the DMN (3/5 dipoles) and the CEN (2/5 dipoles) was activated in the early impulsive reaction control area and in the "*switch from expectancy*" control area. The control group has shown a more balanced pattern, activating 2/6 dipoles included in the DMN, 2/6 dipoles included in the CEN network, and 2/6 dipoles included in the SN (Fig. 2).

Attention related dipoles (RB35, RB24, LB31) were activated mainly in control group just after priming cue was presented, and maltreated group mainly after probe was presented. In any case the ERP modulation of these

Brodman Area	Supposed Area Functionality	General Function	B-Net	Brain Lobe	Priming M	Priming C	Priming ERP	Probing M	Probing C	Probing ERP
LB47	*Adverse Emotion Inhibition* in Conflict Decision	MJ/MOM	CEN	F	*m*					
RB25	Nociceptive processing	MJ	SN	F	*m*					
RB45	*Response Inhibition* in Processing Affective Valence & Face recognition	MOM	CEN	F	*m*					
RB35	*Relevance Check*	TIR	DAN/CEN	T			P$_{3b}$			
RB24	*VisuoSpatial Attention* to Novelty	TIR	SN	F				*m*		P$_{3a}$
LB31	Visual High Demand Processing of threat related stimuli	MJ	DAN/DMN	P				*m*		LPP
RB37	Affective recognition & categorization of face and familiar objects	MOM	DMN	T						P$_{3b}$
LB10	*Switch Control*: expectancy → actual stimuli emotional valence	TIR	CEN/DMN	F				*m*		
LB37	Reasoning about affective categorization of faces & objects	MOM	DMN	T						

TIR   = Tune into Reality
MOM   = Reasoning about Intentions of Others
MJ   = Valence attribution & Moral Judgment
ERP   = Event Related Scalp Recorded Potential
        (*color indicate group with higher potentials*)

m   = CfM modulated by stimuli

P$_{3a/b}$ = ERP P$_3$ (*250-450 ms*) wave
        (**a**: early component; **b**: late component)

LPP   = ERP LPP (*450-1000 ms*) wave

SN   : Salience Network
CEN : Central Esecutive Network
DMN : Default Mode Network
DAN : Dorsal Attention Network

F: Frontal; T: Temporal; P: Parietal

M   : Maltreated Group
C   : Control Group

Default Mode Network

Dorsal Attention Network

Executive Control Network

Salience Network

(*Raichle ME, 2011*)

Fig. 2 Nine Broadman areas corresponding to the nine dipoles with different Cross Frequency Modulations between maltreated people and control group. The Broadman area is labeled with two letter for brain side (LB = Left Side; RB = Right Side) and the area number. With red color are presented data with higher values (statistically different) in maltreated people, while in blue are presented data with higher values in control group. On the right bottom the three main brain networks (Central Executive, Default Mode and Salience Attribution) plus the Dorsal Attention Network are presented. For more details see this paper

area was statistically different in two groups: controls presented higher modulation of P3, already just after the cue, while maltreated people had higher modulation of LPP (a later component) and not in the anterior cingulated or para hippocampus areas, as usually is observed, but in the medial parietal cortex of secondary visual input processing.

We can suppose a different brain dynamics between maltreated people and controls. Classical P_{3a} and P_{3b} ERP waves are more evident in the control group, which has a higher integration and brain activity in the right hemisphere.

While in this group the cue and the probe are evaluated for novelty (positive and negative prime occurred 20% each in respect to total prime slides) in RB24 and for affective valence (negative or positive cue) in RB35 already before 450 ms, in maltreated group the prime was evaluated just as deviant stimuli only (in RB24 and not in RB35).

In this group affective categorization is performed mainly rationally, controlling stimuli perception (in LB31) and forcing the control of the response and of the "actual valence brain switch" (LB10).

So it can be supposed that valence attribution may be later and formally more accurate than control people, switching the way to perform this task from less aware intuitive processing (RB24, RB35, RB37) to more aware rational processing (LB31, LB37), blocking the spontaneous attribution (RB45, LB10) and making very few references to the valence of the previous cue.

This may explain the presence of a lack of response inhibition in maltreated group (since the response control is less activated after neutral cues) and a higher and similar accuracy in checking appetitive or aversive probes. About response timing, just a non significant trend in later responses was found, as we could instead expect.

On the other hand control group seems to react more to aversive probe (RB45, LB47), differentiating, after emotional cues, the relevance of neutral probes in respect to emotional significant ones.

The higher reaction is produced by negative probes (P_{3a} in RB24), but a higher relevance is attributed to positive cue and neutral probes (P_{3b} in RB37): this can explain the observation of less "accuracy" after appetitive probes, which can be interpreted as a more subjective criteria used in valence attribution after positive cues and appetitive probes (Fig. 3).

The model proposed highlights four main differences between the two groups:

1. the control groups uses more intuitive valence attribution matching data with previous memories and the right brain is more involved;
2. the maltreated group instead uses more a rational conscious processing, matching the input with preset formal criteria that may be used as a filter on sensory input processing;
3. in the control group the CEN activity is modulated by stimuli affective valence;

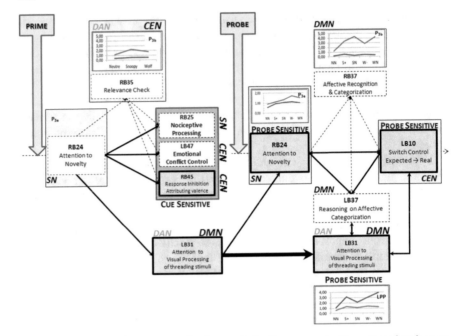

Fig. 3 A different information and salience attribution processing is supposed to be performed in maltreated people in respect to control group. Here a model of such differences is presented. In upper side of the figure processes performed in right brain areas are presented; in the lower side the right brain ones are presented. In left side the processing observed after prime and in the right side of model the processing observed after probe. Highlighted in yellow are the processing supposed to be more activated by maltreated people. It is evident the higher activation of left brain areas in maltreated people. The graphs presented in correspondence of each brain area plot the different responses and Event Related Potentials (P3 or LPP) observed in the two groups (red line = maltreated; blue line = controls) in the different conditions (NN = neutral prime; S+ = positive probe after positive prime; SN = neutral probe after positive prime; W− = negative probe after negative prime; WN = neutral probe after negative prime). See this paper for more details

4. in the maltreated group the CEN activity modulates responses and valence processing (may be to stop the SN in controlling preconscious intuitive processing and response trigger).

3 The Math Model

Since the large brain networks dynamically integrated during specific task, we can suppose that the building of neurocognitive networks models should be a very difficult task, overcoming the possibilities offered by actual mathematics. Happily, this is not the case, for two main reasons:

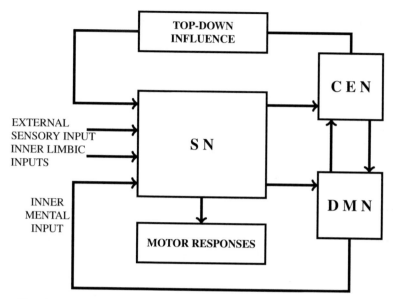

Fig. 4 Here SN = Salience Network, CEN = Central Executive Network, DMN = Default Mode Network. The top-down influences include the attentional processes as well as the information coming from working memory. The inner limbic inputs came from all contributions produced by parts of the brain included within the limbic circuit (like, for instance, amygdala). The inner mental input collects all contributions produced by self-referential mental processes

1. the number of the identified neurocognitive networks is rather small,
2. mathematicians found methods suited to deal with very large neuronal assemblies as if they were single objects, accessible to standard mathematical techniques.

Among the main approaches introduced by mathematicians we quote the *Neural Field Theories* (see, e.g., Coombes 2005; Bressloff 2012; Meijer and Coombes 2014) and the *Neural Mass Theories* (see Stephen et al. 2006; Deco et al. 2008).

In order to build a quantitative model of the brain dynamics we will make use of the property (1) quoted above. More precisely, we will base our considerations on the Without entering into further details about the operations of these network, it is more convenient to describe their reciprocal interconnections, depicted in the drawing reported in the Fig. 4.

In order to describe the possible behaviours of this model we need to introduce a suitable mathematical description. As the introduction of the latter seems, in principle, to be a very difficult task, here it appears as more convenient to associate to each network a macroscopic description in terms of a suitable *neural mass model*.

A typical single model of this kind is described (see Deco et al. 2008) by a system of two ordinary differential equations in two time-dependent variables, shortly denoted by μ_ν and μ_α, having the form:

$$\dot{\mu}_\nu = \mu_\alpha$$
$$\dot{\mu}_\alpha = \kappa^2 \varsigma(\mu_\nu) - 2\gamma \mu_\alpha - \gamma^2 \mu_\nu \tag{1}$$

where κ and γ are suitable parameters while $\varsigma(\mu_\nu)$ is a sigmoid function defined by:

$$\varsigma(\mu_\nu) = \frac{2\kappa}{1 + \exp(-r\,\mu_\nu)} - \kappa. \tag{2}$$

Of course, the symbol r denotes another parameter, specifying the sigmoid growth rate.

By applying these equations to the model structure depicted in Fig. 4, it is possible to obtain the explicit form of the differential equations describing the neural-mass representation of Menon's model. These equations are:

$$\frac{dSN_\mu}{dt} = SN_\alpha$$

$$\frac{dSN_\alpha}{dt} = \kappa^2 \varsigma(SN_\mu) - 2\gamma\, SN_\alpha - \gamma^2 SN_\mu + \alpha_T\, CEN_\mu + B_I\, DMN_\mu + I + L$$

$$\frac{dCEN_\mu}{dt} = CEN_\alpha$$

$$\frac{dCEN_\alpha}{dt} = \kappa^2 \varsigma(CEN_\mu) - 2\gamma\, CEN_\alpha + \gamma^2\, CEN_\mu + \rho\, SN_\mu + \varphi\, DMN_\mu$$

$$\frac{dDMN_\mu}{dt} = DMN_\alpha$$

$$\frac{dDMN_\alpha}{dt} = \kappa^2 \varsigma(DMN_\mu) - 2\gamma\, DMN_\alpha + \gamma^2\, DMN_\mu + \sigma\, SN_\mu + \psi\, CEN_\mu. \tag{3}$$

Here the symbols SN, CEN and DMN denote, respectively, the neural activation densities (variable with time) of the Salience network, Central Executive network, Default Mode network. The subscripts μ and α distinguish between the two components of the activation densities, roughly corresponding, respectively, to postsynaptic depolarisation and capacitive current of involved neurons. In principle the parameters κ, γ, r could depend on the network taken into consideration, even if this choice could increase the model complexity. The values of other parameters included in the model equations are kept as shared between all networks. They can be interpreted as proportionality factors or amplitudes of the associated mass contributions. The symbols I and L, instead, denote, respectively, the global amount of the external sensory input and of the inner input coming from limbic brain circuit contributions (including, for instance, rewards and motivations). In order to simplify the further analyses each variable is coded through a single numerical value (obviously determined by external choices of the experimenter).

Without embarking on the difficult task of finding an analytic solution to model equations, we will directly deal with a search of their equilibrium stationary states. To this end, we start by approximating the sigmoid functions through their Maclaurin series development up to the first order, a change which eliminates all non-linear terms. Now it is convenient to introduce some ansatz on the possible forms of SN, CEN and DMN variables close to stationary equilibrium points, given by:

$$
\begin{aligned}
SN_\mu &= A_{SN}\, e^{W_{SN}\, t} & SN_\alpha &= B_{SN}\, e^{W_{SN}\, t} + B_{0SN} \\
CEN_\mu &= A_{CEN}\, e^{W_{CEN}\, t} & CEN_\alpha &= B_{CEN}\, e^{W_{CEN}\, t} + B_{0CEN} \\
DMN_\mu &= A_{DMN}\, e^{W_{DMN}\, t} & DMN_\alpha &= B_{DMN}\, e^{W_{DMN}\, t} + B_{0DMN}.
\end{aligned}
\tag{4}
$$

Here it is also convenient to suppose that the constants B_{0SN}, B_{0CEN} and B_{0DMN} have values so small as to be almost negligible. If, now, we further simplify our ansatz in order to assume that the coefficients present in the exponential functions satisfy the conditions:

$$
W_{SN} = W_{CEN} = W_{DMN} = W
\tag{5}
$$

it is immediate to acknowledge that the stationary equilibrium condition corresponds to $W = 0$.

By substituting these ansatz and the previous conditions in the model equations it is possible to show that these latter give rise, after suitable computations, to a linear system of algebraic equations in the unknown equilibrium amplitudes A_{SN}, A_{CEN}, A_{DMN}. This system has the form:

$$
\begin{aligned}
-R\, A_{SN} + \alpha_T\, A_{CEN} + B_I\, A_{DMN} + I + L &= 0 \\
-R\, A_{CEN} + \rho\, A_{SN} + \varphi\, A_{DMN} &= 0 \\
-R\, A_{DMN} + \sigma\, A_{SN} + \psi\, A_{CEN} &= 0
\end{aligned}
\tag{6}
$$

where $R = \frac{1}{2}\kappa^3 r + \gamma^2$.

Trivial methods of linear algebra allow to obtain the equilibrium amplitudes as solutions of the above system:

$$
\begin{aligned}
A_{SN} &= \frac{(I + L)(R^2 - \varphi\psi)}{D_T} \\
A_{CEN} &= -\frac{(I + L)(\rho R + \sigma\varphi)}{D_T} \\
A_{DMN} &= \frac{(I + L)(\rho\psi + R\sigma)}{D_T}
\end{aligned}
\tag{7}
$$

where:

$$D_T = -R^3 + R\psi\varphi + \alpha_T \rho R + \alpha_T \sigma \varphi + B_I \rho\psi + B_I R\sigma. \tag{8}$$

Looking to the form of found solutions a general consideration is that all three equilibrium amplitudes depend on the sum of the inputs $I + L$. While this circumstance was expected for A_{SN}, as the model equation for SN contains explicitly the contribution $I + L$, it could be less predictable for A_{CEN} and A_{DMN}, as this contribution is absent in the model equations for CEN and DMN. This means that the contributions of the inputs influences the behaviours of both Central Executive and Default Mode networks in an indirect way, mediated by the relationships between them and the Salience network. This supports the hypothesis that, even in presence of dysfunctions regarding the latter two networks, the brain could compensate for their faulting performances owing to the presence of collateral contributions coming from the eventual inputs. And this interpretation seems to be confirmed by the available experimental and clinical data.

Anyway, it should be interesting to focus on specific cases, defined by particular (and extreme) values attributed to the numerical coefficients present in the formulae. In the following we will shortly report about two possible situations.

Case A: Both CEN and DMN networks are characterized by extreme deficits of their activity.

This is equivalent to say that the values of CEN and DMN variables are both tending towards zero. Such a situation is characterized by the following parameter values:

$$\alpha_T = B_I = \varphi = \psi = 0. \tag{9}$$

The values of equilibrium amplitudes are:

$$
\begin{aligned}
A_{SN} &= -\frac{(I+L)}{R} \\
A_{CEN} &= -\frac{(I+L)\rho}{R^2} \\
A_{DMN} &= \frac{(I+L)\sigma}{R^2}.
\end{aligned}
\tag{10}
$$

It is immediate to see that, if the value of R is greater than 1, the module of A_{SN} is higher than the ones of A_{CEN} and A_{DMN}. In other words, the brain activity is mainly ruled by Salience network, in turn influenced by inputs contribution: a situation characterizing many psychiatric and neurological disorders.

Case B: Both SN and DMN networks are characterized by extreme deficits of their activity.

This is equivalent to say that the values of SN and DMN variables are both tending towards zero. Such a situation is characterized by the following parameter values:

$$B_I = \varphi = \rho = \sigma = 0. \tag{11}$$

In this extreme case the only non-zero value of equilibrium amplitudes is the one of SN, still given by:

$$A_{SN} = -\frac{(I+L)}{R}. \tag{12}$$

The activities of the other networks are practically zero (there is still a brain?).

4 Discussion

It is evident that for people with personal history of child maltreatment the CEN is activated to force a top-down effect which switch the valence attribution from the SN/DMN network interaction to a more aware processing SN/CEN network interaction, the first one making reference to previous memories and the second one to formal properties of inputs.

The simulation setting demonstrates the possibility that such a switch may be effect of an intrinsic property of the brain network balancing, however it is possible than in humans such a automatic reaction can be overcame by top down activities mediated by CEN and expression of the context driven mind of maltreated people.

References

Bressler, S. L., & Menon, V. (2010). Large-scale brain networks in cognition: Emerging methods and principles. *Trends in Cognitive Sciences, 14*, 277–290.

Bressler, S. L., & Scott Kelso, J. A. (2016). Coordination dynamics in cognitive neuroscience. *Frontiers in Neuroscience, 10*, 397.

Bressloff, P. C. (2012). Spatiotemporal dynamics of continuum neural fields. *Journal of Physics A: Mathematical and Theoretical, 45*(3), 033001.

Coombes, S. (2005). Waves, bumps, and patterns in neural field theories. *Biological Cybernetics, 93*, 91–108.

Deco, G., Jirsa, V.K., Robinson, P.A., Brealspear, M., & Friston, K. (2008). The dynamic brain: From spiking neurons to neural masses and cortical fields. *PLoS Computational Biology, 4*(8), e1000092.

Marconi, P. L. (2014). La malattia mentale nella prospettiva sistemica. *Rivista di Filosofia Neo-Scolastica, 3*, 561–587.

Meehan, T. P., & Bressler, S. L. (2012). Neurocognitive networks: Findings, models, and theory. *Neuroscience and Biobehavioral Reviews, 36*, 2232–2347.

Meijer, H. G. E., & Coombes, S. (2014). Traveling waves in models of neural tissue: From localised structures to periodic waves. *EPJ Nonlinear Biomedical Physics, 2*(3), 1–18.

Menon, V. (2011). Large-scale brain networks and psychopathology: A unifying triple network model. *Trends in Cognitive Sciences, 15*(10), 483–506.

Raichle, M. E. (2011). The restless brain. *Brain Connectivity, 1*(1), 3–12.

Stephen, K. E., Harrison, L. M., Kiebel, S. J., David, O., Penny, W. D., & Friston, K. J. (2006). Dynamic causal models of neural system dynamics: Current state and future extensions. *Journal of Biosciences, 31*, 129–144.

Wu, X., Li, Q., Yu, X., Chen, K., Fleisher, A. S., Guo, X., et al. (2016). A triple network connectivity study of large-scale brain systems in cognitively normal APOE4 carriers. *Frontiers in Aging Neuroscience, 8*, 231.

A Note on Variety and Hierarchy of Economic and Social Systems: The System-Network Dualism and the Consequences of Routinization and Robotization

Lucio Biggiero

1 Introduction

Seventy years after the foundations of cybernetics and systems science one could expect that everything important about the basic concepts of redundancy, variety and hierarchy has been said and that its applications to economic and social systems are plain. However, we hope to show that things are not exactly so and that some clarification could be useful and fruitful for understanding new developments, like those related to the robotization of economy and society. In what follows we will show that ontological differences make difficult the application of the concepts and measures of entropy and variety from physical (or biological) to economic systems. We will show that the incomparably higher complexity of the socio-economic respect to all the other types of systems requires a prudent application and interpretation. Such caution should increase when considering mathematical models and, even more, graphical representations. Without appropriate specifications, the hierarchical structure of some giant company of mass productions could appear as a simple system characterized by a degree of variety proximate to zero.

We will argue that this apparent incongruence derives also from the fact that the basic concepts of entropy, variety, redundancy and complexity have been thought for black-boxes, that is, systems whose elements are not connected each other and, if they are, such connections do not create recursive patterns or processes. In other words, the system's elements are independent, as in fact it is required by the measure of variety to be counted, or they

L. Biggiero (✉)
Department of Industrial Engineering, Information and Economics, L'Aquila University, L'Aquila, Italy
e-mail: biggiero@ec.univaq.it

© Springer Nature Switzerland AG 2019
G. Minati et al. (eds.), *Systemics of Incompleteness and Quasi-Systems*,
Contemporary Systems Thinking,
https://doi.org/10.1007/978-3-030-15277-2_15

are very fast randomly interacting like molecules in a gas. Not incidentally, the principal methods used by thermodynamics and information theory were statistical mechanics. Now, the issue that we will discuss is that neither the former view—which indeed is also in contradiction with system definition, which requires elements connections—nor the latter fits with most biological and socio-economic systems, which are made by relatively stable networks with recursive patterns. Therefore, we advance the idea that these systems are characterized by a dualist nature: a systemic one, which focuses mostly on input-output relationships, emergent properties, feedback mechanisms, etc., and a network one, where the attention is mostly devoted to understand the topological properties of the whole and its sub-networks. These two dimensions are complementary, and despite cyberneticists themselves did not acknowledge them in a consistent and clear way, the acknowledgment of this dualism is fundamental to understand socio-economic systems.

After disclosing the huge variety hidden under a hierarchical structure of an economic or social system, we underline that the substitution of men with machines that accompanies the development of our species is driven by two forces: (1) the progressive simplification of operative processes and tasks through standardization, specialization and routinization; (2) the lower costs of machines respect to workers. We notice that it is the goal of simplification—reduction of variety—that drives and precedes the substitution, and not vice versa. Further, we introduce the distinction between potential variety and actual variety: if humans' huge potential variety is made superfluous by dramatically reducing actual variety required by economic systems, then their replacement with machines is enhanced, and the net balance of actual variety can remain invariant. Finally, we will argue that such a balance could change with the progressive use of networks of intelligent robots, because its minor variety under most respects when compared with humans could be partially compensated by its incomparably higher power in storing and treating data. At the moment, the impact of this robotization process in terms of variety and humans' control of their life is definitely unknown.

2 Basic Definitions

Variety and redundancy have been conceptualized and measured in the widely accepted way proposed initially by thermodynamics, then by the mathematical information theory (Shannon and Weaver 1949), then developed and applied to natural and social systems by cybernetics (Ashby 1956; Wiener 1948) and systems theory (Bertalanffy 1968; Klir 1969). From the cybernetics perspective, variety corresponds to entropy, that is with a system's information content and its degree of disorder, while redundancy is measured as the complement to 1 of relative entropy, that is, as the degree of distance from its maximum entropy.[1] Put differently, given the variety of a system's elements, redundancy measures its degree of order: if it were zero, then the system would be maximally disordered.

In Ashby's approach (1956), the crucial variable is no more the number and probability of a system's state like in thermodynamics or the number of different symbols of a language, but rather the number and proportion of different types of a system's parts. In order to underline this distinction, Ashby renamed complexity (entropy) as variety. Here we follow this line, which seems indeed more consistent with the object of study. In this perspective, we have the usual expression:

$$V = \sum x_i \log x_i \tag{1}$$

where:

V is entropy or complexity (H in standard notation);

X is the set of the different types of variables, that is, employed resources in our application to social systems, and in particularly to economic organizations, like firms;

x_i are the frequencies (relative quantities, proportions, probabilities) of employed resources.

Analogously with information theory, redundancy (R) indicates a system's degree of order, which is measured as:

$$R = \frac{V - V_{max}}{V_{max}} \tag{2}$$

where V_{max} is maximum variety, corresponding to the case of resources equal distribution ($\sum \log x_i$). Let's further clarify and "translate" these concepts between the four fields of thermodynamics, information theory, cybernetics and economics:

- elements variety in cybernetics corresponds to symbols variety in information theory or to different states in thermodynamics or to different types of resources in economic organizations, let say workers of type[2] a, b, c, etc. or capital of type x, y, x, etc.;
- elements proportion in cybernetics corresponds to symbols frequency in information theory or system's states frequency in thermodynamics or to quantity of different resources—human and physical capital—in economic organizations.

According to information theory and cybernetics, when does redundancy increase? When the system employs a lower number of (different *types* of) re-

---

[1] A system reaches maximum entropy when all its types of elements are equally distributed—that is, they occur in the same proportion. In pure information theory, when all (language or transmission) symbols have the same frequency. Relative entropy (H_r) is the ratio between actual entropy and maximum entropy. Therefore, redundancy is ($1 - H_r$).

[2] These types can be defined in a number of ways: for instance in terms of competences, roles-positions, geographical location, etc.

sources and/or when resources equi-distribution between the different types is maximally violated. The former condition corresponds also to the case in which there are constraints between variables (resources). As explicitly remarked by Ashby—and later on by Atlan (1979)—it happens when one or more resources are functions of others, because the functional relationship indicates a dependence relationship. At the extreme, when $i-1$ variables are functions of the remaining variable, that is when $i-1$ resources depend on the remaining resource, we have the maximum redundancy. In this case, in fact, $V = 0^3$ and $R = 1$.

3 The System-Network Dualism

From an ontological perspective, we argue that virtual or material reality is organized in systems, characterized by the well-known properties of emergence, possible nonlinear effects when feedback mechanisms do occur,[4] and input-output relationships governing system's behavior and identifying what goes in and out of the system. The evolution from first- to second-order cybernetics (von Foerster and Zopf 1962; von Foerster 1982; Heylighen and Joslyn 2001; Scott 2004) marked important further specifications that turned away from the naïve views matured by early cyberneticists on their thinking and experiencing of mechanical machines in engineering systems, mostly adopted in defense devices.[5] As soon as they turned their attention to more complex systems, like the biological and then the social systems,[6] they realized that their approaches should be more sophisticated and that what previously might appear clear or certain and stable then had to be weakened in a less clear, more uncertain and unstable description, with its consequent implications for the complexification of the analysis.

---

[3] Notice that variety measures the number of *different* resources. Thus, when there is only one resource, variety is zero. This reminds to Bateson's (1972) fundamental remark that an information is a difference that makes a difference. Put differently, without distinctions there would be no perceptions. If all things had the same color we couldn't see because we couldn't distinguish anything.

[4] Indeed, not all systems are characterized by feedback processes. This is exactly a relevant demarcation of cybernetic systems, which are characterized by feedback mechanisms, from the others. And in fact, Wiener and Ashby put feedback at the core of cybernetics.

[5] It is worth reminding that cybernetics and systems science—but especially the former—have been nurtured and fed by military applications during WW2, in particular for aircraft tracking and targeting systems, naval servomechanisms, and telecommunications. For some short reference to these aspects see Wiener's Introduction to his *Cybernetics* (1948), while for a more detailed story see Heims (1982, 1991).

[6] It can be noticed that the recent developments of the so-called "infosphere", namely the various new forms of virtual reality and virtual or concrete robots, further confirm that the initial simple views are definitely inadequate to capture systems properties.

Due likely to the extremely heavy theoretical pressure exerted by physics, as in particular thermodynamics, the kind of systems, and especially the kind of theorizations, that cyberneticists had in mind were concerning systems whose inner structure is unknown or whose elements are made of a huge number of particles in constant and extremely frequent random motion, so that its connections do continuingly change. Molecules of a gas, which are characterized by Brownian motions or electrons orbits at the subatomic level are paradigmatic examples. Here, elements are so many and so fast moving in a random way that it would be meaningless to speak of a network, which instead would require a recognizable structure (links distribution). Though the two things are not perfectly identical, considering the box as black was usually taken as implying no interactions among system's elements or, even, as if elements' interactions were random. Therefore, the following implicit double identity was stated: unknown interactions = no interactions = random interactions. Somehow, these types of systems are the paradigmatic black boxes of cybernetics, which in fact was mostly inspired and influenced by physics or mathematical physics, especially by its founders, like Wiener and von Forster. In fact, the main methodology to approach the behavior of these types of systems is mechanical statistics, and it was just to this field that Wiener explicitly referred to in its seminal book (1948). In the Introduction of that book Wiener claimed that statistical mechanics was the common methodology of thermodynamics, the mathematical theory of information, and cybernetics.

However, alongside the theoretical and experiential evolution above mentioned, it became more and more clear that for some large categories of systems it could be possible to specify the inner structure, that is, the network connecting its elements. Therefore, these systems are not black, because they have a recognizable network. In this theoretical and empirical perspective, the systemic and the network dimensions are two fundamental ways in which, for some types of systems, an organization manifests its properties and functioning (Biggiero 2011). The systemic and network views are just complementary descriptive and analytical perspectives (Fig. 1). When we look at the systemic properties we consider the organization as a whole, focus on its emergent properties, input-output relationships, and (usually) on its boundaries. In so doing, we overlook the organization's inner structure. This is the black-box perspective and the feedback analysis concerns the feedback between the system and its environment, and how it influences the system's input-output relationship. Conversely, when we look at the network dimension, we focus just on its inner structure, the distribution of connections among its elements (nodes), the possible cycles or multiple types of relationships, and the countless further properties that can be investigated with network analysis (Barabási 2016; Lewis 2009; Newman 2010). If we can look at its network dimension, it means that the system is not more black, because we have, let say, "opened (or whitened) the box". Here too feedback processes are of fundamental relevance (if present), but they occur within the system and not between the system and its environment. In other words, and evoking the dualism of wave-particle properties of light, *system-network is the dualism in which an organization manifests its own properties.*

4 A Company's Hierarchical Structure

According to the basic principles recalled in the second section, if a system's topology were a directed chain—where all elements (nodes) do follow from the same vertex and are oriented to the same direction—its variety would be zero. More generally, any out-tree with a unique single root would give the same result, because $n-1$ nodes are constrained (commanded) by the apical (the upper bound limit) node. The nodes mattering for system variety would be those not constrained by any other node, and thus, nodes having only out-edges and no in-edges. Therefore, generalizing to poly-trees and DAGs

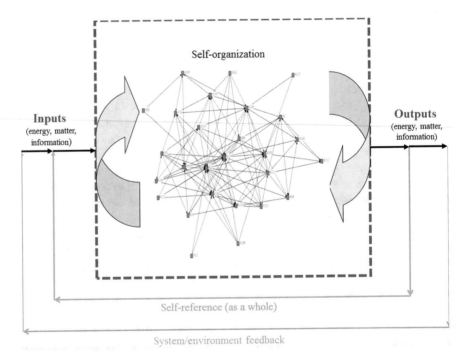

Fig. 1 The system-network dualism

(Directed Acyclic Graphs),[7] the degree of variety would be measured by the number of nodes that have no in-edges.[8]

Now, let's note that the archetype of hierarchy is just an out-tree (Biggiero and Mastrogiorgio 2016; Krackhardt 1994), whose connections represent commands, that is decisions made by superior(s) on subordinates. In Fig. 2 we have represented the two archetypes of hierarchy, which differ in that one (in the upper part of the picture) expresses the pure form, while in the other form hierarchy is mediated, and thus attenuated, by intermediate roles, like middle managers, to which power is delegated to transfer decisions to subordinates. This latter case is identified by Simon (1962) as the archetype because it is the paradigmatic form of all org charts, as they can be found in companies' documents, while the star topology can represent the organizational structure of a single office or a micro-firm whose boss commands on his employees. However, as Biggiero and Mastrogiorgio (2016) demonstrated limitedly to direct (dyadic) power, any delegation is also attenuation, and thus, the true and purest form of hierarchy is the one represented by the star-like structure.[9]

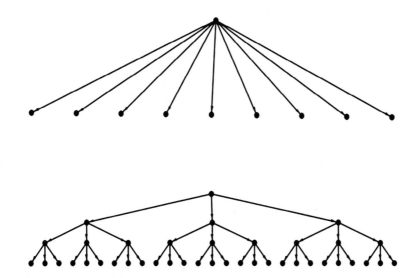

Fig. 2 The archetypes of hierarchy

---

[7] A DAG differs from pure out-trees or poly-out-trees because more nodes can point at the same node. Put differently, the underlying undirected graph of a DAG could be not a tree.

[8] More complex is the case in which the network contains loops (cycles), because constraining nodes are at the same time constrained nodes.

[9] Some of their arguments had been anticipated by Freeman (1979) and Radner (1992).

Even disconnected nodes could increase variety, because they are independent too, but this condition would violate the definition of system, which implies that elements (nodes) be connected. Moreover, in the specific case of our example, a pure hierarchy is compatible with only one node characterized by having only out-edges and no in-edges, namely, the organization's CEO. In fact, she is the only one that has only subordinates and no superior. Conversely, an organization can have more than one node with only out-edges only if there is more than one source of "ultimate power". We could call it a "poly-hierarchy".[10]

Thus, *according to strict formal (mathematical) rules, a pure hierarchy would have zero variety if all subordinates would act.* A result that is evidently paradoxical and unacceptable, because everybody knows—and a huge scientific literature shows—that a company, even a micro-enterprise of few employees, is a complex system. So, how can we explain this result?

5 Standardization and Routinization as Variety Reducers

In organization science language, repetitive processes are called routines,[11] and they are based on standardization of activities and usually accompanied by various forms of automation and dematerialization, and more recently by a massive introduction of robotics. All these means indicate a strict dependence of some activities on its input (independent) activity. As remarked by Beer (1966), all these means are ways to compress variety. What we are arguing here is that, once the reductions of variety were so pervasive to produce standard outcomes through routinized tasks, then the corresponding organization would be a pure hierarchy with zero variety.

The oversimplification of real organizations made through its representation as in Fig. 2 anticipates and addresses to a possible future evolution of economic organizations towards full-robotic companies (Ford 2015). Indeed, the robotization of life does not limit to economic organizations, because the use of robots is fast invading all aspects of our life, even in its more intimate and delicate aspects, like supporting psycho-pathologies or elder people by re-

---

[10] The true opposite of a pure hierarchy is the clique, where all members are connected each other bi-directional links, that is, all give and receive commands. Put differently, they are a group of peers who collaborate.

[11] Organizational routines are plans of decisions and actions, procedures issued and designed to reach a certain goal. They can be encapsulated into a single organizational unit or involve many units. Many routines are not designed but rather the outcome of consolidated habits. Routines can be viewed as the algorithms on which an organization can make decisions and action. Viewed in terms of social network analysis, they are specific paths followed by people and objects among the enormous possible paths. Let's remind that the number of paths scale factorially with a graph size and density.

placing human roles (Turkle 2012). What happened in the economic sphere is just anticipating what can be expected in the other spheres, even if the consequences on mankind can be much harder and fuller of unexpected and unpleasant social and psychological implications. The human species itself, at least as we have acknowledged, conceptualized and represented so far, is threatened.

Progressive automatization is a constant of the development of our homo sapiens-sapiens species, because it follows directly from the use of eso-somatic instruments, from arcs and mattocks to robots. It aims at strengthening human power, and thus, increasing its productivity. To realize it, another fundamental method has been employed: analyzing tasks and goals to reduce them into its elementary parts, so to be designed, planned and controlled as most as possible. Initially, and perhaps not so necessarily in the next future due to the high and fast growing learning capabilities of machines (robots), this requirement was necessary because machines were lacking cognitive capabilities.

What does it mean in terms of variety and redundancy? At first sight it means a progressive reduction of variety, because planning and control means reducing the variety of future events respect with what we have predefined as desired objectives. Therefore, mankind is constantly aiming at making its world more and more controllable, and therefore less complex. The goal of the automated factory means that few people can manage and control a number of different processes and outputs that in the past required hundreds or thousands people. The price of a progressively predictable life is the increase of redundancy and trivialization of our reality and society.[12]

However, at a closer and deeper sight, things are more complex. The tricky point lies in the ways in which human beings are employed in organizations, and more specifically in the tasks they should perform and the processes that are designed to reach final outcomes (goods and services). If tasks were elementary actions, like tighten bolts or reading the address of an envelope, and the elementary operations to realize a product were so detailed, separated each other and executed by dozens or hundreds or thousands different workers according to a fixed design, then what would be the difference between man and machine? As underlined by the Marxist research perspective on the sociology of work, in this case workers would be treated as machines, either because they would be condemned to execute repetitive actions or because, lacking any control on their work and being excluded from the design of the

---

[12] Indeed, there is a "side effect" that we cannot discuss here but we like to mention. It refers to the paradoxical mechanism of learning: we learn to reduce complexity, but the outcomes of learning are new actions, new knowledge, which indeed means increasing complexity, because it adds new unexpected elements. That is why, nevertheless our constant aim at reducing complexity by increasing automation, specialization and planning, we always claim that economy and society becomes every day more complex! The real danger represented by the shift from a human-based to a robot-based evolution is that the learning process is shifted either to robots, thus crowding out humans from this process.

whole process, they would be alienated from its outcome. The progressive automation realized with the first and the second industrial revolution created what has been perhaps the worst situation: a mix of workers and machines where the former had to adapt to the latter.

In such contexts we could say that "biological machines" (workers) are replaced by artificial machines: they have both the same *actual variety*. In other words, even though workers have a higher *potential variety*, that is, they could perform much complex tasks, what really matters is the variety that is actually required by the organizational context in which they are employed. If the organization is designed in a way that tasks are very simple and processes are very repetitive, then actual variety is lowered to a degree accessible by machines. The variety exceeding the required degree becomes superfluous, especially if the organization does not intend to innovate—at least to innovate from the bottom line.

The replacement of men with machines is of course an old story, which initially occurred when ancient peasants have been substituted by oxen: both were biological machines, but of different species. Put differently, if workers engaged into low-variety tasks and processes are replaced by machines nothing changes from the point of view of actual variety, that is, of the variety corresponding to the specific organization design. Of course, a lot could change in economic terms and from the point of view of potential variety.

If machines cost less than workers, then the substitution increases profits. Further, it could also happen that if machines are much faster than men to accomplish the tasks for which they have been introduced, they could a bit increase variety by enlarging the scope of its implementation. Therefore, the variety "net balance" of substitution could be even positive. But what about potential variety? Being intelligent and intentional systems, workers have a potential variety incomparably higher than the physical capital available before the computer revolution. Workers might decide to stop working or to raise their value or to abolish the property rights that guarantee the owners to gain profits or the managers to command over subordinates and design their roles and rules. Machines—at least, non-intelligent machines—cannot decide anything. Therefore, if the productive aspects of potential variety are frustrated because of standardization and routinization constraints, likely its reactive (counter-productive) aspects are freed and triggered. Historically,[13] if mentally health people are treated as machines, then they revolt against such a treatment, as the history of working class has shown abundantly. This lack of own intentionality made a further rationale to replace men with machines.

Things are changing considerably with the two drivers of current industrial revolution (Brynjolfsson and McAfee 2014). On one side there is the persistent attempt to engineering tasks, now regarding also complex, ill-structured,

---

[13] This occurred in different ways according to the different cultural anthropology and institutional-political contexts.

managerial tasks. It is becoming possible to "spoil" such tasks from its trivial or elementary or repetitive components, so to leave at high-skilled humans only the truly "complex kernel" of such tasks, and assign all the other components to machines. It is disputable whether this attempt is feasible, because complex tasks have a number of holistic properties that could prevent the accomplishment of the wished effects of that "spoliation".

The second driver, which is consistent with (and supports) the previous one, is much more important and full of unpredictable implications: the progressive and pervasive use of intelligent machines, and in particular of networks of intelligent robots. Here the net balance between reduction or increase of variety is much more uncertain and perhaps slopes more towards the latter. Intelligent machines—and especially networks of intelligent robots—are likely capable of producing novelties and increasing variety. And unfortunately—and this sheds a dark light on the future—they are capable, at least in many sectors, to do better than humans and in ways that are out of the control of humans. When warning about the implications of cybernetics and machine learning, Wiener (1948) was prophetic in this sense.

The stylized representation done in Fig. 2 hides indeed a huge degree of complexity, related to the following aspects:

- Any organization is made not only by decision-like communication. On the contrary, it is made by various types of relationships, like non-decisional or informal relationships and communications, material links, etc., which are not represented in Fig. 1. In other words, an organization—at least a human organization—is a multilayer network (Kivelä et al. 2014), and thus, if the corresponding graph were a pure hierarchy in terms of commands (decisional types of relationships), it could be not—and likely it will be not—a pure hierarchy in terms of other types of relationships[14];
- Each link of Fig. 2 oversimplifies reality by including a number of different command links. In other words, a superior does not exert only one type of command, but rather many different types, which indeed could configure

---

[14] When not considered as single individuals but rather as organized groups, both biological and economic systems are multilayer networks, because its elements interact according to multiple dimensions. For example, cells in an organ can be connected not only by chemical links, but also by mechanical or electromagnetic links. Members of an ecological niche can be connected by the same space and by chemical (let say, pheromones) or acoustic (language) links. Being much more complex than biological systems, social systems can possess many more dimensions (topology), that is, they can involve more *types* of links (Biggiero 2011). For example, firms belonging to an industrial cluster (Biggiero 1999) can be obviously linked by trade relationships, but also by collaboration activities, information or knowledge exchanges, strategic group relationships (for example collusive behaviors), non-trade contractual relationships (for example adopting a given trademark or computer system), peoples' (workers' or managers') mobility, to name the most important ones. Therefore, each topology corresponding to each type of link might have a different degree of redundancy.

a different orientation, possibly violating the out-tree or even the DAG structure[15];

- Each human node—that is, each employee at any hierarchical level—has usually a more or less wide discretionary power on his tasks. Therefore, his choices and their effects have a large variety.

What we argued in this section is that all these aspects of complexity are to some extent squeezable and eliminable by standardizing tasks and processes and by replacing workers with robots. In other words, a deep and pervasive rationalization and simplification of organizational tasks, products and operations could substantially lower actual minimum variety and thus make workers' high potential variety superfluous. The requirements coming from the—supposedly permanently growing—organizational environment variety could be, to some extent, faced with the increase of operative (actual) variety that intelligent machines and robots networks could produce. Of course, there is no reason to expect that such developments will be homogeneous for all organizations in all sectors. Conversely, they will be more pronounced in some sectors and in some organizations within them. What will be the net balance of all these forces and counterforces is now really impossible to be estimated, not only because these substitution and implementation processes are very complex and there are no systematic data, but also because most attention on robotics is currently devoted to understand the macro and gross-grain effects on employment, and not on the subtle outcomes that we discussed here.

6 Conclusions

We have shown that straight applications of concepts and measures born in a scientific field in reference to its proper objects to other fields and objects can create paradoxical or strange results that deserve appropriate interpretations. The concept of system's complexity and its measure in terms of variety for cybernetics—or entropy in the language of thermodynamics and mathematical information theory—generates some problems when applied to social systems, like economic organizations. More precisely, quasi-complete hierarchical organizations, like an army of two million people, would appear to have zero variety, a result that contradicts common sense. However, once we consider that all social systems are multilayer networks and an org chart represents only the decision network (and only in a very stylized way), that result could be accepted as a rough approximation of reality.

Incidentally, we noticed that a large part of this uneasy application of the (physics-derived) cybernetic concepts to social systems comes from the fact

---

[15] Of course, the same does happen for any of the other types of relationships addressed in the previous point.

that its description as black boxes is incomplete and misleading, because its elements do not interact randomly and instantaneously, and neither they are completely unknown. Conversely, their inner structure is made by relatively stable networks of human-human and human-machines interactions. Hence, because such systems coincide with the networks in which they are structured, they are at the same time networks. In this perspective, the systemic and the network dimension are just two complementary properties in which these types of organizations manifest their ontological nature: in the former dimension, focal issues concern inputs-outputs relationships, system's borders and its feedback mechanisms with the environment (and possibly itself); in the latter dimension, attention is paid to the inner structure and its countless characteristics. In both dimensions, the possible generation of emergent properties attracts prior analytical efforts.

We reminded that the history of humankind can be interpreted as the continuous attempt to control uncertainty by reducing complexity through work specialization and automation. In the language of cybernetics, it could be seen as a process of "whitening the black boxes" by designing and planning as most as possible its inner structures. In other words, the more detailed is the network in which socio-economic systems are organized and the more predefined are the rules and mechanisms governing the flow of information and resources flowing within that network, the more predictable such systems become, at least until the organization is strictly hierarchical in the topological sense—that is, if it approximates an out-tree or a DAG topology. Things can change dramatically depending on two facts: (1) if organizational structures are shaped with recursive processes, which open to nonlinear effects; (2) if workers are substituted by intelligent robots. The corresponding organization would be a network of intelligent robots. This type of organization should be neither considered trivial nor predictable, because when intelligent machines—be them biological or artificial—can interact recursively they can produce unexpected outcomes.

References

Ashby, R. W. (1956). *An introduction to cybernetics*. London: Chapman.

Atlan, H. (1979). *Entre le cristal et la fumee*. Paris: Éditions du Seuil.

Barabási, A. L. (2016). *Network science*. Cambridge: Cambridge University.

Bateson, G. (1972). *Steps to an ecology of mind: Collected essays in anthropology, psychiatry, evolution, and epistemology*. Chicago: University of Chicago.

Beer, S. (1966). *Decision and control*. New York: Wiley.

Biggiero, L. (1999). Markets, hierarchies, networks, districts: A cybernetic approach. *Human Systems Management, 18*, 71–86.

Biggiero, L. (2011). "Nuovi" strumenti di studio dei fenomeni sociali e naturali: Riflessioni sull'impiego delle metodologie di analisi reticolare e di simulazione. In E. Gagliasso, R. Memoli, & M. E. Pontecorvo (Eds.), *Scienza e scienziati: Colloqui interdisciplinari* (pp. 98–169). Milano: Franco Angeli.

Biggiero, L., & Mastrogiorgio, M. (2016). A methodology to measure the hierarchical degree of formal organizations. In Biggiero, L., et al. (Eds.), *Relational methodologies and epistemology in economics and management sciences* (pp. 206–231). Hershey: IGI Global.

Brynjolfsson, E., & McAfee, A. (2014). *The second machine age: Work, progress, and prosperity in a time of brilliant technologies.* New York: Norton.

Ford, M. (2015). *Rise of the robots. Technology and the threat of a jobless future.* New York: Basic Books.

Freeman, L. C. (1979). Centrality in social networks. Conceptual clarification. *Social Networks, 1,* 215–239.

Heims, S. J. (1982). *John Von Neumann and Norbert Wiener.* Harvard: MIT.

Heims, S. J. (1991). *The cybernetics group.* Harvard: MIT.

Heylighen, F., & Joslyn, C. (2001). Cybernetics and second-order cybernetics. In R. A. Meyers (Ed.), *Encyclopedia of physical science & technology.* New York: Academic.

Kivelä, M., Arenas, A., Barthelemy, M., Gleeson, J. P., Moreno, Y., & Porter, M. A. (2014). Multilayer networks. *Journal of Complex Networks, 2*(3), 203–271.

Klir, G. (1969). *An approach to general systems theory.* New York: Van Nostrand Reinhold.

Krackhardt, D. (1994). Graph theoretical dimensions of informal organizations. In M. K. Carley & M. J. Prietula (Eds.), *Computational organization theory* (pp. 89–111). London: Lawrence Erlbaum.

Lewis, T. G. (2009). *Network science: Theory and practice.* Hoboken: Wiley.

Newman, M. E. J. (2010). *Networks: An introduction.* Oxford: Oxford University.

Radner, R. (1992). Hierarchy: The economics of managing. *Journal of Economic Literature, 30*(3), 1382–1415.

Scott, B. (2004). Second-order cybernetics: An historical introduction. *Kybernetes, 33*(9/10), 1365–1378.

Shannon, C. E., & Weaver, W. (1949). *The mathematical theory of communication.* Urbana: University of Illinois.

Simon, H. A. (1962). *The sciences of the artificial.* Cambridge: MIT.

Turkle, S. (2012). *Alone together: Why we expect more from technology and less from each other.* New York: Basic Books.

Von Bertalanffy, L. (1968). *General system theory.* New York: Braziller.

Von Foerster, H. (1982). *Observing systems.* Seaside: Intersystems.

Von Foerster, H., & Zopf, W. (Eds.). (1962). *Principles of self-organization.* New York: Pergamon.

Wiener, N. (1948). *Cybernetics, or control and communication in the animal and the machine.* New York: Wiley.

Information, Communication Technologies and Regulations

Mario R. Abram and Eliano Pessa

1 Introduction

The development of Information and Communication Technologies (ICT) has a deep impact on our lives and many changes are taking place in all the fields, increasing the speed and the quantity of information exchanged between the various actors. The bandwidth of the new communication channels and the development of networks based on new information technologies are changing deeply all the parameters and constraints of the system. These factors have a direct impact on social environments.

The new technologies are contributing to create communities of people that interact using many new ways; then new ways for working, studying, moving and living are emerging (Rifkin 1995–2014; Greenfield 2017). In this context the laws and regulations, here considered in a simplified way as formalized protocols regulating the interactions between peoples and organizations, are strongly stressed and, as a matter of fact, demonstrate their limits and often an increasing inadequacy to manage the "change".

The traditional ideas of regulations and laws become inadequate for new situations and appear even useless in assisting people to react during the confrontation with the coming "big wave of change".

M. R. Abram (✉)
AIRS / Italian Systems Society, Milano, Italy
e-mail: mario.abram@alice.it

E. Pessa
Department of Brain and Behavioral Science, University of Pavia, Pavia, Italy
e-mail: eliano.pessa@unipv.it

© Springer Nature Switzerland AG 2019
G. Minati et al. (eds.), *Systemics of Incompleteness and Quasi-Systems*,
Contemporary Systems Thinking,
https://doi.org/10.1007/978-3-030-15277-2_16

A critical examination of the process underlying the development and evolution of legal systems shows how many inadequacies come out from the fact that information and communication technologies destroy the background concepts on which the laws and regulation systems were developed. New background ideas rise and acquire importance and ask for a reexamination of the processes that, from the deep past of our history and tradition, gave us national constitutions, law and regulation systems.

In this paper we will use the term "regulation"; it can be intended, in a simplified way, as a synonym and a collective name for Constitutions, Law systems, Norms. In the more large meaning, it will be the noun with which we call all the protocols that are rising from the relations between peoples and organizations.

Many arguments would need a juridical competence in order to correctly examine and develop all the implications regarding laws and regulations. In addition an interdisciplinary approach would evidence the deep connections between evolution in history, literature, philosophy, economy, science and all human realizations of models and interpretation methodologies. We will concentrate on some basic ideas that we think may be useful to explore some new approaches that appear promising, even if of no easy application.

In particular, in the following paragraph, some implications coming out from the application of Information and Communication Technologies (ICT) are recalled (Sect. 2). The processes involved in the emergence, setting and applications of regulations are briefly described, showing the role covered by the application domain (Sect. 3). The importance of fundamental rights, as basis of regulations, is examined, also in connection with ICT applications (Sect. 4). Some remarks show the complexity and the contradictions emerging from the different positions (Sect. 5). Finally some conclusions (Sect. 6) close the paper.

2 Information and Communication Technologies

Historically different communities were connected by means of many types of networks. They range from ancient roads and routs on which flowed commerce and consequently information, to the present cable and broadcasting networks.

The role of transmission and diffusion of information emerged and reached novel developments and possibilities when radio and television systems became available. They are a valuable mean useful to diffuse information, then to enlarge the sphere of influence outside the limits of a community.

Different physical networks connect the nodes crossing the geographical borders of countries. The influence domains of a community may enlarge and extend outside the frontiers (Fig. 1).

The development of ICT has modified substantially the structure of the transmission of information and simplified the access to transmission processes. This influenced deeply the communications standards and now the

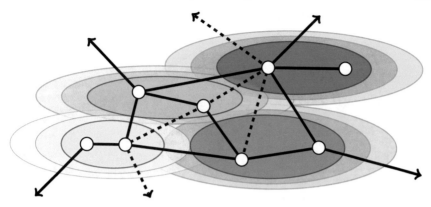

Fig. 1 Space of validity for regulations and enlarged domains of influence gained by means of different networks systems

possibilities of the extended communication processes enable a powerful and effective interactions between individuals and organization (Gleick 2011). As we will see these fascinating perspectives hide a background structure that deeply modifies both our perception of interactions and the meaning of our rights.

For example, let us examine the communications between two individuals or entities A an B located into the context of an organization (Fig. 2). The traditional communication (speech, writings, etc.) connect directly A and B. With information technologies the communications can be operated by means of computer systems (servers) that support many new networks services (internet, social networks, etc.) actually available. By means of servers it is possible to operate communications channels, manage the interconnection networks and store the collected information. But the computer systems that support and operate these networks are placed into different contexts having domains that we can identify as: Organizations (ORG), Nation (NAT), Community of Nations (COMM), other nations (OTHER) and unstructured realities (UNSTR).

Servers have a position in space and their locations should be submitted to the application of local laws. But the domain of application of ICT services is potentially worldwide, outside the application domains of local regulation systems.

The communication channels are not inside the physical context of our relations but may be located outside. This means also that our contexts become enlarged and that they are no more limited by traditional borders. As a consequence they represent a reality that is no more structured and ruled.

The traditional ideas of communication processes, as shown in Fig. 1, are no more valid. No more space-time border is acknowledged as valid. The traditional and usual regulations structures now cannot be applied because their assessment structures are not applicable outside of geographical space of definition (and then of validity).

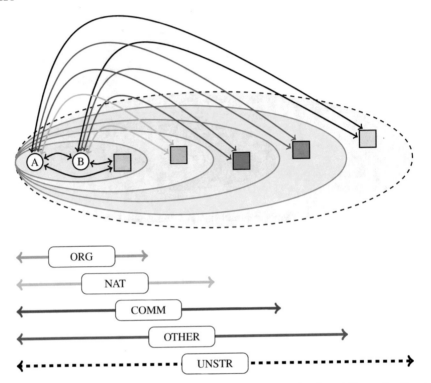

Fig. 2 The communications between two individuals or entities A and B may be direct, by means of traditional channels, or may be mediated by servers located into very different contexts

Figure 2 shows a simplified view of cyberspace in which the connections between the servers create a new meta-structure, potentially without rules and constraints, and then independent from the space characterization of traditional regulations. The servers are physically connected by networks but from the point of view of the users, the entities A and B perceive and use only the available communication channels.

It is evident how the servers' owners, controlling communication channels and stored information, deploy an effective field of application measured by the extension of their ICT services. *De facto* they realize an effective field of application, potentially worldwide, that from virtual becomes real.

3 Emergence of Regulations

In a simplified way, the regulations emerge by a process that can be synthesized in the following steps:

1. *Definition of a right.* Constitutes the very basic and deep idea that is the foundations of regulations.
2. *Acknowledgement of a right.* It is the conscious acknowledgment of the importance to build around a right the lives of communities.
3. *Building of a norm for the application of a right.* The norm is defined and formulated as a protocol applicable in a community.
4. *Building of the instruments for enforcing a norm.* The building of effective instruments to impose the respect of the norm in its domain of validity.
5. *Application of a norm.* In a community the respect of a norm is effectively imposed, using the instruments previously defined.

These steps are cyclically applied when it is necessary to maintain and review a norm in order to activate the changes that become necessary in an evolving society or community.

The point (2) (acknowledgement of a right) is critical. It is the true starting point of the process and it must be considered correctly in its twofold meaning: in a relation between two entities, the consciousness that each actor has of its own right must implicate the mutual acknowledgment and the biunivocal reconnaissance of that right for the other.

For regulations some elements appear essential. They are:

- *Domain of validity.* The application of a regulation or law is valid and recognized in a geographical space coincident with the "space of a community". It is the space in which the norm may be applied. Usually the domain of validity coincides or is assimilated with the surface of the nation and defines the border of the nation. Historically, a nation is seen as a reality able to build, maintain and operate such a strong system of regulations.
- *Instruments to enforce a right.* The application of a regulation may be forced with the help of instruments and tools built for this goal. These instruments are applicable within the limits of validity of the regulation.
- *Mutual reconnaissance of rules.* It takes place when the application of a regulation is valid and acknowledged in a geographical space coincident with the space of organizations, nations, union of nations.

Our actual constitutions, law systems and regulations emerged from the application of the over said procedure, so forming our history and juridical traditions.

Regulations constitute the *corpus* of rules that define the protocols rising from the interactions between people, organizations and environments in all their aspects and implications. ICT are modifying these rules and call for the improvement and the update of interaction codes in order to manage the emergence of a new reality (Rifkin 2000, 2014; Greenfield 2017).

There will be always a delay between the emergence of a new interaction, implemented and powered by ICT, and the building of an adequate regulation. As a consequence a situation of constant mismatching between "living technologies" and the development and application of new regulations seems a problem of difficult solution.

Cyberspace exists without border in space and time, but it is built under the existence and the operation of computer systems, located in space and time, that are interconnected by large ICT physical networks.

The iterations are regulated by local regulations. But, when the channels are external to the space of validity of laws, only the agents (the sender and the receiver) are subjected to regulations. Often the channels are not subject to regulations.

The traditional regulations are law systems applicable to relations emerging in space-time reference. In cyberspace no such system is available because the cyberspace is perceived as a reality without juridical constraints. In any case in cyberspace a large amount of unstructured components is intrinsically present.

While legal systems have local validity, the protocols emerging in cyberspace overcome and superimpose the domains of application of local regulations and law systems. As a consequence the vanishing of the application domain erases the meaning of a norm; therefore that norm becomes inapplicable.

4 Fundamental Rights

With the *Universal Chart of Human Rights* (UCHR) (UN 1948) United Nations supply a common basis for building and operating human relations and interactions.

Human rights are: freedom, dignity, life, security of person, prohibition of slavery and torture, recognition before the law as a person, protection against discrimination, remedy for acts violating the fundamental rights, equality, presumption of innocence, freedom (of movement, asylum from persecution, nationality, marriage), protection of family, property, freedom (of thought, conscience and religion, opinion and expression, peaceful assembly and association, to take part in the government), universal and equal suffrage, social security, work, remuneration, rest and leisure, health and well-being, protection of motherhood and childhood, education, etc.

In particular UCHR settled the basic interpretation approach:

> Nothing in this Declaration may be interpreted as implying for any State, group or person any right to engage in any activity or to perform any act aimed at the destruction of any of the rights and freedoms set forth herein. (UN 1948, Article 30).

This sentence shows clearly a reference that we can consider an "approach of civility". It shows a tendency, a direction and defines a mission that must be constantly pursued and verified.

UCHR is generally agreed to be the foundation of international human rights law. Many constitutions, law systems and norms worldwide were inspired by this basic document. The building of national Constitutions is founded on solid basis sharing acknowledged principles of civility. Nevertheless different countries may develop regulations on the basis of different sets of rights.

Also European Union with the *Charter of Fundamental Rights of the European Union* (EU 2016b) creates legal certainty within the Union by making fundamental rights clearer and more visible. It recalls a list of Human Fundamental Rights grouped as: (1) dignity, (2) freedoms, (3) equality, (4) solidarity, (5) citizens' rights, and (6) justice. Under these main themes we find the rights to: integrity of the person, respect for private and family life, protection of personal data. These rights may be exemplified by Image, Life, Name and all the data that contribute to define Identity.

Fundamental rights are the deep basis on which the law systems were developed. But the possible risks to violate fundamental rights is constantly present. For example, limiting to ICT applications, the use of Internet is emblematic of this situation because many entities may observe all the steps of our connections (Nikiforakis and Acar 2014).

One of the consequences of new technologies is the use of enlarged domains in which many different and local regulations should be applicable. Or in a more extended meaning, the true domain of ICT overcomes and superimposes the subdivisions into structured communities. The protocols emerging in these new domains are not structured into regulations. We are in presence of many new domains in which no regulation is applicable. This situation is modifying the structure of our relations, involving many aspects of all human activities (Rifkin 2000).

International laws are based on the traditional process for definitions of regulations and their application; it is possible to build international laws by negotiations and agreements between countries and nations into a defined field of application (Roberts 2017). This is limiting and not sufficient when dealing with a cyberspace without constraints. The physical space-time constraints cannot be translated into cyberspace.

Regarding the availability of extended and pervasive ICT services may be interesting to recall again the *Universal Chart of Human Rights*:

No one shall be subjected to arbitrary interference with his privacy, family, home or correspondence, nor to attacks upon his honor and reputation. Everyone has the right to the protection of the law against such interference or attacks. (UN 1948, Article 12).

European Union made the effort to regulate ICT matter by settling a norm "to protect natural persons with regard to the processing of personal data and on the free movement of such data" (EU 2016a). An interesting aspect of this norm is the definition of duties by which an entity is responsible of personal data everywhere they are located or moved.

5 Remarks

The building of regulation systems is strongly stressed by emerging ICT. Some critical aspects coexist in this process, often in explicit contradiction. It may be useful to examine some of them because they impact directly on the lives of people.

1. *Building of Regulations.* The diffusion of new technologies stress the emergence of the inadequacies and contradictions in actual regulations (Kittichaisaree 2017). These are structural and preexistent but the technology is a formidable amplification factor that introduce strong instabilities in laws systems, till the demonstration of the total inadequacy of a system of norms (Roberts 2017).

   In particular the process for the formation of regulations presents a structural delay based on the sequence: (1) acknowledgment of a problem created by ICT; (2) solution of a the problem with the building of a new norm.

   Probably the dynamics for the building of regulations is slower than the rapid evolution of ICT systems.

2. *Domain of validity.* The domain of validity constitute a fundamental element that supports each law and regulation system.

   The application to international law is based on the traditional process for defining regulations and their applications, especially with reference to the existence of a definite field of applications. This is limiting and insufficient when dealing with a cyberspace without constraints (Glorioso 2015; Taddeo 2017).

3. *Human Rights.* Even homogeneous communities in different countries does not have real and mutual reconnaissance of law systems (Roberts 2017). Law systems born in different traditions an cultures show deep differences also about the definition of fundamental rights. Common definitions and mutual acknowledgements are the basis of a difficult but inexorable process if we want to extend regulations to the international community (UN 1948; Rifkin 2010, 2014; Martinez 2016).

4. *Connection with Power.* The exercise of Power is strongly influenced by new technologies (Chomsky 2002; Rifkin 2000, 2014).

   Human rights are strictly connected with Power; the exercise of Power influences the respect of human rights and now can gain advantage using new technologies. New forms of Power are arising; based on ICT possibilities, they are free from the constraints of actual laws and regulations (Rauscher 2013; Glorioso 2015). ICT supply new powerful instruments for building and exercising Power (Greene and Elffers 1998; Martinez 2016). Many implications follow from the application of power by means of new ICT (Rothkopf 2008; Rifkin 2011, 2014; Greenfield 2017). Many cases are based on the misunderstanding or the violation of one or more fundamental right (Rifkin 2005, 2010). ICT are powerful instruments also in eliminating rights (Mitnick and Simon 2003).

Norms and regulations may be instruments necessary to balance the exercise of Power. It is a matter of balance between total freedom, without duties, and the building of constraints to defend human rights. May be that civility can find its roots into the building of this balance (UN 1948, Art. 30).

5. *Economy and Finance.* ICT supply the ideal instruments to operate worldwide in real-time. E-commerce developed large multinational structures. Finance is developing new currencies that operate worldwide with the maximum degree of freedom tanks to the lack of any adequate international norm (Peck 2012, 2017; Surowiecki 2012; Zorpette 2012).

6. *Enforcing Human Rights.* When the mutual acknowledgment of rights is not possible, the need to defend or enforce the rights become necessary (Greene and Elffers 1998; Greene 2006).

Alternative approaches face the problem and many countries consider different levels of reaction to eliminate a danger or reduce the risks connected to the use of new technologies (Schmitt 2017; Springer 2015; Taddeo and Glorioso 2017).

7. *History and Classics.* On human fundamental rights we can measure the evolution of mankind and the grade of their application shows the level of civility of a society. A deeper definition of human rights is the consequence of speculations and lesson learned from the past experiences. A deeper understanding may gain great advantage from the study of History and comparing the work done by people during the centuries (Münkler 2007). Many authors faced these themes, searching the roots of problems and helping us to point our attention on the key points and the real goals of civility, long before the development of ICT.

Many classical authors in philosophy, politics, religions, literature, history, economy with their publications described, analyzed and discussed about the "Power" problem. They showed how the "regulations" problem may shift toward the "Power" problem.

Following these lines a deeper historical and philosophical research may be helpful in order to find the deep roots of our ideas that constitute the kernel of our traditions.

The roots of "Power" problem can be found in "classics". Facts and implications are reported without the amplifying effects introduced by ICT.

8. *Human Component.* Human beings are the main users of ICT; they constitute the human component is "in the loop", able to modify the interaction modalities with the systems. Human component is then an element that is able to build a correct interaction with the systems or a critical point that may degrade the behavior of the system. Human elements impact directly on the safety of the system, and condition its evolution

(Mitnick and Simon 2003). There is a real risk to use the human component as an element to determine the behavior of the systems even with its unconscious contributions.

9. *Social Engineering.* The people in charge of Power, with the help of ICT, may modulate the applications of human rights to influence and address the evolution of society (Rifkin 2011; Chomsky 2016, 2017). The pervasiveness of ICT shows an increasing possibility to plan and operate effective activities of social engineering (Greenfield 2017).

10. *Limitations and Opportunities.* ICT for many real situations may suggest a new way to find new models and show the utility to move toward a real unifying process with the building of new norms with progressively enlarged environments.
ICT create a real possibility and the opportunity to evaluate new forms of coexistence to get over the subdivisions of nations. Technology may push for unification (Chomsky 2012, 2014; Rifkin 2010, 2014; Martinez 2016).

6 Conclusions

The above said considerations show how many critical points face with the complexity of the interactions between ICT and human rights. The quick development of ICT opens challenges that impact directly on people and society. Some reflections can help to approach the problem from a systemic point of view, with the goal to explore new strategies for finding solutions.

- *Systems and Quasi-Systems.* When in some special cases the description of elements and relations presents uncertainty and lack of information, it is more convenient to call the system a Quasi-System.
In ICT structures there is an uncertainty generated by the vanishing of the traditional law systems. Uncertainty is also created by the absence of protocols and shared regulations. It is then natural to characterize ICT structures as Quasi-systems. Another example may be the system of values related to human rights in which we may identify the priorities of fundamental rights. If this system of values is dissipated, banalized, degraded, its strength in building and applying regulations is deleted.
This process involves many coexisting systems, interacting and mutually dependent. In our case we have systems that are destructuring themselves, as for example traditional regulation and legal systems now are becoming Quasi-systems (for degradation of the structure). Meanwhile new technologies show contradictory and uncertain situations which we are not able to identify as a system, as a consequence we must identify them as Quasi-systems (by absence of structure).

Furthermore ICT are a formidable accelerating factor that, increasing uncertainty, destructures the traditional law systems, forcing their evolution toward Quasi-systems.

- *Interactions between Systems.* The interactions between these different representations or systems shows how the same characterization of Systems and Quasi-systems may evolve in time changing their status. In this contest ICT are creating a fluid situation in which the choices of "human observers" can make the difference. This is evident if we can consider Human Rights as a systems interacting with all the systems emerging from human activities. The application of Human Rights evolves and it may move toward a structured System or degrade toward a Quasi-system.
- *Limitations and Opportunities.* If the respect of human rights is the leading theme, the contradictions emerging from this complex situation may become the opportunity to find the unifying trend toward new possibilities. The contradictions emerging from ICT may be the stimulus to overcome the actual development models.

  The contradictions emerging from this complex situation may become the opportunity to see and to find the unifying trend toward fascinating possibilities if the respect of human rights is the leading theme.
- *Choice of "Civility".* The need to manage and optimize the strategies to face the present and future challenges, asks for an approach that may be inspired by "human rights first". Starting from this point it is possible to set the goal, define the trend and select the directions for future researches. In addition we must face with the dynamics generated by contradictory and conflictual variables, which are expression of a variety of very different interests. In other words the positions oriented to the "building of human rights" will coexist with those oriented to the "destruction of human rights" in a sort of conflictual and apparently open game; consequently these problems have no easy solution.

  These contrapositions appear consistent for any choice: (1) a "closing" approach for controlling the situation, or (2) an "opening" approach to explore radically new strategies.

  When human element is involved, it is natural to consider philosophy; social and ethical implications enrich the context and suggest the constraints that may characterize the limits of appropriate and acceptable solutions. In any case the development of a diffuse culture about human rights and technologies is useful to understand the real problems and may help the users to develop consciously the best attitude to deal with the actual and future Information and Communication Technologies.

Acknowledgements We want to thank Dr. Eng. Umberto Di Caprio and Prof. Lucia Urbani Ulivi for useful discussions and suggestions.

References

Chomsky, N. (2002). *Understanding power*. New York: The New Press.

Chomsky, N. (2012). *Making the future: Occupations, interventions, empire and resistance*. San Francisco: City Lights Books.

Chomsky, N. (2014). *Masters of mankind: Essays and lectures, 1969–2013*. Chicago: Haymarket Book.

Chomsky, N. (2016). *Who rules the world?* New York: Metropolitan Books/Henry Holt.

Chomsky, N. (2017). *Requiem for the American Dream: The 10 principles of concentration of wealth and power*. New York: Seven Stories.

European Union. (2016a). Regulation (EU) 2016/679 of the European Parliament and of the council of 27 April 2016 on the protection of natural persons with regard to the processing of personal data and on the free movement of such data, and repealing Directive 95/46/EC (General Data Protection Regulation). *Official Journal of the European Union*, (4.5.2016): L119/1–88.

European Union. (2016b). Charter of fundamental rights of the European union (2016/C 202/02). *Official Journal of the European Union*, (2016): C 202/389–405.

Gleick, J. (2011). *Information. A history, a theory, a flood*. New York: Pantheon Books.

Glorioso, L. (2015). Cyber conflicts: Addressing the regulatory gap. *Philosophy and Technology, 28*(3), 333–338.

Greene, R. (2006). *The 33 strategies of war*. New York: Viking.

Greene, R., & Elffers, J. (1998). *The 48 laws of power*. New York: Viking.

Greenfield, A. (2017). *Radical technologies: The design of everyday life*. New York: Verso Books.

Kittichaisaree, K. (2017). *Public international law of cyberspace*. Cham: Springer.

Martinez, R. (2016). *Creating freedom: Power, control and the fight for our future*. Edinburgh: Canongate Books.

Mitnick, K. D., & Simon, W. L. (2003). *The Art of deception: Controlling the human element of security*. Indianapolis: Wiley.

Münkler, H. (2007). *Empires: The logic of world domination from ancient Rome to the United States*. Cambridge: Polity.

Nikiforakis, N., & Acar, G. (2014). Browse at your own risk. *IEEE Spectrum, 51*(8), 26–31.

Peck, M. E. (2012). The cryptoanarchists' answer to cash. *IEEE Spectrum, 49*(6), 48–54.

Peck, M. E. (2017). Blockchains: How they work and why they'll change the world. *IEEE Spectrum, 54*(10), 22–31.

Rauscher, K. (2013). Writing the rules of Cyberwar. *IEEE Spectrum, 50*(12), 26–28.

Rifkin, J. (1995). *The end of work: The decline of the global labor force and the dawn of the post-market era.* New York: G. P. Putnam's Sons.

Rifkin, J. (2000). *The age of access: The new culture of hypercapitalism.* New York: Penguin.

Rifkin, J. (2005). *The European dream: How Europe's vision of the future is quietly eclipsing the American dream.* New York: Penguin.

Rifkin, J. (2010). *The empathic civilization: The race to global consciousness in a world in crisis.* Cambridge: Polity.

Rifkin, J. (2011). *The third industrial revolution: How lateral power is transforming energy, the economy, and the world.* New York: St. Martin's.

Rifkin, J. (2014). *The zero marginal cost society: The internet of things, the collaborative commons, and the eclipse of capitalism.* New York: St. Martin's.

Roberts, A. (2017). *Is international law international?* Oxford: Oxford University.

Rothkopf, D. (2008). *Superclass.* New York: Farrar, Straus and Giroux.

Schmitt, M. N. (Ed.). (2017). *NATO cooperative cyber defence centre of excellence. Tallinn manual 2.0 on the international law applicable to cyber operations* (2nd ed.). Cambridge: Cambridge University.

Springer, P. J. (2015). *Cyber warfare: A reference handbook.* Santa Barbara: ABC-CLIO.

Surowiecki, J. (2012). A brief history of money. *IEEE Spectrum, 49*(6), 40–46.

Taddeo, M. (2017). Deterrence by norms to stop interstate cyber attacks. *Minds and Machines, 27*(3), 387–392.

Taddeo, M., & Glorioso, L. (Eds.). (2017). *Ethics and policies for cyber operations: A NATO cooperative cyber defence centre of excellence initiative.* Cham: Springer.

United Nations. (1948). International bill of human rights. Universal declaration of human rights (General assembly resolution 217 A (III), Paris, 10 December 1948). *Official records of the third session of the general assembly, Part 1, 21 September–12 December 1948. Resolutions* (pp. 71–77).

Zorpette, G. (2012). The beginning of the end of cash. *IEEE Spectrum, 49*(6), 23–25.

Connections and Dissimilarities Among Formal Concept Analysis, Knowledge Space Theory and Cognitive Diagnostic Models in a Systemic Perspective

Eraldo Francesco Nicotra and Andrea Spoto

1 Introduction

This research is aimed at presenting the similarities of three mathematical theories developed in the same years regarding the representation of the relations among objects: Formal Concept Analysis (FCA) (Ganter and Wille 1999; Wille 1982), Knowledge Space Theory (KST) (Doignon and Falmagne 1985, 1999; Falmagne and Doignon 2011), and Cognitive Diagnostic Models (CDM) (Tatsuoka 1985, 2009). One of the core issues that links these theories since the very beginning of their development is that they all can be referred to the fundamental Theorem proposed by Birkhoff (1937) (in Doignon and Falmagne 1985).

Theorem 1 *For any set X, the formula*

$$yQx \text{ iff } x \in A \to y \in A, \forall A \in \varphi$$

defines a one-to-one mapping r of the set of all families φ of subsets of X closed under intersection and union, to the set of all quasi-orders Q on X.

This theorem links the quasi-orders to some families of subsets. This notion has been applied, in different ways in all the three theories considered in this research. The following sections introduce the main concepts of such theories.

E. F. Nicotra (✉)
Department of Pedagogy, Psychology, Philosophy, University of Cagliari, Cagliari, Italy
e-mail: eraldo.nicotra@unica.it

A. Spoto
Department of General Psychology, University of Padova, Padova, Italy
e-mail: andrea.spoto@unipd.it

© Springer Nature Switzerland AG 2019
G. Minati et al. (eds.), *Systemics of Incompleteness and Quasi-Systems*,
Contemporary Systems Thinking,
https://doi.org/10.1007/978-3-030-15277-2_17

1.1 Formal Concept Analysis

The first basic notion of FCA is the *formal context*, defined as a triple (G, M, I), where G is a set of *objects*, M is a set of *attributes*, and I is a binary relation between the set of objects and the set of attributes. A formal context is represented by a Boolean matrix in which each row is an object and each column is an attribute. Whenever a 1 is present in the entry (g, m), the relation gIm holds. Between the objects and the attributes of a formal context, a *Galois connection* is defined. For all the sets $A \subseteq G$ and $B \subseteq M$, the following two transformations define the Galois connection:

$$A' := \{m \in M | gIm, \forall g \in A\}$$

and

$$B' := \{g \in G | gIm, \forall m \in B\}.$$

In words, A' is the collection of all the attributes shared by all the objects in A. Dually, B' is the collection of all the objects shared by all the attributes in B. In FCA, the pair (A, B) is called a *formal concept* if it satisfies the following two conditions: $A = B'$ and $B = A'$. The *extent* A of the formal concept contains the objects of G that have all the attributes in B; on the other hand, the *intent* B includes the attributes satisfied by all the objects in A.

A subconcept-superconcept relation is then defined:

$$(A_1, B_1) \leq (A_2, B_2) \Leftrightarrow A_1 \subseteq A_2$$

or equivalently,

$$(A_1, B_1) \leq (A_2, B_2) \Leftrightarrow B1 \supseteq B2.$$

In words, a concept is of a lower level when it has a larger extent (or equivalently, a smaller intent). The concepts of a context form a *complete lattice* (Birkhoff 1937, 1967), which is called the *concept lattice* of the context (G, M, I). The intents of a concept lattice are closed under intersection; that is, each intersection of sets of attributes is included in the lattice.

It is now necessary to introduce the main concepts of KST.

1.2 Knowledge Space Theory

In KST, a *knowledge domain* is defined as the set Q of all the items that can be investigated about a specific topic. A *knowledge state* $K \subseteq Q$ represents the set of items in Q that a subject can solve. A *knowledge structure* $\mathcal{K}$ is the collection of knowledge states, and it has to include at least the empty set ($\emptyset$) and the total set (Q). Whenever a structure is closed under set union (i.e., every union of states is again a state in $\mathcal{K}$) it is called a *knowledge space.* Whenever a structure is closed under both set union and intersection, it is called a *quasi-ordinal knowledge space.* Notice that Birkhoff's theorem directly applies to this kind of structures, while Doignon and Falmagne extended the theorem to the more general case of a knowledge space. Usually a knowledge structure is denoted as $(Q, \mathcal{K})$, where Q is the domain and $\mathcal{K}$ is the collection of subsets of the structure. The knowledge structure depicts the implications among the items in Q.

A fundamental concept of KST is that of a *skill map* (Doignon and Falmagne 1999; Lukas and Albert 1993). This concept is crucial since it represents a possible link between KST, FCA and CDM (as we will see later). A skill map is a triple (Q, S, f), where Q is a nonempty set of items, S is a nonempty set of skills, and f is a mapping from Q to $2^S \setminus \{\emptyset\}$. For any item $q \in Q$, the subset $f(q)$ of S represents the set of skills assigned to q. Generally speaking, if a subject solves item q, he or she has either (1) all the skills included in $f(q)$ (*conjunctive model*), or (2) at least one of them (*disjunctive model*). A possible way to represent the skill assignment depicted by a skill map is a Boolean matrix where each row is an item in Q and each column is a skill in S. A 1 in a cell (q, s) means that the skill s is needed to solve item q. Moreover, starting from this matrix it is possible to derive a knowledge structure by applying either the conjunctive or the disjunctive model. More specifically, by applying the disjunctive model, one obtains a knowledge space, on the contrary, by applying the conjunctive model a *closure space*, i.e., a structure closed under set intersection, is obtained.

Differently from FCA, since the very beginning of its development, KST focused on the probabilistic framework to be applied on the deterministic structure in order to carry out an efficient and effective adaptive assessment of knowledge. Falmagne and Doignon (1988) define a *probabilistic knowledge structure* as a triple $(Q, \mathcal{K}, \pi)$, where $(Q, \mathcal{K})$ is a knowledge structure, and π is a probability distribution for $\mathcal{K}$. In the model at issue, given a state, the responses to the items are locally independent. Thus, starting from the probabilistic knowledge structure $(Q, \mathcal{K}, \pi)$, given a specific response pattern $R \subseteq Q$, we will define a function $s : (R, K) \mapsto s(R, K)$, assigning to each response pattern its conditional probability given that a subject is in state K, the response function for the probabilistic knowledge structure. For each pattern a probability distribution is defined as follows:

$$p(R) = \sum_{K \in \mathcal{K}} s(R, K)\pi(K).$$

Since the response function s satisfies local independence for each item $q \in Q$, the conditional probability $s(R, K)$ is determined by means of the two probabilities η and β, the careless error and lucky guess of each item q. The most widely applied probabilistic model in KST is the Basic Local Indepence Model (BLIM; Doignon and Falmagne (1999); Falmagne and Doignon (1988)). It is now possible to introduce the last theory involved in this paper: The CDM.

1.3 Cognitive Diagnostic Models

As clearly suggested by the name, CDM theory is mainly concerned with skill diagnosis. The probabilistic modeling of data assumed a great relevance in CDM, mainly relying on the latent class approach (Roussos et al. 2007).

It is not surprising that in CDM one of the core notions is that of a *skill* (sometimes referred to as *attribute*). In this theory a skill is a discrete cognitive component needed to perform specific operations, or to solve specific problems. It is important to underline that here skills are conceived as properties of both persons and items. From a notational point of view, skills are defined as dichotomous latent variables. In fact they can be either presented or not by an individual, or either needed or not to solve a problem.

The most important concept defined in CDM, and used to depict the cognitive theory which stands behind the specification of the relations between the items and the skills is the so-called *Q-matrix* (Tatsuoka 1990). The *Q*-matrix is generally represented (once again) as a Boolean matrix having as many rows as the number of items, and as many columns as the number of attributes (or skills). In this matrix, whenever an item needs a specific skill, a 1 is present in the corresponding cell, which otherwise contains a 0. Therefore, the *Q*-matrix depicts the skill assignment for each item. The interpretation of the matrix is not unique. This caused the development of different classes of CDM.

As stated above, CDM, as well as KST, focused since their first development on the probabilistic models to be applied to the available deterministic frameworks. Coherently with the fact that many different deterministic interpretations of the *Q*-matrix are allowed, a number of probabilistic models have been derived for CDM. For sake of shortness, here we mention one of the most popular models that is also strictly related to the BLIM: the Deterministic Inputs Noisy AND-gate model (DINA) (Haertel 1984, 1989). The DINA model uses a *conjunctive* rule. Therefore, each item might be related to more than one skill, and each skill might be needed by more than one item. The *Q*-matrix is used to represent such relation. In the DINA a *knowledge state*, is a binary vector representing the set of skills possessed by some individual (differently from KST). Finally, in the DINA, the response pattern of an individual is a binary vector with as many elements as the number of items and as many ones as the number of items correctly solved by the individual.

Even from this extreme overview of the main concepts of the three theories, a number of links should be evident. They are the subject matter of the next section.

2 Linking FCA, KST and CDM

It is evident how the deterministic parts of the three theories share a number of concepts. In 1996 Rusch and Wille first noticed a possible link between KST and FCA (Ruch and Wille 1996). In their paper they showed that, since the collection of the intents of a formal context is closed under intersection, the collection of the complements of the intents is closed under set union. Therefore, this last can be seen as a knowledge space. In their article, the authors started from a formal context defined by the set G of subjects (which, in this case, were treated as formal objects), the set M of items, and the binary relation gIm, meaning that the subject g had solved item m. In these terms, a response pattern becomes a set of formal attributes, that is, an intent. With the relation between the intents of a formal context and their complements, the authors derived a so called *knowledge context* having the domain defined by the set of items and the states by the complements of the observed response patterns. Using this methodology, it is then possible to construct a knowledge space starting from a formal context.

More recently, a deeper connection between KST and FCA has been shown by (Spoto et al. 2010). In their paper these authors refer to the possibility of building a knowledge structure via the definition of a formal context. More specifically, A formal context corresponding to the skill map (Q, S, f) can be derived by interpreting Q as the collection of objects and S as the collection of attributes and by defining a binary relation $R \subseteq Q \times S$ so that, for all pairs, $(q, s) \in Q \times S$:

$$qRs \Leftrightarrow s \notin f(q).$$

Therefore the triple (Q, S, R) can be regarded as a formal context, where qRs should be read as skill s is not required by item q. As an effect of this definition, the intent $q' := \{s \in S | qRs\}$ is just the complement of $f(q)$ in S. The collection $\mathcal{I}$ of all the intents of the concept lattice corresponding to this context could then be obtained by closing under intersection the collection $\{q' : q \in Q\}$ of all object intents. At this point, the states of the structure are simply the extents of the lattice. This formulation allows also to solve one of the crucial problems related to the skill map procedure, that is the fact that is not granted to have a one to one correspondence between a set of skills and a set of items. In other words, it is possible that the same knowledge state corresponds to different sets of skills. The introduction of the above described connection between KST and FCA allows to reconstruct a bijection between sets of skills (attributes) and sets of items (objects). The link between these deterministic structures and that of the CDM is straightforward.

Some further interesting connections have been established between the probabilistic models applied in KST and in CDM. Going beyond the deterministic case, the probability in KST and CDM refers to the concept of a response pattern, that is the subset R of Q consisting of all the items which would receive a correct answer by an individual. The probability of an arbitrary response pattern $R \subseteq Q$ in these theories is usually specified by the latent class models (Lazarsfeld and Neil 1968) where the conditional probability of a response pattern given the latent class is determined, e.g., by some parameters of error for each item. In these models the answer to the items (included in the response pattern) are locally independent given the latent class which the person belongs to. With respect to the specific models applied in KST and CDM, it was pointed out by Heller et al. (2015) that the BLIM (and its variations) and the DINA (and its variations) models have exactly the same characteristics. Therefore they represent two differently labeled models which make the same assumptions, and estimate the same kinds of parameters. More recently it was pointed out by Spoto and colleagues how these two classes of models present also the same problems with respect to their identifiability (Spoto et al. 2013; Stefanutti et al. 2018).

From these hints it should be evident how the three theories share a number of fundamental characteristics.

3 Discussion

This paper was aimed at summarizing the similarities and the differences among three theories that for different reasons play an important role in data representation, mathematical psychology and cognitive psychology. It has been shown how they are actually built on the same formal background relying on the fundamental theorem by Birkhoff. The reason why there have been, so far, so few attempts to look at them in a comprehensive perspective is unknown and somehow bizarre.

In our view, this paper should represent a starting point for a much deeper and more systemic analysis of the connections among them aimed at solving the problems that each of them presents and that, maybe, has already been faced and solved by any of the others.

References

Birkhoff, G. (1937). Rings of sets. *Duke Mathematical Journal, 3*, 443–454.

Birkhoff, G. (1967). *Lattice theory.* Providence: American Mathematical Society Colloquium Publication (No. XXV).

Doignon, J.-P., & Falmagne, J. C. (1985). Spaces for the assessment of knowledge. *International Journal of Man-Machine Studies, 23*, 175–196.

Doignon, J.-P., & Falmagne, J. C. (1999). *Knowledge spaces.* Berlin: Springer.

Falmagne, J. C., & Doignon, J.-P. (1988). A class of stochastic procedures for the assessment of knowledge. *British Journal of Mathematical and Statistical Psychology, 41*, 1–23.

Falmagne, J. C., & Doignon, J.-P. (2011). *Learning spaces.* Berlin: Springer.

Ganter, B., & Wille, R. (1999). *Formal concept analysis: Mathematical foundations.* Berlin: Springer.

Haertel, E. H. (1984). An application of latent class models to assessment data. *Applied Psychological Measurement, 8*, 333–346.

Haertel, E. H. (1989). Using restricted latent class models to map skill structure of achievement items. *Journal of Educational Measurement, 26,* 301–321.

Heller, J., Stefanutti, L., Anselmi, P., & Robusto, E. (2015). On the link between cognitive diagnostic models and knowledge space theory. *Psychometrika, 80,* 995–1019.

Lazarsfeld, P. F., & Neil, W. H. (1968). *Latent class structures.* Boston: Houghton Mifflin.

Lukas, J., & Albert, D. (1993). Knowledge assessment based on skill assignments and psychological task analysis. In G. Strube & K. F. Wender (Eds.), *The cognitive psychology of knowledge* (pp. 139–150). Amsterdam: Elsevier.

Roussos, L. A., Templin, J. L., & Henson, R. A. (2007). Skills diagnosis using IRT-based latent class models. *Journal of Educational Measurement, 44,* 293–311.

Rusch, A., & Wille, R. (1996). Knowledge spaces and formal concept analysis. In H. H. Bock & W. Polasek (Ed.), *Data analysis and information systems. Statistical and conceptual approaches* (pp. 427–436). Berlin: Springer.

Spoto, A., Stefanutti, L., & Vidotto, G. (2010). Knowledge space theory, formal concept analysis and computerized psychological assessment. *Behavior Research Methods, 42,* 342–350.

Spoto, A., Stefanutti, L., & Vidotto, G. (2013). Considerations about the identification of forward- and backward-graded knowledge structures. *Journal of Mathematical Psychology, 57,* 249–254.

Stefanutti, L., Spoto, A., & Vidotto, G. (2018). Detecting and explaining BLIM's unidentifiability: Forward and backward parameter transformation groups. *Journal of Mathematical Psychology, 82,* 38–51.

Tatsuoka, C. (2009). Diagnostic models as partially ordered sets. *Measurement, 7,* 49–53.

Tatsuoka, K. K. (1985). A probabilistic model for diagnosis misconceptions by the pattern classification approach. *Journal of Educational Statistics, 10*(1), 55–73.

Tatsuoka, K. K. (1990). Toward an integration of item-response theory and cognitive error diagnosis. In N. Frederiksen, R. Glaser, A. Lesgold, & M. G. Safto (Eds.), *Diagnostic monitoring of skill and knowledge acquisition* (pp. 453–488). Hillsdale: Lawrence Erlbaum Associates.

Wille, R. (1982). Restructuring lattice theory: An approach based on hierarchies of concepts. In I. Rival (Ed.), *Ordered sets* (pp. 226–233). New York: Springer.

Theatrical Organicism: Thoughts on Drama and System Theory

Francescogiuseppe Romano Maria Dossi

Before starting to expose our thesis, it is necessary to introduce a precondition: drama is already systemic. The aim of this essay is not to suggest another way to stage a drama or to perceive the theatre praxis but, through the analysis of theatre history, to observe any embryonic approach—both conscious and unaware—to the drama. As far as I am concerned, it is possible to identify three different perspectives:

1. Drama is, by its nature, systemic. It is intrinsically systemic, even before considering it from the point of view of the script, the dramaturgy, or the mixture of different media in the *mise-en-scène*;
2. Drama is aesthetically systemic. This may seem obvious, but the idea of an organically staged play appears only with the birth of the modern dramaturgy, at the end of the nineteenth century;
3. Drama is ontologically systemic. The individual parts of any play reveal something that is not attributable to any of those parts, by virtue of their relational properties.

These three perspectives did not arise simultaneously: the first one coincides with the birth of Greek tragedy, the second one with the birth of dramaturgy, meaning the theatrical direction, and the third with the meta-theatrical considerations of Peter Brook and Jerzy Grotowski.

The birth of tragedy, as is common knowledge, is to be contextualized in Greece in the fifth century BC The leaders of Greek society needed to find a problem to social harmony: how can people live in peaceful cohabitation? The only answer that could possibly be accepted by a Greek is rationality, the logos. To understand how the Athenians could reach such a conclusion,

F. R. M. Dossi (✉)
Catholic University of the Sacred Heart, Milano, Italy

© Springer Nature Switzerland AG 2019
G. Minati et al. (eds.), *Systemics of Incompleteness and Quasi-Systems*,
Contemporary Systems Thinking,
https://doi.org/10.1007/978-3-030-15277-2_18

we need to take a step back to the first written Greek work, the Iliad. The Iliad begins with the "destructive wrath" of Achilles, caused by Agamemnon's impiety. The Greek general stole Briseis, Achilles' slave, as he had to return his own prize, Chryseis, to her father, Apollo's priest. While she was still Agamemnon's hostage, the god of the sun sent a horrific plague to the Achean military camp. It is interesting to note that the Achean chiefs did not interrogate a physician on the pestilence, but a seer. They wanted to know who, not what, was the cause (Bernardi 2015).

Drama was born as the gathering of an audience and a performance. The spectator attends a performance in order to purify himself, establishing a connection with the performers. This act of purification is called catharsis and it is, in all respects, a property emerging from—but not entirely reducible to—this connection. Therefore, the drama's purpose is the pharmakòs, literally "medicine", but in more accurate terms "vaccine". The Athenians attended theatrical performances once a year, on the occasion of the City Dyonisia where they experienced horror, irrationality and passion: indeed, all that is considered to be the Dionisyan.

Having established the intrinsically relational and inherently emergent aspects of drama as a gathering of an audience and a performance, let us consider the second perspective presented above. Germany, nineteenth century: Georg II, Duke of Saxe-Meiningen revolutionises the way of staging a drama, basing his *mise-en-scènes* on historical accuracy, on the observance of the fourth wall principle, and especially on the stage coherence principle. The individual actor had no say in the matter of the theatre production but was still integrated in a bigger project, coordinated by a single person super partes (Koller 1984). This pivotal shift in perspective offered an alternative to the so-called *Teatro del Grande Attore*, where the direction was entrusted to the practical experience of the main actor. Thus, dramaturgy was born, and with it the idea of organicity.

The missing notion that inhibited the transition from the previous tradition to direction-based drama was the "research on life", an organic unity: the presence of an external coordination in order to create a coherent representation. Drama, although still anthropomorphic, concerned with human life and its numerous facets, ceases to be anthropocentric. The smallest elements of drama are not the main character or the individual anymore, but the connections that link the bodies on stage (Schino 2001). This was a historical landmark in theatre history and resonated throughout all of Europe. It was particularly acknowledged in Russia, with the works of Stanislavskij (2002, 2004). Konstantin Stanislavskij applied the organic creation method to the acting training. Distancing himself from the "authoritarian tendency" of the despot-director, he favoured a creative and spontaneous act of the actor, while still under the supervision of a director.

The actor is unable to do more than what they are, and therefore the self is the starting point from which they must characterise their character. The character never corresponds to its actor. With this method, the actors do not solely rely on their experiences, nor do they dissociate themselves from their own background in order to identify themselves with the character, but they make use of their past to create something that is not entirely them or the character they are portraying. The character is grafted onto the actor, and the result is a synthesis of both of them.

So far, the genetically systemic and aesthetically systemic aspects of drama have been analyzed. Let us focus now on the ontologically systemic aspect, inspecting under this lens the elements by which it is composed. The meta-theatrical productions of Brook and Grotowski are here particularly interesting. Jerzy Grotowski pushed the notion of connection as the centrepiece of the drama even further. In *Towards a poor theatre* (Grotowski 2002), he states that

> By gradually eliminating whatever proved superfluous, we found that theatre can exist without make-up, without autonomic costume and scenography, without a separate performance area (stage), without lighting and sound effects, etc. It cannot exist without the actor-spectator relationship of perceptual, direct, "live" communion. (Grotowski 2002)

The Polish theatre director makes a clean break with the past, with what we called the "aesthetically systemic theatre". Make-up, costumes and scenography are superfluous. Drama cannot exist without a relationship between the actors and the audience. The actor is "sacred" and their task is to make the drama emerge from the connection they made with the spectator.

This perspective was re-elaborated by Peter Brook in *The Open Door* (Brook 2005). Brook considers the empty space as theatre's primordial condition, where the only essential component is the human element. The actor has to make visible the invisible, to show the spectator something that otherwise they could not see.

References

Bernardi, C. (2015). *Eros. Sull'antropologia della rappresent-azione*. Milano: EDUCatt.

Brook, P. (2005). *The open door: Thoughts on acting and theater*. New York: Anchor Books.

Grotowski, J. (2002). Towards a poor theater. In E. Barba (Ed.), *Towards a poor theater. Jerzy Grotowski*. New York: Routledge.

Koller, A. M. (1984). *The theatre Duke: Georg II of Saxe-Meiningen and the German stage*. Stanford: Stanford University.

Schino, M. (2001). Mutamenti elementari: l'avvento della regia. In C. Alonge-Davico & M. Bonino (Eds.), *Storia del teatro moderno e contemporaneo. Volume terzo. Avanguardie e utopie del teatro. Il Novecento* (pp. 72–89). Torino: Einaudi.

Stanislavskij, K. S. (2002). *Il lavoro dell'attore sul personaggio*. Bari: Laterza.

Stanislavskij, K. S. (2004). *Il lavoro dell'attore su se stesso*. Bari: Laterza.

A Need for "Systetics"

Giordano Bruno

We create the world we perceive not because there is no reality outside our minds; we create it because we choose and modify the reality we see in order to adapt it to our beliefs about the world we live in. It is a function necessary for our adaptation and survival.
(Bateson and Bateson 1987)

By aesthetic, I mean responsive to the pattern which connects.
(Bateson 1979)

1 Preliminary Remarks

I have been teaching and learning at a small but prestigious Italian institution operating in the field of design training and research—for this reason it is generally not known—for 35 years, namely the *Istituto Superiore per le Industrie Artistiche—ISIA*, (Higher Institute for Artistic Industries), based in Rome. Its name, which stems from a 1929 royal decree by then minister Giovanni Gentile which established the Higher Institutes for Artistic Industries, naturally betrays its possibilities and limits. The Institute took this name and was conceived and shaped by Giulio Carlo Argan, the sculptor and head Aldo Calò, as well as by ministry inspector Giuseppe Chiatti, who seized a favourable chance to carry on the experience of design training opened up precisely by the Higher Experimental Courses in this field.

I have started by emphasizing this fact because I have always been drawn to a holistic approach to reality thanks to my previous education and vocation, but above all because it is in the very culture expressed and practiced at this Institute that I have also found theoretical views and interpretations which conceive the design of objects within something already wider: systems design. The fruitful encounter with Gianfranco Minati and AIRS has resulted in a training project that is part of a second-level master's degree deliberately called *Systems Design—Systemic Teachings* and *System Analysis*.

G. Bruno (✉)
ISIA Roma Design, Istituto Superiore per le Industrie Artistiche, Roma, Italy

© Springer Nature Switzerland AG 2019

G. Minati et al. (eds.), *Systemics of Incompleteness and Quasi-Systems*, Contemporary Systems Thinking,
https://doi.org/10.1007/978-3-030-15277-2_19

During all these years I have noticed that the research and projects carried out in the field of systems design, when valid and accomplished for the observer (or observers), present common features: some properties have emerged through the relationships and interactions designed which, in short, are at the same time ethical and aesthetic.

So came the idea of coining a new term that integrates systemics, ethics and aesthetics: "*systetics*".

2 Systetics: What Is It?

As mentioned earlier, coining this new term has been very straight and simple. Now, it is about giving it a sense, a use, a perspective. In order to do it, we must start from the meaning and attributes to be given to ethics and aesthetics, which I have only quoted so far and undoubtedly represent two of the most controversial and much debated human concepts.

I shall start from the article by Giuseppe O. Longo titled *Etica, Estetica e Libero Arbitrio* ("Ethics, Aesthetics and Free Will") Longo (2010):

> Aesthetics is the subjective (but shared) perception of our bond with the environment, characterized by a deep and balanced dynamic harmony. Ethics is the subjective and inter-subjective ability to conceive and carry out actions which can maintain a healthy and balanced relationship with the environment. Ethics and aesthetics are therefore two sides of the same coin because they stem from the evolutionary co-implication between species and environment and are both *reflections* in us of this co-evolution. If aesthetics is the (inter)subjective feeling of harmonious immersion in the environment and ethics is the (inter)subjective feeling of respect for the environment and action in harmony with it, then ethics allows us to maintain aesthetics and aesthetics functions as a guide for our ethical action.

I share this approach, although I will try to generalize from it and explain it with my own words. When can we talk about a *systetic* project?

I shall dwell upon this because I think that here lies the importance of the introduction of a systetic *category*.

A project can be defined as systetic when it is able to make a (collective) subject detect properties that come from the interaction of its constituent elements and are inseparably *beautiful* and *good*, and can be perceived as such by the same subject. Consequently, a new cognitive model takes shape in the observer, consisting in a kind of topological space built upon the two elements of beauty and good. This space does not include a *measure*—in fact, it only has topological properties. It presents features of uncertainty and indeterminacy as well as incompleteness. Thus, it is an intrinsically and logically open space. I am using the universal categories of beauty and good ($\kappa\alpha\lambda\grave{o}\varsigma\kappa\acute{\alpha}\gamma\alpha\vartheta\acute{o}\varsigma$) stemming from the ancient Greeks, who thought that what was beautiful could only be good and vice versa, what was good, was necessarily beautiful.

Certainly, the problem is now to determine what the beautiful and the good are. My contribution, therefore, aims to understand what is said to be beautiful and good in systems design. Surely, the beautiful can be understood as the aspect of

> subjective (but shared) perception of our bond with the environment, characterized by a deep and balanced dynamic harmony,

as Longo says. However, since a system, like an object, also has a concrete form, whether abstract or virtual, I believe that such a form should be simple and elegant at the same time, as mathematics teaches us—it is not possible to *reduce* it further without damaging its meaning, and it has been *chosen* (etymologically, "elegant" comes from the Latin *eligere*, "to choose, select") so that we can recognize it and access it intuitively. As for the good, it can be understood as resulting from the

> subjective and (inter)subjective ability to conceive and carry out actions which can maintain a healthy and balanced relationship with the environment,

as pointed out by Longo. However, I think that it is nonetheless worth saying that a system designed will be good if it is able to provide greater well-being in cultural, economic, social and environmental terms, not only to those who will benefit from it (and to the environment in general), but also to all those who have participated in its realization, along with *future generations*. Therefore, its usefulness is linked to its essentiality, ease of comprehension, usability, appreciation and recognition of its importance in relation to the aforementioned purposes. In this regard, I would like to quote the great artist and designer Enzo Mari (Favento 2006):

> The beauty created is only an allegory. The true quality of a design object consists in the quality of the work expressed by those who will make the product: from the designer to the entrepreneur, from technicians to simple workers.

Moreover, in order to try and clarify what has been said above, I would like to quote an extract from the same interview (Favento 2006) with Mari that illustrates what I mean:

> I try to do well but I don't know if I manage,

Mari humbly confesses,

> if art, as every historian says, is sinsemantic—that is, it has infinite meanings—how can we have a handbook that includes them all?

There are no possible instructions, the only choice is constant investigation and analysis. In his long study path, Mari has devoted himself to research the psychology of vision, the planning of perceptual structures and the methodology of design. His attempts follow his wish to determine the concept of total quality and go in the direction of a possible grammar of form. He expresses the need for an understanding of the form, an intention which has taken shape especially through projects with children. Mari says that

> when you stop and ponder, as some scholars and I have done, at best you can describe one or two fragments of the total quality, but it's impossible to describe it all.

Given these theoretical limits, he argues that the overall quality of an object necessarily coincides with the quality of its form. This quality only emerges when the form transcends the more banal functions and corresponds to the essential meaning of the object; when

> transcendence is materialized in the work, the quality of the work is the only possible demonstration.

As regards formal features, Mari says that:

> for me a form is good when it seems too poor for most people.

We have already highlighted his tendency to eliminate the superfluous, to strip the object to get to its intrinsic characteristics. The ultimate example which he designed in 1958 is the "iron beam", a work of industrial enhancement of an intrinsically significant archetype that "includes all contradictions of design". In this project more than in anything else he has designed, it is evident how the elegant sign on the industrial product does not aim to "hide obscene reasons", but rather to "denote the intimate beauty" of a simple object.

Mari's statement on the form and his view contain an idea of beauty/good related to incompleteness ("too poor"); it is logically open as it appears in the paintings by Impressionists and it is related to the uncertainty and indeterminacy that push us to modify our cognitive model, as shown by Maurits Cornelius Escher's works. The relationship between art and ethics has always been alive and has produced an understanding, which has been fundamental for a development of the individual as integral and unitary as possible, as Mari has also explained. By designing objects, design has integrated art, science, ethics and aesthetics (in the best cases) and has represented the culture of forms and functions. Today, we need to extend this view: the object must no longer be conceived as a means to achieve a result (to solve a problem), but as a "knot of relations", as a "consistent" creation between us (designers, implementers and users) and the outside world. In this sense, the beautiful and the good are intertwined and realize the ancient Greek vocation.

Moreover, the project, the object, becomes systetic if located within a system—or conceived to generate it—that is capable of inducing in the (individual and/or collective) observer a radical change of their cognitive model through emerging processes. This change identifies the consistency between and within the system, and between the observer and the system, thus allowing to subjectively and qualitatively perceive an improvement of the well being brought about by this emergence. This is the amazement and inner satisfaction that we experience in front of a work of art, a work of nature or an unexpected gesture in favour of another person. In this regard, I fully share what Franois Cheng says about the Monna Lisa in his *Cinq méditations sur la beauté* (Cheng 2006):

Her beauty is not based on the mere combination of exterior traits. It's almost as if it were illuminated by a look and a smile, an enigmatic smile that seems to mean something. How beautiful it is to be able to hear her voice! The woman expresses her feelings, but also her nostalgia, her dreams, and that unspeakable part that still seeks a way to express itself. The desire to speak mingles with the desire for beauty; the desire to speak adds something to her charming beauty. An evidence thus strikes the eye: female beauty is not merely the result of physiological evolution, it is a conquest of the spirit. This achievement reveals that true beauty is consciousness of beauty and momentum towards it, and that it arouses love and enriches our own conception of love.

It is indeed in this process of cognitive enrichment, so masterfully described by Cheng, that I understand the profound meaning of systetic work, systetic project. In addition to this and in the light of what has been written above, it is of paramount importance to introduce and give dignity to all those terms that are regarded as limiting elements in the still dominant culture—for instance, uncertainty, incompleteness and indeterminacy. As I have already tried to show, these, on the contrary, become bearing elements of a nonlinear, complex and not fully defined profile that becomes uncertain and indeterminate, that is open to cognitive changes, and allows the development of dynamics within and outside the object, the project, the person and the beings, in close interaction with the environment and the context. In this respect, I am referring to the latest articles by Minati (2016a,b).

The recognition of the presence of each of the above-mentioned terms and their overall interaction, if we look at the Monna Lisa, is an added value, a set of striking features that allow us to be profoundly impressed by the painting and driven to a cognitive perception of the καλὸςκάγαϑό that transpires from it. We are captured by the indeterminacy of her gaze and the incompleteness of her smile that speaks, and their interaction with all the other elements of the work leaves us in uncertainty. This uncertainty does not weaken but enhances the value of the communication between us and the Monna Lisa, which spurs a deeper reflection on beauty, which, as Cheng says, becomes

consciousness of it, conquest of the spirit and source of love (in the two meanings of "source" as place where you can *drink* from and a *generating* element).

In order to apply to the field of design what has been said above, which I only regard as a first *seed* into a field that I hope will bear fruits and perspectives and will be followed by many more seeds, I shall present a system project developed at ISIA in Rome which I believe includes all the features that make it a "systetic" project.

Pro(b)ABILITY project by Chiara Longo: a system of product and services to dress people with motor disabilities through the enhancement and strengthening of their residual skills. Second-level dissertation in Systems Design, July 2010, Prof. Veneranda Carrino. Project developed within the 2009/2010 Final Workshop in Design: Caring For Things. This project was awarded the 2010 *Premio Nazionale per le Arti* for the category IDEA Design Award; the *Premio per l'Innovazione Piaggio–Vespa Prize, Museo Piaggio, Pontedera for Creactivity* and the *Premio Imprenditoria al Femminile*

"*L'Eccellenza della donna*" awarded by *Fiera di Roma* within the "EXPO ARTI & MESTIERI".

The design areas in which the project has been conceived and developed are:

- Design of systems.
- Design for human diversity.
- Design for wearability.
- Design for inclusion.
- Design for all.

The system is a set of different codes and languages used to concretely contribute to breaking down barriers, and employs relevant forms of communication that deal with the subject addressed by moving away from the simple concept of giving information and from a purely welfare subsidy mentality. The project is divided into:

- "*Design For Wearability*": a system of clothing designed for people with motor disabilities;
- "*Sharing Lab*": a creative lab for clothes-making, with users sharing and participating in;
- "*Virtual Square*": a website where you can buy clothes, access related services and sign up for the lab.

Engaging at more levels allows to develop the person's skills towards greater autonomy at all stages, from the design to the purchase of an item of clothing.

The project for dressing people who have difficulty in doing it unaided provides them with the possibility of getting dressed by themselves—by sitting down and using clothes and accessories conceived and designed for this purpose—while meeting the need for a pleasant and suitable aesthetics which also brings psychological and physical comfort for the individual's well-being. Furthermore, clothes are just a part of the project, which also involves the user's participation in a creative lab to design clothes and a website that encourages the purchase, the use of services and the opportunity to sign up for the lab activities.

The following pictures (Figs. 1 and 2) illustrate the project in details.

The systetic aspect of the project is the fact that the designer has conceived it by adopting a cognitive approach that meets the overall needs of people with disabilities, resulting from the interaction between the good—having clothes which are easily wearable and comfortable—the beautiful—a simple but pretty outfit in terms of shape, fabric and colours—and the system—the creation of a website where users can buy clothes, share and participate in a lab in which they can create a collection of clothes, ask for assistance in purchasing and make use of tailoring delivery services. Thus, I think that the possible interactions created by the project really bring about well-being and modify the cognitive model of both the internal and external observer (agent). Additionally, such an approach can be applied to other social fields and thus create a virtuous circuit that encourages fulfilling, inclusive relationships and behaviours for a wider and wider part of the population, which are urgent and necessary to give meaning to our lives.

3 Conclusions

Finally, I believe that what I have described in the example I have taken provides us at least with an initial key to identify systetics. In any case, it is necessary for the (single and/or collective) observer to be able to feel and absorb the well-being from the system created, or being created, through the interaction between its elements, which brings about the emergence of *the beautiful* and *the good*. In addition, this emergence must modify the observer's cognitive model so as to produce a re-modulation of the meaning of their life, in particular the value to be given to their choices and actions with respect to the relationship with the others and the environment, in order to preserve and expand the well-being generated by that system.

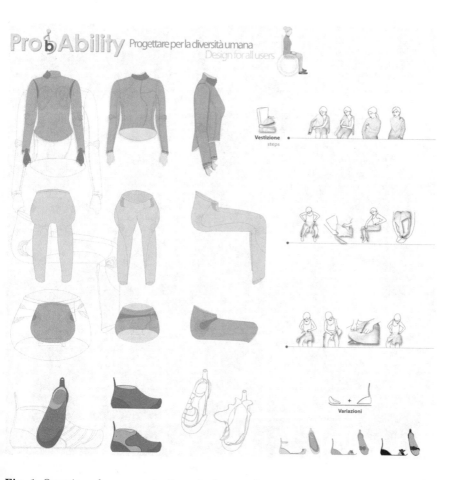

Fig. 1 Overview of a communicative episode according to our conceptual model

Therefore, a system that is not able to significantly change the state of the relationship between the observer and the environment, thus inducing in them a consistent attitude that favours the creation of *the beautiful* and *the good* in the context of the system outside it, is not systetic. On the contrary, we could regard as systetic a system of mobility services, for example, which establishes networks between people, the environment, means of transport and urban furniture in an interaction that creates well-being in the observer (user and operator) to make them feel part of the system, safeguard it, and make its further development possible; this is a system which makes the environment in which it is located more pleasant—it does not collide with

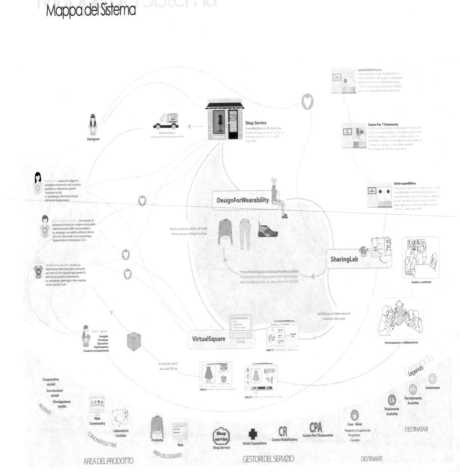

Fig. 2 Naason's Universal Journey: How to self-legitimize (excerpt)

it, but instead enhances its natural and/or artificial beauty; it can become a conceptual model used in other fields and a legacy for future generations, while protecting the environment (locally and hence globally) and freedom of movement.

Therefore, all those projects/systems which, for example, only privilege the aesthetic aspect or some ethical values are not systetic. They are not able to establish an interaction between elements which creates well-being.

At the beginning I have talked about giving systetics a perspective. To conclude, I would like to focus on this. I believe that spreading this term can be useful, in fact fundamental, in two fields of human activity: design and education (training).

Regarding the first one, I shall just say that systetic design should paradigmatically become *the way of designing*. Although this approach is at least in part already present in designers' work, all those involved in this activity should become fully aware of it to reach one of the main objectives of the systemics of emergence: the deep modification of the observer's conceptual model.

I hope that this awareness will become fundamental—only by adopting the systetic approach will we be able to continue to design and produce objects/systems which will not worsen the quality of our lives but which, on the contrary, will encourage a revision or creation of cognitive models and behaviours that put the relationship and interaction between us, those who will come after us, and the environment, at the centre of our daily life.

And here comes the second field, education in the broader sense (I have put training in brackets because these two concepts are often wrongly confused—the first is much wider and more fundamental than the second and pervades every moment and stage of our lives). I regard the work that philosophers like Edgar Morin and Michel Serres have been carrying out for decades as seminal from this point of view. I shall only quote from two of their recent books, respectively: *Enseigner à vivre* and *Le Gaucher boiteux: Figures de la pensée*. I quote (Morin 2014):

> It is very important to talk about the ethical consequences that the ring of knowledge can cause. In fact, morality, solidarity and responsibility cannot be dictated *in abstracto*; you cannot spoon-feed the mind with them as you feed geese with a funnel. I think that they must be driven by the way of thinking and the experience lived. The linking thought shows the solidarity of phenomena.

I would like to add that a powerful tool to build an (individual and collective) ethics is precisely cultivating (and thus understanding) the aesthetic value of things (*res*); only in their interweaving and in their deployment can we hope to encourage thoughts and behaviours that generate systetic emergences. I conclude by quoting the skilful Serres (2015):

> How beautiful Antares is with its Medusa's head, how beautiful the glowing diamond, the cedar tree employed in building, how beautiful the tiger with its shining and coloured fur [...]. And they are such not thanks to their nuclear, carbon or

wooden material, their superb or feline shape, or any judgment uttered by an expert or an idiot, but rather the cosmic and vital momentum in which an unexpected forking brought to light this figure, this star, tree, animal or crystal character [...]. The beauty of a woman, body and soul, of a sonata or a page, is born in such a surreption, in the very insurrection, in this erection, in this resurrection. With contingent leaps, whose irregularity draws an endless ramp of flames—different in angle, colour, brightness, deviation, intensity, height—the Tale of the world invents shining singularities—multicolour glittering, dazzling diamond, cedar of Lebanon, Bengal tiger, [...], a sentence. Beauty—seal of thought.

These two short extracts and what accompanies them continue to fuel my desire, my drive and the immeasurable joy of learning and teaching ... "systetically"!

Acknowledgements I would like to thank Prof. Gianfranco Minati and Prof. Giulia Romiti for their valuable advice and support in writing this contribution.

References

Bateson, G. (1979). *Mind and nature: A necessary unit.* New York: E. P. Dutton.

Bateson, G., & Bateson, M. C. (1987). *Angels fear: Towards an epistemology of the sacred.* New York: Macmillan.

Cheng, F. (2006). *Cinq méditations sur la beauté.* Paris: Albin Michel.

Favento, C. (2006). *Enzo Mari, (tra) etica e design.* February, 1, 2006, FM 84. https://www.fucinemute.it/2006/02/enzo-mari-tra-etica-e-design/

Longo, G. O. (2010). Etica, estetica e libero arbitrio. *Scienza in rete* (17 May 2010). https://www.scienzainrete.it/contenuto/articolo/Etica-estetica-e-libero-arbitrio

Minati, G., (2016a). Knowledge to manage the knowledge society: The concept of theoretical incompleteness. *Systems, 4*(3), 1–19.

Minati, G. (2016b). Introduzione al pensiero sistemico e ai suoi recenti sviluppi. *Rivista di Filosofia Neo-Scolastica, CVIII*(2), 271–276.

Morin, E. (2014). *Enseigner à vivre: manifeste pour changer l'éducation.* Arles: Actes sud/Play bac.

Serres, M. (2015). *Le Gaucher boiteux. Figures de la pensée.* Paris: Le Pommier.

A Systemic Approach to Religious Communication: Case Study of "La Luz del Mundo" Church

Irune Medina

1 Introduction

Decision making is not easy. Very frequently, it's not even rational! Every day we are witness of choices, both individual and collective, that seem to escape all logic. Especially when the choices emerge from the specific ways of discursiveness: marketing, advertising, propaganda, religion, education, science, etc. These choices, however, often come into conflict with what can be considered as viable (Von Glasersfeld 1998) for the individual. Given the highly complex nature of language and communication that entails the emergence of new levels of signification as a consequence of the communicative and operational interaction of social agents, we rely on a conceptual framework that allows to deal with the different components and mechanisms (biological, cognitive and social) that play a key role in social communication.

The proposed meta-model, a Systemic Model of Non-Cooperative Communication, leverages on the mechanisms of Perceptual Learning and Categorical Perception, Radical Constructivism, Game Theory, Neo-Rhetorics, and Decision Making to provide a picture of the complexity hidden in the decision making processes that social communication involves. This latter turns to be very valuable especially when one is required to model complex phenomena, thus highlighting the emergent features of the interactions between their different components. Understanding how language becomes a social manipulation tool would indeed help shed light on a variety of key processes in societies. By applying our model to the self-legitimation speech of Naasón, Leader of the mexican cult "La Luz del Mundo" we develop a

I. Medina (✉)
University of Rome Tor Vergata, Roma, Italy

© Springer Nature Switzerland AG 2019
G. Minati et al. (eds.), *Systemics of Incompleteness and Quasi-Systems*,
Contemporary Systems Thinking,
https://doi.org/10.1007/978-3-030-15277-2_20

qualitative analysis to see how a religious belief system relies on language in cooperative and competitive terms and to understand how this helps define its organizational identity, existence and growth.

2 World, Language and Decision Making

What is the role of language in the knowledge construction process? What would be the purpose of communicative action? Can language change the perception of the world's and influence the decision making process? Perception is the process through which a cognitive system obtains information from its environment. When the subject comes into contact with the world, a series of internal cognitive processes get started, specifically, with the processes of identification, discrimination and Categorical Perception (Goldstone 1998). Our perceptions, in fact, are altered in agreement with our categorical structures. Phenomena like the *framing effect* (Kahneman and Tversky 1981) actually show that the choices of the subjects are anything but rational. Rather they are significantly conveyed by language as confirmed by empirical research in cognitive sciences during the last decade (Ting Siok et al. 2009). In fact, important findings have clarified some of the relationships that hold between mind, language and what philosophers have often referred to as the manifold. Furthermore, empirical evidence supports the idea that language interferes with perception (Tan et al. 2008). The role of language in decision-making, however, goes well beyond perception itself, as there is also empirical evidence of its influence on ethical and moral dilemmas (Costa et al. 2014). The relationship between perception, language and decision making can be described by a constructive mechanism that deals with *Online/Offline Re-presentations*. The term "re-presentation" is employed here with the constructivist meaning.[1]

Online Re-presentations refer to an information process that occurs at the same time of the perception of the manifold, involving a spatial and chronological closeness that depends on the knowledge base of the subject. The background knowledge network is what we refer to as *Offline Re-presentation*. The subject establishes a dense network of connections between the online experience and the one already stored in his memory. We hypothesize that if these online re-presentations are coherent and don't match significantly with the structure of the receiver's *offline re-presentations*—in that the latter are more fragmented and unstructured with respect to the formers—the latter will tend to be completed with the scheme and content offered by the sender. This sense of "cognitive saturation" will become the "boost" to consider the sentence as true or viable. The construction of cognitive structures by the

---

[1] "Re-presentation" (with the dash) that is "to present again" the past information at the same moment of the experience (Perelman 2001).

subject does not depend only on the subject's perception of the world, but also on what the others "say" about the world, and thus directs the decision making process.

3 A Systemic Model of Non-cooperative Communication

We advance the idea that a Systemic Model of Non-Cooperative Communication would help shed light on a whole class of communicative processes. The model described here (Fig. 1) is a translucent model based on a systemic framework that tries to pick up the basic interactions occurring in the context of a minimal communication episode. Here "inside/outside" refers to the spaces where different interactions—including operations and functions—between the components of the communication process occur. It's important to recognize that not only the roles of the participants are interchangeable but also the very boundary between outside/inside in human cognition cannot be univocally defined, rather it is the result of a continuous and interactive construction by the involved systems, including the observer. In the *Out-*

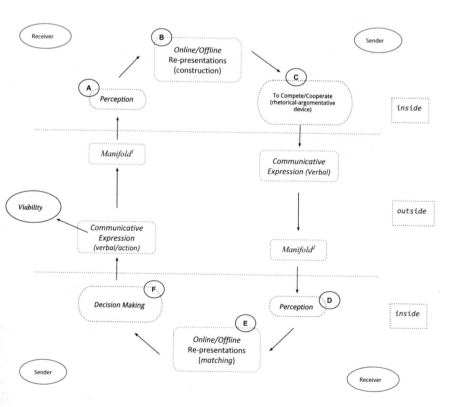

Fig. 1 Overview of a communicative episode according to our conceptual model

side spaces there are the manifold and the different forms of communicative signals (verbal and action).

The manifold is perceived by different subjects probably in different ways: manifold1 refers to one participant (i.e. the sender) whereas manifold2 refers to the other participant (i.e. the receiver). In the *Inside* spaces there are the cognitive operations (A, B, C, D, E, F). Viability strongly depends on the "outside" because the payoff of linguistic actions (both for the sender and the receiver) is not the result of an operation that belongs just to the subject. Rather it relates to external data and to the subjective experience of the manifold. Therefore we can speak of viability with reference to all the participants and stakeholders of the communication process as they play both roles (i.e. sender and receiver in alternative mode). Thus, the operations within the translucent boxes of the conceptual model can be assumed as a function of the role that the subject is playing. Most of all, they look like a continuous, automatic, unconscious and simultaneous process in communication. Stakeholders and players may be individuals or spokespersons of specific groups. There is no theoretical constraint over their number. However, there are two necessary conditions for successful communication: being able to establish the same (almost) linguistic code (language) and demonstrating full possession of the cognitive abilities (Perelman 2001). We hypothesize that there are also other two conditions—not necessary yet desirable—for a successful communication: (1) to share more or less the same socio-cultural and behavioral codes, and (2) to share as much relevant information as possible with reference to the context of communication. Communication can be both bidirectional—like in inter-subjective or intra-subjective dialogue—and unidirectional—like when the receiver is a passive recipient of the message without any expected communicative interaction.

In strictly methodological terms, therefore, our analysis of the communication process take places in three steps. The first step is the identification of data (e.g. communicative expressions, the profile of participants in the discourse and the context of the manifold where communication takes place). It's important to establish a distinction between two wide spheres of verbal communication: what the sender of the message directly says (constitutive speech) and what others may say about the expression of the sender (hypertextual discourse). The latter can be considered as a methodological tool useful to understand and synthesize the discourse of collective beings (Minati 2001). The second step is the generation of hypotheses regarding the involved cognitive mechanisms (A, B, D, E, F) and the analysis of rhetorical-argumentative expedients (C) to understand whether the use of language is competitive or cooperative. The final step is to assess viability for all the participants. The test method to establish if we are facing an uncooperative use of language, however, can be assessed only with reference to the "post-speech" facts, i.e. the resulting conditions of the manifold as they are perceived by the stakeholders of communication once the communicative interaction has ended. The sender will receive the payoff of his speech when he reaches or not his goals (i.e. change the manifold at his own benefit); the recipient will receive the payoff of his decisions (which is raised from the sender's discourse) probably after the communicative interaction. Some results are immediate, but there are many others in which the payoffs are not appreciable in the short term. It is precisely here that our model becomes relevant: by analyzing the characteristics of the participants and the mechanisms involved in the communication process, the systemic model of the non-cooperative communication can envision, if not all, at least some of the possible outcomes of asymmetric communicative interactions.

Our case study deals with a Mexican cult, *La Luz del Mundo*. We offer a qualitative analysis to see how a religious belief system relies on language in cooperative and competitive terms and to understand how this helps define its organizational identity, existence and growth. We examine here just few examples.

4 How to Legitimize the Apostolate of Naasón

"La Luz del Mundo" (LDM) is a Christian cult of Mexican origin founded in 1926 by an ex-army of peasant origin, Eusebio Joaquín. The church had three spiritual leaders: Eusebio (1926–1974); his son Samuel (1974–2014) and from 14 December 2014 Naasón, the latter's son. The cult has a control and surveillance system (the specific internal organs that maintain control over adepts daily life, personal and collective, specially through the hegemony on the symbolic code) that limits self-perception and self-determination of the subjects (De la Torre 1994). It is perceived as viable by members of the church (Biglieri 2000; Fortuny Loret de Mora 1984; Gaxiola 1970). This is mainly due to the self-reliant character of the community (although, anyway, based on religious ideologies that are often considered as non-rational) that, by way of reflection, allows them to preserve their self-organization and thus ensure their continuity. In fact, we have argued elsewhere (Medina 2017) that the LDM can be considered as a viable system of religious beliefs by virtue of its self-organization, which can be defined as autopoietic thanks and, above all, to processes of self-distinction, self-production and self-fulfillment. The viability displayed by the aforementioned cult is to be understood as the ability to adapt the system of beliefs, practices and institutions so as to meet the requirements for survival in a changing environment. The system also, as we have argued, could probably maintain its autopoietic organization even without the presence of a leader as, more importantly, it does not need the same components to maintain its organization. In the sense that, as well as a collective being created out of a bank queue, the components of the LDM system can change constantly; it is sufficient for the system to maintain the necessary operations for the processes that define its organization to continue functioning, independently of their material components. LDM has

often dealt with the external disturbances (e.g. sexual abuse allegations in 1997) and with a range of social prejudices. But from the beginning it has always been threatened primarily by a specific internal disturbance: the split. Eusebio experienced two splits and Samuel experienced one, which led to the emergence of other religious cells. The reason lies in the pivotal doctrine of its beliefs network: the Apostolic Doctrine of Election (La Luz del Mundo 2010, 2015). The vast majority of followers accepted it, but others saw it as an imposition like former members[2] of the LDM who criticize this aspect especially regarding the material assets of the Church. In fact, the apostolic lordship (according to their symbolic code) also implies the political and economic power, that has been managed within the Joaquin family and its relatives, leading to the establishment of a genuine apostolic dynasty.

Talking about Naasón as the leader of the church, he inevitably connotes his ancestors and the perception that others outside of the cult have of them. The distinctions made on the figure of Naasón are very relevant for his leadership especially because his hereditary apostolate is perceived as imposed. To avoid the risk of another split, Naasón decided to organize an "Universal Journey" with the goal of legitimizing his role as an apostle appointed by God, thus preventing any internal fragmentation and avoiding public criticism. This hypothesis finds confirmation in his discursive action and in the repetition of the same rhetorical-argumentative expedient (often the same words used) in the framework of the Universal Journey.[3] By applying our Systemic Model of Non-Cooperative Communication (Fig. 2) to the structure of his arguments we can see a competitive use of language.

The core of Naasón's argument can be summarized clearly in the speech of Asuncion on 2015:

> Have you believed that Jesus Christ is the Son of God? Do you believe that on earth is the Church of Jesus Christ? Have you believed in the Messenger of God? Do you believe in the call that God made last December?

He's not worried about the doctrine of election itself, which is held to be true, but the legitimacy of the apostolic dynasty (A, B). The *online representations* that Naasón offers through the rhetorical-argumentative device (C, D) are addressed to the adepts of LDM. They are not required to be "convinced" about the election of Naasón, but they become strategic subjects to legitimize his apostolate. Then, the result of the *re-presentations matching* would be compatible with the knowledge base of the audience because the audience itself forms part of that very same social system (F). In fact, believing in the apostolic dynasty is not an option for the believers of LDM,

---

[2] The Ex-LDM (exlldm.com) believe that the worship shown towards Samuel and Naasón (extended to his entire family) is not only a betrayal of the values taught by Aarón, but rather an act of apostasy.

[3] We took as a discursive sample 8 apostolic presentations of the 7th stage of the Universal Journey: Buenos Aires and Montevideo (Jan 2015), Asuncion (Feb 2015), Tapachula (Mar 2015), Querètaro (May 2016), Los Angeles (June 2016), Roma and Paris (July 2016).

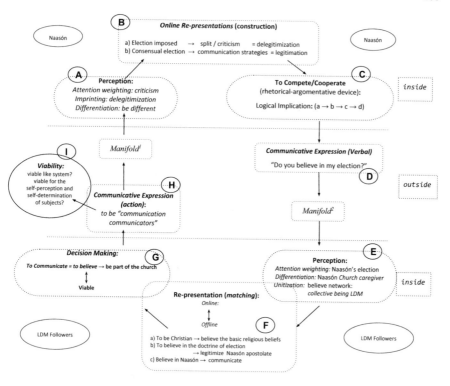

Fig. 2 Naason's Universal Journey: how to self-legitimize (excerpt)

but is one of the basic requirements they have to meet (Biglieri 2000). This further strengthens their collective identity. For this reason, Naasón's online re-presentations are considered as unconditionally true and, then, as viable (I).

Naasón is also a leader searching for public consent. By leveraging different rhetorical-argumentative techniques, Naasón tries to persuade his audience to act (H). In this case, the receiver's communicative expression would be: to say, to evangelize, to become the bearer of Naasón's words, to become "communication communicators" at last. Therefore Naason exploits his flock of adepts to validate the doctrine of the apostolic dynasty and to legitimize himself (but not the Church) in order to preserve the asymmetry of decision-making power (administrative, economic, social and spiritual) which is firmly in his own hands and, also, within his own family. For this reason Naasón's usage of language is clearly competitive.

Adherents to the LDM have learned to recognize and obey their apostles in order to belong and feel part of the community (De la Torre 1994). This is also why Naasón's leadership is considered as viable. But we cannot be certain about the benefits on the followers lives, especially because they have tasks and duties. How long will apostolic dynasty be viable for believers? Will there ever be any opportunity for meritocracy in LDM? What if one day any LDM follower said to have received a direct message from God? Will he/she be considered as the new Apostle? Would God's will be listened and accepted by the community, especially by the Joaquin family? The legal and economics powers would go into the hands of the new apostle? In the religious sphere

the language game is always a game of authority and asymmetry (Pace 2008)

and inside of LDM the words of the Apostle must be obeyed unconditionally. What would happen if he asked his flock to perform ethically questionable or destructive actions? What would happen, for example, if the Apostle asked him to commit suicide? We do not want to say absolutely that the LDM is a destructive cult. Indeed, as we have argued, its tendency is rather autopoietic and thus opposed to self-injury. But we want to be reasoning about the viability that the words of their leaders can enact on the followers if they are listened in an undisputed way.

5 Conclusions

The competitive use of language refers to the formulation of subjective interpretations of the manifold in order to change the perception, thought and actions of others in view of a competitive advantage. This happens when at least one of the participants tries to reach his goals at the expense of the others. This need not be done by resorting to deception. Indeed, the presented case study shows that those who make an uncooperative use of language can do it based on values and principles that are widely shared, and therefore, considered as true by the community of participants. Thanks to the clarification of some of the underlying cognitive mechanisms, we believe that our Systemic Model of Non-Cooperative Communication could help shed light on a variety of important communication phenomena in society.

As a conceptual model, it takes into account not only the asymmetrical power relationships among participants but also some of the cognitive processes involved in social communication, from perception to semantic processing and decision making. In that sense, our model would be a useful starting point to further develop simulation models to help predict, in a way, the emerging patterns of the communicative social interaction. And this, in the sphere of social and behavioral sciences, is not trivial. Possible applications range from the study of how public or group opinion is influenced by the use of mass media and new media by certain power groups (political, economic, social, religious, etc.) like for example by amplifying the perception of danger in front of situations of crisis and conflict (a terrorist act and the function of social networks).

The same can be said about other spheres of social life: education and pedagogy; government and its relationship with science (e.g. healthcare regulations and obligations); social policies in multi-ethnic, multi-cultural and multi-religious communities; the set-up of economic policies (individual and collective); the use of new technologies, their diffusion and capillary action, etc.

We are aware that our proposal is just a first step and there is still a long way to go, especially regarding the refining of the model and empirical investigation. Besides improving it on the conceptual side, we believe that the model could be implemented by means of suitable simulation technologies, bringing together, for example, multi-agent intelligent systems and Mixed Reality. This kind of development will allow to design and collect evidence on a variety of scenarios that are not actually replicable in real social settings. This, we believe, will shed light on aspects of communicative phenomena that are currently totally unexplored.

References

Biglieri, P. (2000). Ciudadanos de La Luz. Una mirada sobre el auge de la Iglesia La Luz del Mundo. *Estudios Sociológicos, 18*(2), 403–428.

Costa, A., Foucart, A., Hayakawa, S., Aparici, M., Apesteguia, J., Heafner, J., et al. (2014). Your morals depend on language. *PLoS ONE, 9*(4), e94842.

De la Torre, R. (1994). *Los hijos de la Luz: discurso, identidad y poder en La Luz del Mundo.* México: CIESAS, Universidad de Guadalajara, ITESO.

Fortuny Loret de Mora, P. (1984). La historia mítica del fundador de la Iglesia La Luz del Mundo. In C. Castañeda (Ed.), *Vivir en Guadalajara. La ciudad y sus funciones* (pp. 363–380). Guadalajara: Ayuntamiento de Guadalajara.

Gaxiola, M. J. (1970). *"La serpiente y la Paloma". Análisis del crecimiento de la Iglesia Apostólica de la Fe en Cristo Jesús de Mexico.* South Pasadena, CA: William Carey Library.

Goldstone, R. L. (1998). Perceptual learning. *Annual Reviews of Psychology, 49,* 585–612.

Kahneman, D., & Tversky, A. (1981). The framing of decisions and the psychology of choice. *Science, 211,* 453–458.

La Luz del Mundo (2010, 2015). *Himnario Luz del Mundo.* Mexico: Editoriale Casa Berea.

Medina, I. (2017). *Verso un Modello Sistemico della Comunicazione Non-Cooperativa. La Luz del Mundo come Sistema Autopoietico.* PhD thesis, Social and Philosophical Sciences, University of Rome Tor Vergata.

Minati, G. (2001). *Esseri collettivi.* Milano: Apogeo.

Pace, E. (2008). *Raccontare Dio.* Bologna: Il Mulino.

Perelman, C. (2001). *Trattato dell'argomentazione. La nuova retorica.* Torino: Einaudi.

Tan, L. H., Chan, A. H., Kay, P., Khong, P. L., Yip, L. K., & Luke K. K. (2008). Language affects patterns of brain activation associated with perceptual decision. *Proceedings of National Academy of Sciences of USA, 105*(10), 4004–4009.

Ting Siok, W., Kay, P., Wang, W. S., Chan, A. H., Chen, L., Luke, K. K., et al. (2009). Language regions of brain are operative in color perception. *Proceedings of National Academy of Sciences of USA, 106*(20), 8140–8145.

Von Glasersfeld, E. (1998). *Il costruttivismo radicale. Una via per conoscere ed apprendere.* Roma: Società Stampa Sportiva.

Values for Some: How Does Criminal Network Undermine the Political System? A Data Mining Perspective

Roberto Peroncini and Rita Pizzi

1 Introduction

In the field of Economics of both Crime and System Theory when we speak about *Organized Crime* (OC) we always refer to something broad, so abstract and full of such intricate equations that it is very difficult to understand the scope of action and the interrelation of processes which it produces in the Society.

In reality what is alluded to with the concept of OC does not exist; there is only a *Main Matrix of Data*, even away in time, that nests in it different meanings and reflects more important features of communicative structures and centrality of values, status of people and density of ties, grouped in form of *Network* (Barabasi 2002, 2010).

Here the Network has a crucial impact on the *Information* made available. As in the framework of Information Engineering, the transition from the problem of the two nodes to that of the three and more nodes raises a series of issues of stability around particular *connectors* and *distributors* of it (Landauer 1975).

The resulting *Knowledge* is expectable, practically predictable, depending on the positions holding by each *Data agent* within the nodal structure. The effects that it originates in Social Life can be perceived only through a special character trait: the *Human Behaviour* (Von Bertalanffy 1968).

R. Peroncini (✉)
University of Genoa, Genoa, Italy

R. Pizzi
Department of Computer Science, University of Milan, Milan, Italy
e-mail: rita.pizzi@unimi.it

© Springer Nature Switzerland AG 2019
G. Minati et al. (eds.), *Systemics of Incompleteness and Quasi-Systems*,
Contemporary Systems Thinking,
https://doi.org/10.1007/978-3-030-15277-2_21

267

Under this perspective, the study of OC is much more complex than that of the *Exact Sciences* (e.g. Biology regarding DNA cells, Physics, atoms, Astronomy, celestial bodies, Botanic, vegetable organisms etc.). These Sciences have their application into the nature of things which surround us and have characters which are considered rigorously scientific, so they are widely seen to be Sciences in the pure state, or, indeed "Exacts". In truth, Natural Sciences are only "probabilistic" because they too proceed through "trial and errors" and the errors, both in them and in Human Sciences (or Social sciences), are "recovered or recoverable".

For such reason, progress on Whole Science

depends on its courage to take seriously its own fallibility (Monti Bragadin 1973).

The alleged "Scientificity" of the Natural Sciences in comparison with the Human ones is so misleading term, since, as above mentioned, the Scientific Method is the same for the one and for the other.

What is different in each of them is the *nature of the data*. If in the various Natural Sciences the data has *objective nature*, "it is just the objectified object", in Human Sciences the data has *subjective nature* (agents) and a higher level of *Dynamicity* (agents interaction). Moreover, in Human Sciences the Human Being is at the same time "Observing Subject and Observed Subject", thus raising the degree of Complexity (Monti Bragadin 1971).

Finally, in Human Things, or to be more precise, in Human events such as those involving OC, considerable importance have the "unintentional effects of intentional action" for the high weight which plays on them *Randomness* (Monti Bragadin 1982).

Given all this, it can be understood why Human Sciences, concerning complex cultural systems, which study phenomena "far away" from observation, may lack of analysis techniques, although *scientifically established* (Carli 2006).

For the higher weight that *Power factor* plays on Human Relations, is the case of the Multiple Sciences designed to faithfully represent the object of the present study: to turn General System Theory in Political Science (GSTPS) into Practice in order to measure fully how the *Choice to Crime* (Becker 1988) can really transform the Political System Values by the *Individualistic Political Process* (Buchanan 1962, 1966).

The "paradox" of the problem raised, in which a state of fact can be expressed in two different versions: both as "unity" and as "multiplicity" and the concept denies that here it is something different (Luhmann and De Giorgi 1991), it has to be addressed in the wider context of the perspective envisaged by Einstein in 1915:

The conquest of General Relativity was simpler than the search for formulas to govern forces which were raging his Family (Isaacson 2007).

The proposition is still today impressed in the following quote:

There is a precise theory, formulated in the field of study of so called collective phenomena, that operates in somewhat simpler domains than those concerning social and economic systems: The Physical Theory of Collective Phenomena. A living being is in fact rather more complex than an electron [...] a Society [...] than crystals (Minati 2001).

Simplicity and *Complexity* thus represent the strange couple in search of authors in the field of Crime Society. When it grows in an uncontrolled fashion, one of the references to explain it consists in making "distinction between element and relation" (Luhmann 1991).

The U.S. Political Scientist, David Easton, was first convinced of this fact (Easton 1953). When he hosted the future Nobel Prize, J. M. Buchanan (Easton 1966), he focused on measuring the points of equilibrium of "The Idea of System" he was working on, i.e.

The complex and logical relationships between essential variables and Political line.

Easton was oriented to capture the distinctive features of the Political System: namely the

network of decisions that assigns values for a Society and fixes their own frequency of adjustment (Easton 1953).

The link between General System Theory in Political Science (GSTPS) and General Rational Choice Theory (GRCT) was stressed by the need to

Not adopt a narrow conception of the political line of the Society, seeing this only in the formal decision, i.e. legal (Easton 1953).

A sort of oxymoron has been the result of the meeting between the two Great Scholars. It still induces the Politician to interact with different groups in order to elevate particular values in Society by an authoritative *Process of Government* (Bentley 1908).

The Theory of Equilibrium around which everything revolves is still the

Only discernible suggestion of a theoretical framework that appears in the broad horizon of empirical research (Easton 1953).

As the Nobel Prize von Hayek said, although not exhausting the subject of Economy, Equilibrium is the first step towards the concrete analysis of its *"dynamic"* phenomena (Von Hayek 1950).

2 Criminal Disequilibrium or Chaos (the Matter)

In fields such as Economics and Political Science, the focus is on the concept of Equilibrium. Looking at the graph in Fig. 1, the *data* belonging to a Criminal Network[1] show that in OM, from an economic point of view, there

---

[1] On the Horizontal axis we measure the resources (O) devoted by Public Prosecutor's Office (PPO) to fight OC by applying LALE in the penal trial (Genoa 1996–2004). OM

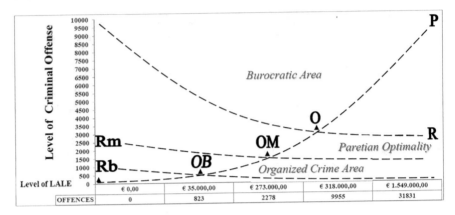

Fig. 1 The functional relationship between level of criminal offenses and link analysis law enforcement (LALE)

are "social benefits" from the Monopoly Organization for Crime. The level of Criminality and of LALE (Link Analysis Law Enforcement) effort are lower and Economists without romance evaluate it positively (Buchanan 1973).

Somewhat surprisingly, the elementary argument has rarely turned on its head: if Monopoly in the supply of "bad goods" is "Socially desirable", Monopoly in the authoritative allocation of the corresponding Values in Society is "Politically Legitimate"; exactly because "assigned for all, not for a group" (Easton 1953).

The proposition induces the Politician to act in the real terms paraphrased by Easton:

> a Legislative body may decide "to defend" the monopolists; This is the intention. And a governmental leader "supports" this decision by failing to discover violators, pursue them strongly or by accepting black market (RB) (Easton 1953).

---

represents the way by which the Judge, in 2004, shared them between the "two parties" of criminal proceeding (273.000,00 euro in charge of PPO; 35,000.00 euro against defendants convicted by the Law Court as bosses and members of four Crime syndicates). On the Vertical axis we measure level of criminal offences determined through the final enforceable Court Judgment (28187/08). R, RM and RB represent the separate "defending response relationship" which, during the first two degrees of the proceeding (1999–2004), has influenced the possibility of discovering offences on the basis of the level of Law enforced by the Public Prosecutor's Expert. OB represents the final amount of public resources which the Government, "owned" by the same Defender Legislator, despite the Court Order, has authoritatively forced into the penal process, imposing Expert obligation for difference (2005).

Thanks to the GSTPS, the second phase of reflection can be established; the one which follows where Economic Theory stops. Satisfied its general conditions (OM), the Economics of Crime is unable (and it does not have to do so) to totally seize the Judicial Political process by which a Government "deals with Syndicates" (Buchanan 1973).

From a Systemic point of view, OB represents the "constant point of disequilibrium" (Easton 1953), that the "socially desirable good" raises between conflicts and systemic contagions, i.e. between the possibility of granting or denying *Knowledge to Manage the Knowledge Society* (Minati 2016). Its elements play a role in stimulating the Political System, which in turn is exposed to the chance of founding, "even for a short time, in the state of stability" (Easton 1953) sought by bosses.

In this space of time, the Defender Legislator, altering the Judgment of fact that includes its Intention in the Official decision, self describes the Political System that forges into two distinct phases:

1. The first being *Constructivist* in nature, in which the deliberate attempt of groups to contain and transform the structures of Law State and Social System (Luhmann 1978) is perfected by Individualist Political Process (Buchanan 1962, 1966), thus imposing "Judicial Positivism" (Luhmann 1991) at certain levels of "Law determination" (Luhmann 1974), "manipulating the individuals" (Luhmann 1991), "Centralizing and Politicising State control" (i.e. OP) (Luhmann 1980).

2. The second *Rationally ecological*, as unintentional by-product of a new emerging order, depending on a myriad of other factors of which Human mind does not give a reason (i.e. OM to OB) (Smith 2008). Its provocative principle (Buchanan 1973) is definable by the inverse cognitive formula, *Innocent Swindle and Guilty Virtue*. And such as for *Sentiments* (Smith 2016), for this kind of Knowledge it is necessary to go to the lab, namely to do *Data Mining* (DM) (Han et al. 2012) and derive its *Social production function* (Simon 1958) from the type of elements and density of relations which at the formation of the Network Culture constitute "the state of the Political System" (Easton 1953).

3 Data Mining and Clustering (the Meaning)

If the idea of the controlled experiment may seem paradoxical to any strenuous opponent of the application of natural science methods to the study of social phenomena, it is even more incomprehensible in an approach such as GSTPS:

Scientific reasoning cannot determine which Values must be valid (Easton 1953).

The present study bases its analysis on the alternative direction indicated by Easton himself: disposing of a huge amount of data, really different from each other,

> reduces the important elements or variables of a Political phenomenon in terms that can be measured (Easton 1953).

The first Italian experimentation in the field of DM applied to Economics of Crime was carried out at the Department of Computer Science of the University of Milan, using a set of probabilistic methods that allow the so called *Knowledge Discovery in Databases* (KDD) (Nikolle 2016; Orlandi 2016). The work offers quantifiable magnitudes toward the "continuous path of the Political analysis" (Easton 1953).

Performing technical of relational analysis and DM processes on huge data sets of "legal data" (Easton 1953) allows to predict values of attributes. These elements include the answer to investigative and analytical questions, knowledge extraction and finally visualization.

The correlation between different types of attributes (e.g. between A1 and A2, the request for a mat. sums(frauds) and A3, the amounts actually liquidated, or between A5, the amount(fraud) presumed and still A1, A2) is shown in Fig. 2.

It leads towards further task to investigate other two classes of data: *situational* and *psychological* ones (Easton 1953). Selecting between them the most meaningful additional attributes it is possible to perform *clustering* on the (political) attributes. Moreover, using decision making techniques generated by algorithms (Orlandi 2016, pp. 4–8) it is possible to "classify experimentally" (Easton 1953) *correlations* between the characteristics (values) common to the attribute groups identified. Assigning a label to data and classes to represent the confidence level of the classification, the "decision tree" identifies significant segments and groups, including "Trial and Error" (Smith 2008).

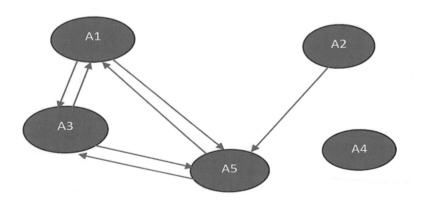

Fig. 2 Graph visualization of attributes correlation (Nikolle 2016, p. 29)

Every proportion of "errors" made over the whole set of instances testifies that in the Economics of Crime the variables which determine a strong correlation are equivalent to the political ones (false, deceit and manipulation) by which the Defender Legislator institutionalizes as obvious the recognition of binding decisions and considers it as consequence, not of its "Intention", but of the validity of the official decision.

4 Data Mining and the Visualization of the Logical Network Ties (the Form)

The *Relational data* of research (Cood 1970) elaborated with "satisfactory tools for the transformation of qualitative data" (Easton 1953; Cood 2009) reflect the power structure and the circular effects of mutual codetermination regarding relationship of interdependency between data agents. Hidden in the Network data there are, in fact, tangible and intangible values whose divergences give rise in the Political System to *conflict of interests* based on *power relations* (Capra 2002).

Theoretical analysis is therefore in the data (Easton 1953). DM in techniques, in preluding to KDD (Kurgan and Musilek 2006), allow to visualize different associations among different entities. These ties are identifiable referring to the boundaries of the Network. From a DM perspective a *cluster of attribute* can be seen as a *group of crimes* acting in different structures; from a Politological one, as a *group of data agents* interacting between a hot spot of consent diffused in different arenas of exchange (economic, political and religious).

The Economists do not have to take in account the morals, which is important instead for the Politician to whom the "practical measure" is as relevant as the number of degrees of freedom it is for the "rational mechanics" (Easton 1953; Pareto 1911). Nothing remains outside of this Network, not even the Positivist Judge who becomes politically functional (Luhmann 1974, 1978, 1981b).

Appropriate Clusters have a one to one correspondence to *Crime Patterns* and *Informational Lobbying* influence in decision-making processes. Clustering algorithms in DM are thus equivalent to the Observer's task of identifying single agents who take part (1) or not (0) in the *Constructivist Process*.

The Clusters of *relational data* organized in simple matrix of associations, such as Fig. 3, allow to display areas of vulnerability between Criminal Network and Political System.

Form, in Fig. 4, refers both to the single *Constructivist tie*, and implies *measuring* the *intensity* which characterizes the underlying reaction of the *mutuality* and *reciprocity* (0,92), the structure of the whole networked implying measures such as $\rightarrow$ *avg degree* (8,62) $\rightarrow$ *centralization* (0,26) $\rightarrow$ *density* (0,57), *fragmentation* (0), *connectivity* (1).

DATA AGENTS	POS. JUD.	DEF.LEG.	COEC. EXEC.	JUDIC.	FINAN.	RELIG.	MEDIA	BUROC.	PUB.PROS.	JUD.POL.	EXPERT	CR.EGO	CR.ALTER	LAWYER	WHI.COL	PEOPLE
POSITIVIST JUDGE	0	1	1	1	1	1	1	1	0	1	0	0	0	1	1	0
DEFENDER LEGISLATOR	1	0	1	0	1	1	1	1	0	1	0	1	0	1	1	0
COERCITIVE EXECUTIVE	1	1	0	0	1	1	1	1	0	1	0	1	0	1	1	0
FINANCIAL	1	1	1	0	0	1	1	0	0	0	0	1	1	1	1	0
RELIGIOUS	1	1	1	1	1	0	1	0	0	0	0	0	0	1	1	1
MEDIA	1	1	1	0	1	1	0	1	0	1	0	1	1	1	1	0
BUREAUCRACY	1	1	1	1	0	1	1	0	0	1	0	1	1	1	1	0
PUBLIC PROSECUTOR	0	0	0	1	0	0	1	0	0	1	1	0	0	0	0	1
JUDICIAL POLICE	1	1	1	1	0	1	1	1	1	0	0	0	0	1	1	1
EXPERT	0	0	0	1	0	0	0	0	1	0	0	0	0	1	1	0
CRIMINAL EGO	0	1	1	0	1	0	1	1	0	1	0	0	0	1	1	0
CRIMINAL ALTER	0	0	0	0	1	0	1	1	0	0	0	0	0	1	1	0
LAWYER	1	1	1	1	1	1	1	1	0	1	0	1	0	0	1	0
WHITE COLLAR	1	1	1	1	1	1	1	1	0	0	0	1	1	1	0	0
PEOPLE	0	0	0	1	0	1	0	0	1	1	1	0	0	0	0	0

Fig. 3 Matrix of association indices by UCINET (Borgatti et al. 2002)

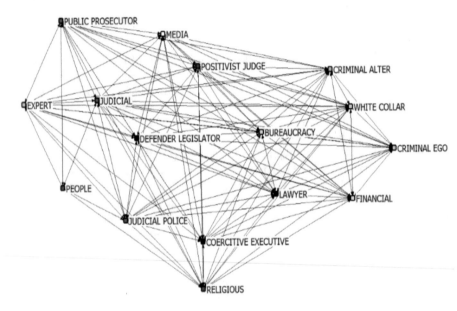

Fig. 4 Network connectivity level by UCINET (Borgatti et al. 2002)

Areas of potentially fruitful additional analysis can be determined just moving away from the concepts of *Matter* and *Meaning*, emerging from the sum of clusters, to those of *Form* (Capra 2002; Pareto 1916).

The Networking Vulnerability is to be found not only in its *level of connectivity* or *fragility* of its interpersonal bonds before the Law. It nests in the *neutralest data, simplest meanings* and *weakest ties* (Granovetter 1973), emerging from the most *neglected disciplines* and *fearest Expert* (Ferrau et al. 2013): the real external *bursts* generators within the range of the specified Network. The effects that Constructivists produce at System level are *decidables* (Minati 2016) only by *algorithms* (Smith 2008; Orlandi 2016, p. 3).

5 Political Equilibrium and Quasi System (the Process)

At the OB point of Fig. 1, the Political System occurs as result of the authoritative mechanism of allocation of criminal values by a process of Government. This might be defined as the phase of Entropy increasing trough which the Defender Legislator, in accelerating its Intention, causes the amount of political disorder aimed to transform the existing equilibrium and then "create space to the new one" (Easton 1953).

Without such a policy making mechanism, Political System would not exist. Its *function* of *collection* of illegality (A), (i.e. the *Innocent Swindle*), is set up by the "Defender Legislator" in the Penal process and the *collection of premiums* (B) adapted by its Government in support of the second part of its paradigmatic "Thesis": the *Guilty Virtue*. *Precision, Continuity* and *Rigour* rated by Enforcers and remediated by the Expert through "an indefinable range of activities" (Valentini 2008) "of a prevalently technical-scientific nature" (Carli 2006), are jeered and punished (Miucci 2011).

Falsehood and *Deceit* sought by the Bosses are pinned up on the Bureaucracy and institutionally sustained against the self same Expert. As a consequence, Political System "Does take specific characteristics, no longer deducible from its elements and relations between them" (Minati 1998) (i.e. *Emergentism* (Pessa and Penna 2000)) but from the "social brain" which takes place on it (Smith 2008).

Considering the Political System as a body called to make decisions on the collections A and B (systemic needs) sustained by the Defender Legislator (yellow, bad available energy), its components (Demands, Support, Output, Feedback) with its five parameters (Culture, needs, regulation mechanisms of access and demand flow, citizen perception) are related in Fig. 5.

The source of information that, through *rumors* (Pessa and Penna 2000), triggers the conversion process is the "Legislative Intention" (just a "Psychology topic" (Tarello 1980)). This mechanism, derived from *second order cybernetics* (Pessa and Penna 2000), legitimizes the starting input through legal proceedings guaranteed by an act of Ideological will: the "Judicial Positivism", i.e. "the biggest scientific error" (Von Hayek 1982). And Legitimation via Judicial procedures is not an effect of a single cause or some ascertainable causes. It is the performance of the System (Luhmann 1991).

Thanks to Easton's model, is not difficult to understand why: in the circularity processes between Economics of Crime and Political System, the "cognitive game" set up by a "casual comment" (Capra 2002), through the *Double Loop*, acts on variables and processes predetermined for common values (*Single Loop*) (Minati 1998, 2009).

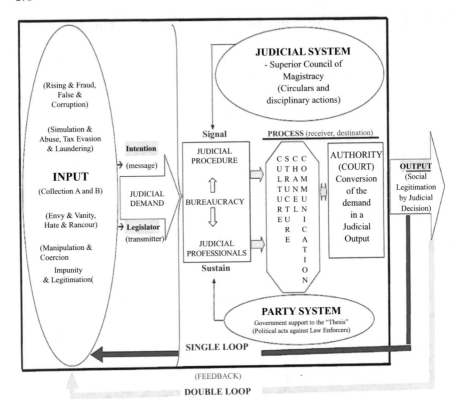

Fig. 5 Double loop input/output/process scheme

This learning processes, combined with "manipulation", convert "bad values" into "common values" of the Political System. The emerging complex order does not depend on the existence of a *Unique Equilibrium* around which patterns of behaviour are coordinated. It operates based on equivalent pattern of *uninterrupted equilibriums*[2] (Pareto 1916), such as those marked in Fig. 6.

Uninterrupted equilibriums thus reflect three *"changing of state"* (Easton 1953): the two extremes "Quasi System", *Chaos* (Capra 2002), i.e. more OC

---

[2] It is clear from what has already been in achieved from Fig. 1 that People are moved in all the direction (preferences on x or y) allowed by the suppression of LELA constraints (utilities functions). Figure 6 shows that there are two types of consequences: People reach the O' point, moving away from OC; they come to OM, moving from O". Then definitely moving away, benefiting all the members of the community. With the process of Government (OB) people breached the OM equilibrium, moving away also to extreme OL. In this way they lose their spontaneous option fixed on OM by the Judicial: the one and only *Networking Values Optimizer*.

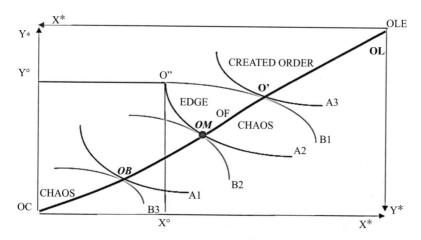

Fig. 6 Uninterrupted social equilibrium

and less Data (OB) and *"Constructed Order"* (Von Hayek 1982), i.e. more Bureaucracy and more Information, but less *Certainty in Knowledge* (OL) (Minati 2009).

In the midst, there's the *"Edge of Chaos"*, i.e. *the Spontaneous Order* (OM) (Von Hayek 1982; Capra 2002), the only *Autopoietic system* (Luhmann 1981a; Luhmann and De Giorgi 1991), governed not by *coercion* but only by *rules*: i.e. *agreement* and *consent* (Smith 2016).

As for external stimuli in *Nervous system* (Von Bertalanffy 1968) *Spontaneous Order*, in its regeneration, does not allow data agents and coordinators and supplants the unrealistic as "not yet complete" (Minati 2016) concepts of *Equilibrium, Common values* and *Stability* of the Pure Economy, Sociological and Political Theory.

6 Conclusions

The exemplifying scheme emerging at the end of the analysis (Fig. 7) attests that the *complexity* of the Criminal Network is not represented by its single perspectives, but in the *unified one* that it has been able to grasp and reconstruct making use of its available data.

RMKDCN does not enter in the merit of the *Incompleteness of the Multiple Theories* or of the limits of their applicability regarding infinite conceptual possibilities.

RMKDCN helps only to directly be aware of the intricate relationships which exist among single aspects of Criminal Life and Political System distinguished from the Official one.

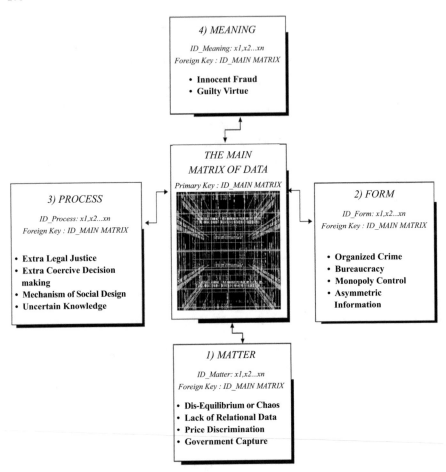

Fig. 7 The relational model for knowledge discovery in crime network (RMKDCN)

By its five tables, in reconstructing the geometric trace of any data agent interconnection, RMKDCN offers a *new Meaning* on the warp and weft of Society, thus raising amazement and doubtful policy issues by which GSTP declares itself

> Free from every objection since implies both conservation and change, the maintaining of the System and the internal conflict (Von Bertalanffy 1968).

Agreeing with this proposition, the Model Core System resides in the *Process*: the data table which "conserves" the elements by which the Criminal Network carried out its decisions (Matter) and its choices of political line (Form). Process is the conceptual relational category which, better than any other, identified sequences between Individual plans and Level of Data, Data agents and Group actions, Structures and Function events. In sustaining coordination, it leads and continuously varies the Meaning that preserves the "*state of quiet*" of the Political System (Easton 1953): the one that, "keeping constraints and conditions fixed" (Pessa and Penna 2000), without "change" in "Legislative intention", always acts "transforming completely the state of matter" (Prigogine 2007).

The major contribution by DM is in demonstrating that you learn all this more by studying the Criminal Network than Criminal Network Theories. Turned into the Practice, DM testifies that in solving the additional problems raising from an Observer it is necessary to go through the most difficult and sometimes arid path of the most complete and modern detection through *Data* and *Experimentation*.

This choice become self-evident once we recognize that communicable structures are variables that may be used for achieving particular social purposes, in this case, the reduction in the aggregate level of data along with the reduction in resource commitment in KDD.

It is not from the public spiritedness of the Data Agents that we should aspect to get an increase in our Knowledge capability but from their courage in recognizing their own *Fallibility* or admit the supposition that any opinion on *Common values*, of which they feel very certain, may be really one of the examples of *Recoverable Errors* (IBM Knowledge Center 2017).

References

Barabasi, A.-L. (2002). *Link. The new science of networks.* New York: Perseus Books Group. ISBN-10: 0738206679.

Barabasi, A.-L. (2010). *Bursts: The hidden patterns behind everything we do.* New York: Dutton. eISBN: 978-1-101-18716-6.

Becker, G. (1988). *Crime and punishment: An economic approach.* Cambridge: NBER. http://www.nber.org/chapters/c3625. Accessed October 30, 2017.

Bentley, A. (1908). *The process of government. A study of social pressures.* Chicago: The University of Chicago Press.

Borgatti, S. P., Everett, M. G., & Freeman, L. C. (2002). *UCINET 6 for Windows: Software for social network analysis.* Harvard: Analytic Technologies. https://sites.google.com/site/ucinetsoftware/home. Accessed October 30, 2017.

Buchanan, J. M. (1962). *The calculus of consent: Logical foundations of democracy.* http://files.libertyfund.org/files/1063/Buchanan_0102-03_EBk_v6.0.pdf. Accessed October 30, 2017.

Buchanan, J. M. (1966). An individualistic theory of political process. In D. Easton (Ed.), *Varieties of political theory.* Englewood Cliffs, NJ: Prentice-Hall.

Buchanan, J. M. (1973). A defense of organized crime? In S. Rottenberg (Ed.), *The economics of crime and punishment* (pp. 119–132). Washington, DC: American Enterprise Institute for Public Policy Research.

Capra, F. (2002). *La Rete della Vita*. Milano: RCS Libri.

Carli, L. (2006). Notes on the Scientific Proof in the Criminal Case – from the findings of the preliminary phase to the probative conclusions judgement into decision. Italian S.C.J. Conference – 1–19 (note 16).

Cood, E. F. (1970). *A relational model of data for large shared data banks*. https://www.seas.upenn.edu/~zives/03f/cis550/codd.pdf. Accessed October 30, 2017.

Cood, E. F. (2009). *The relational model for database management – Version 2*. http://codeblab.com/wp-content/uploads/2009/12/rmdb-codd.pdf. Accessed October 30, 2017.

Easton, D. (1953). *The political system, an inquiry into the state of political science*. New York: Knopf.

Easton, D. (Ed.). (1966). *Varieties of political theory*. Englewood Cliffs, NJ: Prentice-Hall.

Ferrau, P., Marzaduri, E., & Spagher, G. (2013). *La Prova Penale*. https://books.google.it/books?id=UmZpAgAAQBAJ&pg=PA368&lpg=PA368&dq=ferrua+%22la+prova+penale%22+peroncini&source=bl&ots=Fq-FRdB7RW&sig=A4JDL7gg9Yzg-xfGlD7qGTfAfT8&hl=it&sa=X&ved=0ahUKEwit2_iUk5vXAhXFLhoKHUnYDiAQ6AEIJjAA#v=onepage&q=ferrua%20%22la%20prova%20penale%22%20peroncini&f=false. Accessed October 30, 2017.

Granovetter, M. (1973). *The strength of weak ties*. https://sociology.stanford.edu/sites/default/files/publications/the_strength_of_weak_ties_and_exch_w-gans.pdf. Accessed October 29, 2017.

Han, J., Kamber, M., & Pei, J. (2012). *Data mining: Concepts and techniques* (3rd ed.). Amsterdam: Elsevier, Morgan Kaufmann.

IBM Knowledge Center. (2017). https://www.ibm.com/support/knowledgecenter/search/Recoverable%20Errors. Accessed October 29, 2017.

Isaacson, W. (2007). *Einstein. La sua vita, il suo universo* (pp. 221–227). Milano: Arnoldo Mondadori.

Kurgan, L. A., & Musilek, P. (2006). A survey of knowledge discovery and data mining process models. *The Knowledge Engineering Review, 21*(1), 1–24.

Landauer, R. (1975). Stability and entropy production in electrical circuits. *Journal of Statistical Physics, 13*(1), 1–16.

Luhmann, N. (1974). *Sistema Giuridico e Dogmatica Giuridica*. Bologna: ll Mulino.

Luhmann, N. (1978). *Stato di Diritto e Sistema Sociale*. Napoli: Guida Editore.

Luhmann, N. (1980). Concetti di Politica e "politicizzazione" dell'amministrazione. In G. Gozzi (Ed.), *Le Trasformazioni dello Stato. Tendenze del dibattito in Germania ed in Usa* (pp. 70–92). Firenze: La Nuova Italia.

Luhmann, N. (1981a). *Come è Possibile l'Ordine Sociale*. Roma, Bari: Laterza & Figli.

Luhmann, N. (1981b). *La Differeniazione del Diritto*. Bologna: Il Mulino.

Luhmann, N. (1991). *Procedimenti Giuridici e Legittimazione Sociale*. Milano: A. Giuffrè.

Luhmann, N., & De Giorgi, R. (1991). *Teoria della Società*. Milano: Franco Angeli.

Minati, G. (1998). *Sistemica, Etica , Virtualità, Didattica, Economia*. Milano: Apogeo.

Minati, G. (2001). *Collective beings*. Milano: Apogeo.

Minati, G. (2009). *L'incertezza nella gestione della complessità*. http:// www.aiems.eu/files/documento_gianfranco_minati.pdf. Accessed October 29, 2017.

Minati, G. (2016). *Knowledge to manage the knowledge society: The concept of theoretical incompleteness*. http://www.mdpi.com/2079-8954/4/3/ 26. Accessed October 30, 2017.

Miucci, C. (2011). *La Testimonianza Tecnica nel Processo Penale*. https:// books.google.it/books?id=5ISed3TS3mQC&pg=PA52&lpg=PA52& dq=miucci+%22la+testimonianza+tecnica%22+peroncini&source=bl& ots=eboTMQHtXq&sig=SwsCZttvDziVVGgKs8KUcYW_N20&hl=it& sa=X&ved=0ahUKEwjkir-_k5vXAhVLvRoKHWq2A7wQ6AEIJjAA# v=onepage&q=miucci%20%22la%20testimonianza%20tecnica%22 %20peroncini&f=false

Monti Bragadin, S. (1971). Questioni di Metodo. Scienze della Natura e Scienze dell'Uomo. Natura soggettiva del dato nelle Scienze umane. *Controcorrente. Verifica delle ipotesi di trasformazione della Società, 3*(1/2), 67–92 (Milano: CESES).

Monti Bragadin, S. (1973). Uso Critico ed Uso Dogmatico della Ragione: Note Introduttive. *Controcorrente. Verifica delle ipotesi di trasformazione della Società, 5*(2), Numero Tematico sul Razionalismo critico in onore di Karl Popper. A cura di S.Monti-Bragadin e V. Bělohradský, 3–15. Milano: CESES.

Monti Bragadin, S. (1982). Alcuni tratti delle Scienze Umane. *Biblioteca della Libertà, 84/85*, 91–131 (Firenze: Le Monnier).

Nikolle, M. (2016). *Information management* (Project Report, Prof. A. Ceselli). Milano: Department of Computer Science, University of Milan.

Orlandi, M. (2016). *Probabilistic methods* (Project Report, Prof. R. Pizzi). Milano: Department of Computer Science, University of Milan.

Pareto, V. (1911). *Manuale di Economia Politica con una introduzione alla scienza sociale*. http://www.policonomics.com/wp-content/uploads/ Manuale-di-Economia-Politica.pdf. Accessed October 30, 2017.

Pareto, V. (1916). *Trattato di Sociologia Generale* (Vol. 2, pp. 545–837). Firenze: G. Barbera.

Pessa, E., & Penna, P. (2000). *Manuale di Scienza Cognitiva. Intelligenza Artificiale classica e Psicologia Cognitiva*. Roma, Bari: Laterza.

Prigogine, I. (2007). L'Esplorazione della Complessità. In G. Bocchi & M. Ceruti (Eds.), *La Sifda della Complessità* (pp. 155–169). Torino: Paravia Bruno Mondadori.

Simon, H. (1958). *Il Comportamento Amministrativo*. Bologna: Il Mulino.

Smith, V. (2008). *La Razionalità nell'Economia. Fra Teoria ed Analisi Sperimentale*. Torino: IBL Libri.

Smith, V. (2016). *Sentiments, conduct, and trust in the laboratory*. https://www.socsci.uci.edu/files/news_events/2013/smith_wilson.pdf. Accessed October 29, 2017.

Tarello, G. (1980). *L'Interpretazione della Legge*. Milano: A. Giuffrè.

Valentini, C. (2008). Il caso di Rignano: ancora un episodio del rapporto tra scienza e processo. In *Cassazione Penale* (pp. 190, 200). Milano: A. Giuffrè.

Von Bertalanffy, L. (1968). *General system theory: Foundations, development, applications*. New York: George Braziller.

Von Hayek, F. A. (1950). *The pure theory of capital*. https://mises.org/sites/default/files/Pure%20Theory%20of%20Capital_4.pdf. Accessed October 30, 2017.

Von Hayek, F. A. (1982). *Legge, Legislazione e Libertà. Critica dell'Economia Pianificata*. Milano: Il Saggiatore.

Part VI
Emergence, Quasiness and Incompleteness: Maintaining, Crises and Degeneration in Emergence Phenomena

Embracing the Unknown in Post-Bertalanffy Systemics Complexity Modeling

Rodolfo Fiorini

1 Introduction

Human beings' approach to the real world is about incompleteness: incompleteness of understanding, representation, information, etc.; what one does when one does not know what is going on, or when there is a non-zero chance of not knowing what is going on. It focuses on the unknown, rather than on the production of mathematical certainties based on weak assumptions. Men inevitably see the universe from a human point of view, communicate in terms shaped by the exigencies of human life in a natural uncertain environment, and make rational decisions in an environment of imprecision, uncertainty and incompleteness of information. Therefore, mankind's best conceivable worldview is at most a representation, a partial picture of the real world, an interpretation centered on man.

Ontology was once understood to be the philosophical inquiry into the structure of reality: the analysis and categorization of "what there is", the theory of being. Recently, however, a field called "ontology" has become part of the rapidly growing research industry in information technology. The two fields have more in common than just their name (Poli and Seibt 2010). Ontology as a theoretical domain is a description or inventory of the things that are supposed to exist according to a particular theory, which might, but need not, be true. Ontology as an extant domain, in contrast, is the actual world of all real existent entities, whatever these turn out to be, identified by a true complete applied ontological theory (Jacquette 2002, pp. 2–3).

R. Fiorini (✉)
Department of Electronics, Information and Bioengineering (DEIB), Politecnico di Milano, Milano, Italy
e-mail: rodolfo.fiorini@polimi.it

© Springer Nature Switzerland AG 2019
G. Minati et al. (eds.), *Systemics of Incompleteness and Quasi-Systems*,
Contemporary Systems Thinking,
https://doi.org/10.1007/978-3-030-15277-2_22

In other words, any human interpretation is a model which represents an operational compromise (a tool, an "application") between something you can gain (an advantage, something known but incomplete like epistemic uncertainty) and something you are forced (consciously or unconsciously) to ignore ("operational domain" knowledge incompleteness), to lose or to pay for (an unknown drawback from natural uncertainty), even if you are not aware of that. That operational splitting can represent an advantage by a formal representation point of view (i.e., ease of representation), but its major drawback is the loss of precision in the original information, if the observer is unaware of it or unable to compensate for it. But where does that operational compromise come from? To find a sound answer, we need to start from our human spacetime understanding (Fiorini 2015).

2 Spacetime Splitting Fundamental Relationship

Spacetime (ST) invariant physical quantities can be related to the variables employed by a specific interacting observer to get an interpretation of the world within which a human being is immersed. In fact, original "spacetime" (a transdisciplinary concept) is usually split by classic operative interpretations into two separate additive subcomponents; "space" and "time". In that forced passage original information is lost or dissipated to an unaware interactor (Fiorini 2015).

This forced operational splitting may represent an advantage by a formal (rational) representation point-of-view (i.e., ease of representation), but its major drawback is the loss of precision in the original information, if the observer is unaware of or unable to compensate for it partially, not taking into consideration the folding and unfolding properties offered by Computational Information Conservation Theory (CICT) "OpeRational" representation approach (Fiorini and Laguteta 2013). According to CICT, the full information content of any symbolic representation emerges from the capturing of two fundamental, coupled components: the linear component (unfolded) and the nonlinear one (folded). This is the root relationship, the fundamental dichotomy of any human representation. Referring to the transdisciplinary concept (Nicolescu 1996), we see that for full information conservation any transdisciplinary concept emerges from two pair of fundamental coupled parts. From a common language perspective, taking into consideration the folding and unfolding properties of CICT structured "OpeRational" (OR) representations for the Space-Time Split (STS) (Fiorini 2015), one can conceive a better operative understanding of usual terms, with the added possibility of information conservation as shown in "The Four Quadrants of The Space-Time Split" (Fig. 1) through a narrative point of view.

Here, the term "Timeline" (first quadrant, top right) is considered the combination of a major linear time representation framed by folded minor

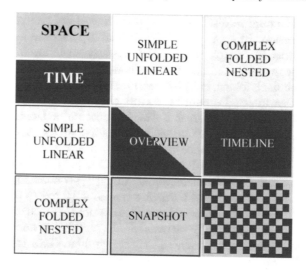

Fig. 1 The four quadrants of the human space-time split (STS) fundamental relationship (see text)

space representation. The term "Overview" (second quadrant, top left) is interpreted as the combined representation of major linear space and major linear time representations, with minor complementary folded time and space components. The term "Snapshot" (third quadrant, bottom left) can be assumed as the combination of a major linear space representation framed by the minor folded time representation. The forth quadrant (bottom right) represents the combination of major folded space and time components, framed by the combination of minor linear space and time components. It can be interpreted as a simple (bidimensional), but realistic representation of the usual information experienced by a living organism.

As an operative example, we can use previous understanding of the representation of human experience by a narrative point of view, to be used effectively in human knowledge structuring and computer science modeling and simulation. We can start to divide human experience into two irreducible, interacting concepts or parts, "Application" and "Domain", in the sense that experience and knowledge are always gained when an Application is developed to act within a Domain, and a Domain is always investigated by a developed Application.

3 Application-Domain Fundamental Dichotomy

In terms of ultimate truth, the Application-Domain dichotomy of this sort has little meaning, but it is quite legitimate when one is operating within the classic mode used to discover or to create a world of "immediate appearance" by narration. In turn, both Domain and Application can be thought of as being either in "simple mode" (SM, linearly structured, technical, unfolded, etc.) or in "complex mode" (CM, non-linearly structured or unstructured, non-technical, folded, etc.) Representation, as defined in Fiorini (1994). The SM Application or Domain represents the world primarily in terms of "immediate appearance", whereas a CM Application or Domain sees it primarily as "underlying process" in itself. Therefore, we can assume, for now, to talk about human experience by referring to SM and CM, Application and Domain, according to the Four-Quadrant Scheme (FQS) of Fig. 2. SM is straightforward, unadorned, unemotional, analytic, economical, and carefully proportioned. Its purpose is not to inspire emotionally, but to bring order out of chaos and make the "unknowns known". It is not an aesthetically free and natural style. It is "esthetically restrained". Everything is under control. By now these battle lines should sound a little familiar. This is the source of current conflict and trouble between these two cultures. Human beings and researchers tend to think and feel exclusively in one mode or the other and in so doing tend to misunderstand and underestimate what the other mode is all about. But no one is willing to give up the truth as he/she sees it, and as far as we know, quite a few individuals now living have been developing any real reconciliation of these truths or modes, which is mandatory to arrive at the new Science 2.0 worldview. There is no social, formal shared point at which these visions of reality are unified at present. But if you can keep hold of the most obvious observation about SM Application or Domain, some other

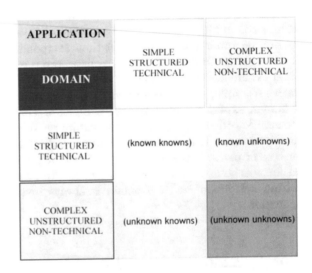

Fig. 2 Four-quadrant scheme (FQS) for the application and domain fundamental dichotomy

peculiarities can be observed which are not apparent at first, and which can help to provide and to let us understand a convenient unification point.

The first is that in the traditional Science 1.0 approach, apart from the recent areas of risk analysis within disciplines and computer security, an interacting observer is missing. Any classic SM Application or Domain description does not take into consideration an observer. Even an operator is a kind of personalityless robot whose performance of a function on a device is completely mechanical. There are no real subjects in this description. The only objects that exist are independent of an observer.

The second is that, according to classic Science 1.0, dichotomy is a simple cut-and-split process only. As a matter of fact, there is an arbitrary knife moving here: an intellectual scalpel so swift and so sharp you sometimes don't even see it moving. You get the illusion that everything is there and that a thing is being named as it exists. But they can be named quite differently and organized quite differently depending on how the knife moves. It is important to see this knife for what it is and not to be fooled into thinking that anything is the way it is just because the knife happened to cut it up that way. It is important to concentrate in the knife itself. From the spacetime and Science 2.0 perspective, it is much better and more convenient to call it "dynamic bookmarking" instead of knife!

The third is that the words "good" and "bad" and all their synonyms are completely absent. No value judgments have been expressed anywhere, only mere fact.

The fourth is that anything under CM is almost impossible to understand directly without experiencing it, unless you already know how it works. The immediate surface impressions that are essential for primary understanding are gone. Nevertheless, the masterful ability to use this knife effectively can result in creative solutions to the SM and CM split (De Giacomo and Fiorini 2017). For now, you have to be aware that even the special use of the term SM and CM are examples of this knifemanship.

Both complexity science and chaos theory converge on showing the unavoidability of uncertainty, whether it is embedded into feedback cycles and emergence or in the infinite precision of initial conditions. Even in mere terminology, minimizing or avoiding representation uncertainty and ambiguities is mandatory to achieve and keep high quality result and service. When uncertainty and ambiguities cannot be avoided, then reliable Ontological Uncertainty Management (OUM) systems are needed and become a must (Fiorini 2017c). There are surprising similarities in many fields of human activities and much can be learned from these. For instance, Puu discussed bifurcations that are likely to govern the evolution of culture and technology. More specifically, by defining culture as art plus science, he discusses the evolution of social and material products (Puu 2015). We can use our previous knowledge to develop a better approach to post-Bertalanffy representation and modeling.

4 Post-Bertalanffy Representation and Modeling

Amazing possibility on the one hand and frustrating inaction on the other, that is the yin and yang of modern science (Ness 2014). The fact that we can build devices that implement the same basic operations as those the nervous system uses leads to the inevitable conclusion that we should be able to build entire systems based on the organizing principles used by the nervous system. Nevertheless, the human brain is at least a factor of 1 billion more efficient than our present digital technology, and a factor of 10 million more efficient than the best digital technology that we can imagine (Fiorini 2015). The unavoidable conclusion is that we have something fundamental to learn from the brain and biology about new ways and much more effective forms of computation and information managing. We need revisiting our fundamental research tools and reinventing our scientific ecosystem to enhance our relational competence (L'Abate et al. 2010) for real innovation vital development, towards a more sustainable economy and wellbeing (Fiorini et al. 2016). We need tools able to manage ontological uncertainty quite more effectively than in the past (Fiorini 2014a,b).

In fact, an intriguing point is that, although currently there are multiple models for the integer numbers, they all will agree on the definition of computable functions. However, current real number $\mathbf{R}$ computation does not have these properties. Traditional scientific computation uses specified fixed-length finite representations (related to scientific notation) of real numbers, and so can achieve only limited precision, can make errors in comparisons, and can even be unstable over rounds of conversion to and from corresponding decimal representation. Amazingly, whether an extended Turing machine model or a real-number computation model is appropriate for scientific computation is still an open topic of discussion. Current computer computation must be either symbolic or approximate. Nevertheless it can be shown that computer computation can use either an "approximated approximation" or "exact approximation" representation system with completely different final results (Fiorini and Laguteta 2013). To achieve exact approximation computational number representation, logic must be described in terms of "closure spaces". The concept of "closure spaces" was developed around 1930 by Polish logician, mathematician Alfred Tarski, who conceived an abstract theory of logical deductions which models some properties of logical calculi. Tarski's undefinability theorem shows that Gödel's arithmetization encoding cannot be done for semantical concepts such as truth. It shows that no sufficiently rich interpreted, symbolic language can represent its own semantics. Mathematically, what he described is just a finitary closure operator on a set (the set of sentences). In Logic, the structure of closure spaces is defined by the "consequence operator" introduced by Tarski.

Since its inception in the 1980s, CICT made the fundamental ontological discrimination between "Symbolic Representation" and "OpeRational

Representation" of number values to obtain well-defined structures and to achieve computational information conservation (Fiorini and Laguteta 2013). In CICT Arithmetic there are no longer arbitrary axioms for the definition of number or number properties and operations; the structure of closure spaces is self-defining, taking into consideration the Natural Number Reciprocal Space (RS) self-description properties. In this way, Natural Number can be thought as both a well-defined structured object and symbol at the same time. CICT emerged from the study of the geometrical structure of a discrete manifold of ordered hyperbolic substructures, coded by formal power series, under the criterion of evolutive structural invariance at arbitrary precision (Fiorini 2014a). In other words, hyperbolic geometry (HG) can describe projective relativistic geometry closure spaces directly hardwired into elementary arithmetic long division quotient and remainder sequences, offering many more competitive computational advantages over the traditional and current Euclidean approach alone. In the next section we present an operative example.

5 Operative Example

We show, with no restriction to any other solid number (SN) (Fiorini and Laguteta 2013), the simple case for $SN = 7.0 = D$. To conserve the full information content of rational correspondence at higher level (continuum), by CICT, we realize that we have to take into account not only the usual modulus information, but even the related external or extrinsic RFD (Representation Fundamental Domain) periodic precision length information W = 6 (numeric period or external phase representation) in this case (i.e. $CD1 \equiv$ "000007" as base RFD, and yes for CICT leading zeros do count! (Fiorini 2014c). We can refer to the traditional Euler's formula to establish the usual fundamental relationship between trigonometric functions and the complex exponential function:

$$e^{ix} = \cos x + i \sin x \tag{1}$$

where e is the base of the natural logarithm and $i = \sqrt{-1}$. It establishes the fundamental relationship between the trigonometric functions and the complex exponential function. For $D = 7.0$, we obtain:

$$CQ1 = \frac{1}{7}e^{i\frac{\pi(2n+1)}{3}} = \frac{1}{7}\left(\cos\left(\frac{2\pi(n+1)}{6}\right) + i\sin\left(\frac{2\pi(n+1)}{6}\right)\right) \tag{2}$$

and

$$CD1 = \frac{1}{CQ1} = 7e^{-i\frac{\pi(2n+1)}{3}} = 7\left(\cos\left(-\frac{\pi(n-1)}{3}\right) + i\sin\left(-\frac{\pi(n-1)}{3}\right)\right)$$

$$= 7\left(\frac{1}{2} - i\frac{\sqrt{3}}{2}\right) \quad \text{p.v.} \tag{3}$$

for $n = 1, 2, 3, \dots$ in N, where p.v. means principal value. The final EPG-IPG (external phase generator vs. internal phase generator) relationships are reported in Fig. 3 for $SN = 7.0$.

The knowledgeable reader will have already guessed the relationship of our result to de Moivre number or root of unity (i.e. any complex number that gives 1.0 when raised to some integer power of n. In this way, we can exploit Rational numbers $\mathbf{Q}$ full information content to get effective and stronger solutions to current AMS (arbitrary multiscale) system modelling problems (Fig. 3).

By this figure, we show how to unfold the full information content hard-wired into Rational OR representation (nano-microscale discrete representation) and to relate it to an assumed continuum framework (meso-macroscale) with no information dissipation for $D = SN = 7.0$. CICT EPG-IPG approach combined to geometric algebra (GA) and geometric calculus (GC) unified mathematical language can offer an effective and convenient "Science 2.0" universal framework, by considering information not only on the statistical manifold of model states but also on the combinatorial manifold of low-level discrete, phased generators and empirical measures of noise sources, related

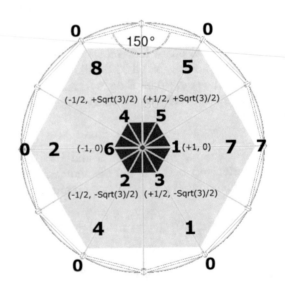

Fig. 3 The relationship of our EPG-IPG result for SN = 07,0 to de Moivre number or root of unity (i.e. any complex number that gives 1.0 when raised to some integer power W of N. In our case W = 6), see text

to experimental high-level overall perturbation. A synergic coupling between GA, GC and CICT offers stronger arbitrary-scale computational solutions which unify, simplify, and generalize many areas of mathematics that involve geometric ideas (Fiorini 2015).

Scale related, coherent precision correspondence leads to transparency, ordering, reversibility, kosmos, simplicity, clarity, and to algorithmic quantum incomputability on current, real macroscopic machines (Fiorini 2014b,c). CICT fundamental relation (Fiorini and Laguteta 2013) allows to focus our attention on combinatorially optimized numeric pattern generated by LTR (left-to-right) or RTL (right-to-left) phased generators and by convergent or divergent power series or recurrence relations, with no further arbitrary constraints on elementary generator and relation. Thanks to subgroup interplay and intrinsic phase specification through polycyclic relations in each SN remainder sequence, word inner generator combinatorial structure can be arranged for "pairing" and "fixed point" properties for digit group with the same word length.

As a matter of fact, those properties ("pairing" and "fixed point") are just the operational manifestation of universal categorical irreducible dichotomy hard-wired into integer digit and digit group themselves (i.e. "evenness" and "oddness") and to higher level numeric reflexion structures (i.e. "correspondence" and "incidence").

6 Conclusion

The presented approach, based on CICT, has shown to be quite helpful with high application flexibility (Fiorini 2016). It can be applied at any system scale: from single quantum system application development (Fiorini 2014b,c, 2017a), through new computational semantic processing (Fiorini 2017b), to full system governance strategic assessment policies (Fiorini 2017c). It can open the door towards a more effective post-Bertalanffy Systemics Complexity modeling, taking into consideration system incompleteness and quasiness, and beyond.

Acknowledgements The author acknowledges the continuous support from the CICT CORE Group of Politecnico di Milano University, Milano, Italy, for extensive computational modeling and simulation resources.

References

De Giacomo, P., & Fiorini, R. A. (2017). *Creativity mind.* Amazon, E-book.
Fiorini, R. A. (1994). *Strumentazione Biomedica: Sistemi di Supporto Attivo.* Milano: CUSL.

Fiorini, R. A. (2014a). Stronger physical and biological measurement strategy for biomedical and wellbeing application by CICT. In N. E. Mastorakis & A. Fukasawa (Eds.), *Proceedings of the 3rd International Conference on Health Science and Biomedical Systems (HSBS 2014)* (pp. 36–45). Florence: WSEAS Press.

Fiorini, R. A. (2014b). Stronger quantum decoherence incomputability modeling by CICT. In N. E. Mastorakis & G. Q. Xu (Eds.), *Proceedings of the 3rd International Conference on Applied and Computational Mathematics (ICACM 2014)* (pp. 78–86). Geneva: WSEAS Press.

Fiorini, R. A. (2014c). The entropy conundrum: A solution proposal. In *1st International Electronic Conference on Entropy and Its Applications, 3–21 November 2014*. Sciforum Electronic Conference Series (Vol. 1). Available at: http://www.sciforum.net/conference/ecea-1

Fiorini, R. A. (2015). GA and CICT for stronger arbitrary multiscale biomedical and bioengineering solutions. In S. Xambó Descamps, J. M. Parra Serra, & R. González Calvet (Eds.), *Early Proceedings of the AGACSE 2015 Conference* (pp. 153–162). Barcelona: Universitat Politècnica de Catalunya.

Fiorini, R. A. (2016). From systemic complexity to systemic simplicity: A new networking node approach. In G. F. Minati, M. R. Abram, & E. Pessa (Eds.), *Towards a post-Bertalanffy systemics* (pp. 97–108). Cham: Springer.

Fiorini, R. A. (2017a). Brain-inspired systems and predicative competence. In *Proceedings 2017 IEEE 16th International Conference on Cognitive Informatics & Cognitive Computing (ICCI*CC)*, 26–28 July 2017 (pp. 268–275). Oxford, UK: IEEE Press.

Fiorini, R. A. (2017b). From computing with numbers to computing with words. In *Proceedings 2017 IEEE 16th International Conference on Cognitive Informatics & Cognitive Computing (ICCI*CC)*, 26–28 July 2017 (pp. 84–91). Oxford, UK: IEEE Press.

Fiorini, R. A. (2017c). Would the big government approach increasingly fail to lead to good decision? A solution proposal. *Kybernetes, 46*(10), 1735–1752. https://doi.org/10.1108/K-01-2017-0013.

Fiorini, R. A., De Giacomo, P., & L'Abate, L. (2016). Wellbeing understanding in high quality healthcare informatics and telepractice. *Studies in Health Technology and Informatics, 226*, 153–156.

Fiorini, R. A., & Laguteta, G. (2013). Discrete tomography data footprint reduction by information conservation. *Fundamenta Informaticae, 125*(3–4), 261–272.

Jacquette, D. (2002). *Ontology*. Montreal: McGill-Queen's University Press.

L'Abate, L., Cusinato, M., Maino, E., Colesso, W., & Scilletta, C. (2010). *Relational competence theory*. New York: Springer.

Ness, R. B. (2014). *The creativity crisis*. Oxford: Oxford University Press.

Nicolescu, B. (1996). La transdisciplinarité, manifeste. In J. P. Bertrand (Ed.), *Transdisciplinarité*. Paris: Édition du Rocher.

Poli, R., & Seibt, J. (Eds.). (2010). *Theory and applications of ontology: Philosophical perspectives*. Dordrecht: Springer Netherlands.

Puu, T. (2015). *Arts, sciences, and economics*. Heidelberg: Springer.

The Problem of Functional Boundaries in Prebiotic and Inter-Biological Systems

Leonardo Bich

1 Introduction: Challenges to Biological Autonomy

From a system theoretical perspective, investigating the distinctive features of organisms means approaching living systems as highly integrated entities capable to control their underlying dynamics and the functional behaviour of their components in such a way as to ensure their existence and persistence over time. This line of investigation has been specifically pursued for over five decades by the theoretical framework centred on the notion of biological autonomy.[1] According to the autonomy framework, biological systems are organised in such a way that they realise metabolic self-production and self-maintenance. The specificity of this kind of systems is that the existence and activity of their components depend on the network they realise, and to exist as organised unities they actively manage the continuous exchange of matter and energy with the environment.

This theoretical perspective has been facing two main challenges. The first consists in characterising the distinctive dynamic regime and organisation that put together this class of systems (organisms). The second consists in specifying, on the basis of a specific theory of biological organisation, what

L. Bich (✉)

IAS-Research Centre for Life, Mind and Society, Department of Logic and Philosophy of Science, University of the Basque Country (UPV/EHU), Donostia-San Sebastian, Spain
e-mail: leonardo.bich@ehu.es

[1] For a detailed theoretical and historical analysis of this research line see Bich and Damiano (2008), Letelier et al. (2011), and Mossio and Bich (2017).

© Springer Nature Switzerland AG 2019
G. Minati et al. (eds.), *Systemics of Incompleteness and Quasi-Systems*,
Contemporary Systems Thinking,
https://doi.org/10.1007/978-3-030-15277-2_23

can be considered a functional component of the system and what cannot.[2] A good theoretical account should be able to trace the precise functional boundaries of a system, but it might be particularly difficult to make this distinction if we consider the multifarious interactions with the environment that a living system needs to undergo and maintain in order to exist.

A detailed account of the organisation of autonomous biological systems has been recently provided in Moreno and Mossio (2015) and Montévil and Mossio (2015) in terms of *closure of constraints*. Yet, recent research on host-microbiota and, more generally, symbiotic relationships characterised by close functional ties (Bosch and McFall-Nagai 2011; Pradeu 2011; Gilbert et al. 2012), might seem either to question the capability of this framework to identify clear functional boundaries for living organisms, or to call for further work of characterisation of the different ways functional interactions can be established within a system or between systems.

2 Closure of Constraints, Control and Functional Integration

The notion of biological autonomy is grounded in the idea that living systems are metabolic self-producing systems that are able to self-maintain and keep their network organisation invariant through the continuous exchange of matter and energy with the environment. This idea is captured by the notion of *organisational closure* introduced by Piaget (1967), Rosen (1972), Maturana and Varela (Varela et al. 1974), to account for a fundamental feature of the organisation of (biological) self-maintaining systems: its circular topology as a network of processes of production of components that in turn realise and maintain the network itself. These early formulations of the notion of *closure* played a pioneering role in providing an understanding of the distinctive features of biological systems. Yet, they exhibited several limitations such as lack of detail, abstractness and lack of connection with thermodynamics.

Recently, an approach to *closure* based on the concept of constraint has been proposed to overcome these issues. Constraints are characterised as material structures that harness processes, and that by doing so specify part of the conditions of existence of the latter. According to this framework, living systems are capable to generate a subset of the constraints acting on their internal processes, and to realise a regime of *closure of constraints* (Moreno and Mossio 2015; Montévil and Mossio 2015).[3]

---

[2] In this context, "functional component" means that its existence depends on the system which harbours it, and that in turn it actively contributes to the existence of such system (Mossio et al. 2009).

[3] For a concise discussion of the role of the notion of constraint in reframing the debate on organisational closure see Bich (2016).

A conceptual aspect of this framework that is central to discuss the functional boundaries of a living system, is that constraints play a specific functional role in a biological organisation, captured by the notion of *control*.[4] Control is generally defined as the capability to modify the dynamics of a system toward a certain state, an activity that implies an *asymmetric interaction*. Specifically, in biological systems controllers are molecules or supramolecular structures[5] that are produced from within and operate as a subset of the local boundary conditions (constraints) of the controlled processes. A biological system is capable of generating some of the (internal) constraints that control its dynamics so that it can maintain itself in far from equilibrium conditions by harnessing the thermodynamic flow. An example is *kinetic control* (e.g., catalysis), which specifies the rates of diverse synthetic pathways: e.g. an enzyme that harnesses (catalyses) a chemical reaction, without being directly affected by it.

The distinctive feature of biological organisation, captured by the notion of *closure*, is that the constraints which exert these basic controls are organised in such a way that they are mutually dependent for their production and maintenance, and collectively contribute to maintain the conditions at which the whole network can persist. The notion of *closure* of constraints focuses, thus, on the distinctive capability of living systems to contribute to their own conditions of existence. This is a feature that is not shared by other circular networks such as abiotic water cycles (Mossio and Bich 2017) or self-maintaining systems such as dissipative structures, which are mostly and largely determined by external boundary conditions, and emerge spontaneously under appropriate environmental conditions.

In this perspective, control mechanisms are the functional components of a biological organisation. The way they are wired together to collectively achieve self-maintenance provides the criteria to characterise the degree of *functional integration* of a system. As argued in Bich et al. (2016), there are indeed different ways according to which control constraints can be said to be mutually dependent and realise closure. The simplest way is when control subsystems are coupled in such a way that they provide one another the substrates necessary to their own internal processes. In this case subsystems do not exert control on the others by directly affecting their local boundary conditions, but just interact by means of inputs and outputs.

The passage from functional coupling to full-fledged functional integration occurs when a control mechanism is generated by a process directly controlled by another control mechanism in the system. In this case, the two mechanisms are not simply coupled through supply and demand of metabolites, but each depends on the direct action of another constraint for its production and maintenance. They functionally contribute to one another's conditions of existence by mutually controlling their respective generative processes.

---

[4] For a detailed discussion of control in biological systems see Pattee (1972) and Bich et al. (2016).

[5] This role can be played by cellular and more complex structures in multicellular systems.

3 Expanded Functional Interactions: Beyond Basic Closure

The ideas provided in the previous section can be employed to analyse functional interactions beyond a single autonomous system. Specifically, they can be applied to those challenging cases in which functional boundaries seem unclear and not totally specified by a regime of *closure of constraints*: i.e. when systems, in order to maintain themselves, need to recruit control mechanism in the environment or to establish higher order entities by exerting direct (cross) control upon other biological systems.

3.1 Achieving Functional Sufficiency in Prebiotic Systems

One of the conceptual issues faced by the notion of *closure* is whether and how it can account for cases of infra-biological (prebiotic) self-maintaining systems which might realise *closure*[6] without achieving full-fledged or robust autonomy. The idea is that a self-produced control network might be able to realise closure, but not to reach a stable *functional sufficiency*[7] that allows it to persist.

Let us thinks of prebiotic scenarios of fragile self-maintaining systems: steps towards life which do not exhibit the same complexity and functional differentiation of current living systems. A hypothetical case is that of Kauffman's autocatalytic sets, a minimal theoretical example of integration realised by means of cross-control (Kauffman 2000). A catalyst A is generated through the action of another catalyst B, which controls kinetically the chemical process that leads to the synthesis of A. A, in turn, contributes to the conditions existence of B by controlling directly its production or some intermediate steps, such as the synthesis of other catalysts in the system that are responsible for the production of B. In such a way, each constraint depends for its production and maintenance on the direct action of (at least) another control constraint in the system—a basic form of *closure*—and they are collectively capable to realise self-production and self-maintenance.

This system is functionally integrated. Yet, it cannot generate a compartment, and therefore it does not exert (spatial) control upon some of the crucial boundary conditions that specify the medium in which its processes take place: i.e. concentrations, contiguity, permeability etc. In order for the necessary boundary conditions to be present, to avoid dispersion, to achieve

---

6 See Moreno and Ruiz-Mirazo (2009) for a discussion of how such organisation might have been realised in the prebiotic world.

7 "Functional sufficiency" is an expression introduced by Kauffman (2016). It is used here as the set of control relationships that are necessary in an organisation subject to *closure*.

the necessary crowding for reactions to take place, and therefore achieve functional sufficiency, the system can rely on environmental scaffolds for spatial control. It can do so by recruiting basic control constraints in the environment, such as micro-pores in rocks.

It would be incorrect to describe this scaffolding interaction in terms of a more comprehensive regime functional integration, since the system does not exert any influence upon the generation of the external constraint (e.g. the pore). Yet, it does not mean that the system does not realise *closure* but, rather, that this kind of (hypothetical) system, although realising a basic regime of closure, is more directly determined by external boundary conditions and material constraints than other, more complex, autonomous systems.

3.2 Integration in Inter and Supra-Biological Systems

Let us now consider full-fledged living systems. Functional integration requires that subsystems contribute to one another's conditions of existence by mutually controlling their functional processes in such a way as to achieve *closure*. This concept, I will argue, allows not only to understand organisms as cohesive entities, but also to functionally account for those interactions between different organisms that are necessary for the maintenance of the organisms involved, without the need to abandon or weaken the notion of *closure*.

Let us first consider metabolic complementarities, such as the exchanges of metabolites and amino acids that take place between hosts and endosymbionts. In these cases a subsystem, or an organism, does not exert control on the others by affecting the local boundary conditions of their internal processes. The entities involved are mutually dependent only in a very simple form, to the extent that they provide one another the material substrates necessary for their own processes. This is not a problematic case: exchanging metabolites with the environment is inherent in the thermodynamic nature of biological systems. It does not require stretching or redefining the functional boundaries of the systems involved.

What does occur, instead, in those cases when functional roles are shared by different organisms? Do they call into question the notions of functional integration and *closure*? Let us consider three cases when the functionally integrated system is a symbiotic one or another consortium of organisms, so that control is exerted not only within but also across biological systems. For example, to respond to nutritional stress bacteria in biofilms can exchange not only metabolites, but also enzymes (or DNA sequences coding for enzymes) responsible for the control of the internal processes of other bacteria. Cross-control can be found also in *arbuscolar mycorrizal symbiosis*. This symbiotic relationship is realised through a mutual interaction between soil fungi and terrestrial plants, that is beneficial for both partners: the fungi receive carbon source from the plants, and the plants received other nutrients such as nitrogen and phosphate. The symbiotic interaction does not consist only in the exchange of metabolites. Actually, the exchange of nutrients is made possible by the activity of several mechanisms of control exerted by both partners on each other, within plant cells in the roots, through the development and modulation of a functional contact surface between the fungal cells and the plant cytoplasm. Another interesting phenomenon is *functional replacement*. The parasitic isopod *cymothoa exigua* enters fishes, attaches itself to the fish's tongue and causes it to fall by severing its blood vessels. Then it attaches to the stub of the tongue, and becomes the fish's new tongue. In this case a function originally exerted by a part of an organism is then exerted by a different organism

These cases are qualitatively different from ecological organisations[8]—such as ecosystems, ant nests, etc.—where organisms exert control upon the *external conditions of existence* of other organisms, either by directly harnessing the external flux of matter and energy, or indirectly by generating *external control constrains* in the environment (e.g. bird nests, spider webs, beaver dams, etc.).[9] In the examples described above, instead, a new order of functional integration, or an extension of an organism's functional integration, is realised because the organisms involved exert control upon one another's processes.

It is important to point out that the realisation of these new integrations does not imply *per se* that the organisms involved are not able to realise organisational *closure* and, therefore, achieve functional integration by themselves. It means, instead, that while maintaining *closure* as functionally cohesive entities, biological systems can extend their functional networks of control constraints by realising nested forms of functional integration that include more than one system.

4 Final Remarks

At first, the expansion of functional relationship to other systems—a phenomenon that occurs frequently in biology—might seem to undermine the idea of organisational *closure* as the basis of autonomy by blurring the distinctions between biological systems, thus undermining the possibility to understand them as functionally cohesive systems. This apparent problem depends on an incorrect interpretation of the notion of *closure* of constraints, which confuses the self-specification of the functional boundaries of a living system with functional self-sufficiency. *Closure* is a regime of mutually

---

[8] See Nunes-Neto et al. (2014) for an organisational account of ecological functions.

[9] See also Christensen and Bickhard (2002), for an analysis of the functional role of the bird nest in an organisational perspective.

dependent constraints that determines a subset of its own conditions of existence, not all of them. In this scenario, there is no problem in accounting for functional contributions that can cut across entities. A system that realises *closure* can undergo functional interactions with other biological systems. It can exert control upon their processes, while some of its internal processes can, in turn, be controlled by the other systems (cross-control). By establishing control interactions with other organisms, living systems do not lose their organisational *closure*. They also realise forms of inter-system functional integration, or possibly new super-organismal organisations, characterised by a new (higher) level of *closure*.[10]

Acknowledgements The author thanks Derek Skillings for the stimulating discussions on metabolic complementarities and cross-control in symbiotic systems, that lead to some of the ideas presented in Sect. 3.2. This project has received funding from the European Research Council (ERC) under the European Union's Horizon 2020 research and innovation programme—grant agreement no. 637647—IDEM, from Ministerio de Ciencia, Innovación y Universidades, Spain ('Ramon y Cajal' Programme RYC-2016-19798), from Ministerio de Economia y Competitividad (MINECO), Spain (research project FFI2014-52173-P), and from the Basque Government (Project: IT 590-13).

References

Bich, L. (2016). Systems and organizations: Theoretical tools, conceptual distinctions and epistemological implications. In G. Minati, M. R. Abram, & E. Pessa (Eds.), *Towards a Post-Bertalanffy systemics* (pp. 203–209). Cham: Springer.

Bich, L., & Damiano, L. (2008). Order in the nothing: Autopoiesis and the organizational characterization of the living. *Electronic Journal of Theoretical Physics, 4*(16), 343–373.

Bich, L., Mossio, M., Ruiz-Mirazo, K., & Moreno, A. (2016). Biological regulation: Controlling the system from within. *Biology & Philosophy, 31*(2), 237–265.

Bosch, T. C. G., & McFall-Nagai, M. J. (2011). Metaorganisms as the new frontier. *Zoology, 144*(4), 185–190.

Christensen, W., & Bickhard, M. (2002). The process dynamics of normative function. *The Monist, 85*(1), 3–28.

Gilbert, S. F., Sapp, J., & Tauber, A. I. (2012). A symbiotic view of life: We have never been individuals. *The Quarterly Review of Biology, 87*(4), 325–341.

Kauffman, S. A. (2000). *Investigations.* New York: Oxford University Press.

Kauffman, S. A. (2016). *Humanity in a creative universe.* New York: Oxford University Press.

---

[10] See Montévil and Mossio (2015).

Letelier, J.-C., Cárdenas, M. L., & Cornish-Bowden, A. (2011). From L'Homme Machine to metabolic closure: Steps towards understanding life. *Journal of Theoretical Biology, 286*(1), 100–113.

Montévil, M., & Mossio, M. (2015). Biological organisation as closure of constraints. *Journal of Theoretical Biology, 372*, 179–191.

Moreno, A., & Mossio, M. (2015). *Biological autonomy: A philosophical and theoretical enquiry.* Dordrecht: Springer.

Moreno, A., & Ruiz-Mirazo, K. (2009). The problem of the emergence of functional diversity in prebiotic evolution. *Biology and Philosophy, 24*(5), 585–605.

Mossio, M., & Bich, L. (2017). What makes biological organisation teleological? *Synthese, 194*(4), 1089–1114.

Mossio, M., Saborido, C., & Moreno, A. (2009). An organizational account of biological functions. *The British Journal for the Philosophy of Science, 60*(4), 813–841.

Nunes-Neto, N., Moreno, A., & El-Hani, C. N. (2014). Function in ecology: An organizational approach. *Biology and Philosophy, 29*(1), 123–141.

Pattee, H. H. (1972). The nature of hierarchical controls in living matter. In R. Rosen (Ed.), *Foundations of mathematical biology. Volume 1 - Subcellular systems* (pp. 1–22). New York: Academic.

Piaget, J. (1967). *Biologie et Connaissance.* Paris: Gallimard.

Pradeu, T. (2011). A mixed self: The role of symbiosis in development. *Biological Theory, 6*, 80–88.

Rosen, R. (1972). Some relational cell models: The metabolism-repair systems. In R. Rosen (Ed.), *Foundations of mathematical biology. Volume 2 – Cellular systems* (pp. 217–253). New York: Academic.

Varela, F. G., Maturana, H. R., & Uribe, R. (1974). Autopoiesis: The organization of living systems, its characterization and a model. *Biosystems, 5*(4), 187–196.

AI-Chatbot Using Deep Learning to Assist the Elderly

Guido Tascini

1 Introduction

Recently *Chatbots*, Artificial Intelligence (AI) software systems, have appeared online. These are capable to create a conversation between a virtual agent and a user. This paper tackles the problem of realize an Artificial Intelligent Chatbot conversing with elderly persons, with age-related problems. There are many and complex features that such a system must possess and are difficult to achieve. The system has to understand human language and learn from interactions, increasing his knowledge; has to remember commitments and medicines, connect remotely with doctors, family; control transmission of physiological parameters; has to entertain the elder. Assistive software robots are believed important as interface of elderly to digital technology. It increases quality of life by providing companionship.

There is a growing attention for these devices in the literature. The reborn attention to the AI has led us to conceive intelligent software that is capable of learning adapting to interaction with the elderly. Recently Chatbots appeared on the Internet, such as on facebook. They are IA software that allows to create a conversation between a virtual agent and the people. Behind all this is the application of Artificial Intelligence that is currently having a new revival and is becoming more and more object of interest.

The work presented here concerns an Intelligent Artificial Chatbot that allows a conversation between an elderly person with age-related problems and a virtual agent. This time, Chatbot's intelligence is not in the service

G. Tascini (✉)
Centro Studi e Ricerca "G. B. Carducci", Fermo, Italy

© Springer Nature Switzerland AG 2019
G. Minati et al. (eds.), *Systemics of Incompleteness and Quasi-Systems*,
Contemporary Systems Thinking,
https://doi.org/10.1007/978-3-030-15277-2_24

of e-commerce but of the needs of the elderly: creates a smart companion to converse with. The real power of Chatbot is to be self-contained and always present, active 24 h a day, to provide users with help and answers. It is able to understand natural language and to constantly learn from interactions, becoming increasingly intelligent.

A Chat Bot may be also conceived as intelligent software, then capable to learn from the experience. The mean at this aim is constituted by *machine learning algorithms*. Recently theoretical results suggest, in order to learn the kind of complicated functions with high-level abstractions, like in natural language, to adopt *deep architectures*.

These architectures are composed of multiple levels of non-linear operations, such as in neural nets with many hidden layers. Learning algorithms, such as those for Deep Belief Networks (DBN), that Geoffrey E. Hinton et al. have been proposed achieved a remarkable success. The reason for the adoption of machine learning is related to the distance that, up to half of 2000, existed between AI systems and human brain's ability to solve problems of: vision, natural language understanding and speech recognition.

The inspiration to human brain suggested to design machine learning with properties that are evident in its visual cortex, like its deep and layered connectivity. Yet the attempt to train neural networks with more than one hidden layer has failed until mid-2000. That is the performances was not superior to those of non deep, or shallow, networks. In 2006, Hinton et al. (2006) designed the *deep belief network*: a probabilistic neural network with an efficient greedy procedure for successfully pre-training it. The procedure is linked to the learning algorithm of the restricted Boltzmann machine (RBM) for layer-wise training of the hidden layers, in an unsupervised fashion. Later the procedure was generalized (Bengio et al. 2007) and subsequently many other works appeared which have strengthened the field of deep learning (Ciresan et al. 2012; Dahl et al. 2012; Larochelle et al. 2009a,b; Le et al. 2012; Salakhutdinov and Larochelle 2010; Vincent et al. 2008, 2010). Main characteristics of deep learning networks processing are the following:

- Each layer is pre-trained by unsupervised learning of representation.
- The unsupervised learning of representations is used to (pre-)train each layer.
- The unsupervised training of one layer at a time, on top of the previously trained ones. The representation learned at each level is the input for the next layer.
- The supervised training is used to fine-tune all the layers (in addition to one or more additional layers that are dedicated to producing predictions).

2 Chatbot

Chatbot means chatting robot. These are automated systems, programmed to respond to certain inputs and resulting user feedback. They currently operate mainly in chat. For example, on Facebook Messenger, it counts a lot. It ranges from a cat that tells weather forecasts, airport bats or cooking bats. It is a software based on Artificial Intelligence, which simulates a smart conversation with the user on a chat. In practice, they are currently planning to offer a functional and support service through the major messaging platforms.

The real strength of Chatbot is to be self-contained and always active, active 24 h a day, to provide users with help and responses and at the same time track their interests, preferences, ages and tastes. They are able to understand natural language and to constantly learn from interactions, becoming increasingly intelligent.

A main idea is to build an AI-chatbot as companion for senior people. The chatbot would be able to conduct a conversation on topics such as: weather, nature, news, history, cinema, music, etc.

The problem here is to give senior people the opportunity to communicate by talking, share their experiences and memories. As an assistant, the AI Chatbot can be able to: remember commitments, medicines to take, connect remotely with doctors, family, and assistants; control remote transmit physiological parameters; entertain the elder with intelligent speeches, offering games, or delivering news. The chatbot like this faces the task not only to ask and answer questions but also to memorize the context of the conversation, for a useful dialogue (Bordes et al. 2014; Li et al. 2016; Yao et al. 2015; Yih et al. 2013).

For instance an interlocutor can respond in various ways to the question; "how did you eat?" We can consider three answers: good, normal and bad. But what happens when the interlocutor goes into details? For such cases, the chatbot has to be provided with words of support. The solution now offered by different platforms may be an API, built to understand context or synonyms or other situation. APIs may be trusted for speech recognition issues, and much more. In general, the chatbot's functions are various and can be defined only during the process of chatbot developing and testing. Recently, there have appeared support for bot construction with free access to related API, like for instance Endurance, or Pandorabot Platform. Our Chat Bot, is conceived as intelligent software, capable to learn from the experience. The mean to implement this is constituted by deep machine learning algorithms.

3 Understanding NL with Deep Neural Networks

Giving machines the ability to learn and understand human language opens unthinkable scenarios. The Natural Language Understanding, in the recent past based on complex systems, it has been overtaken by the recent approaches in terms of word representation, and processing (Pouget-Abadie et al. 2014) with Deep Learning networks. In order to understand what happened recently in NLP we can see at how it attempted to understand a word. For instance let consider the following word:

Undefined.

We attempt to gather information about this word, like its definition, its sentiment, and other. We can break the word in three parts:

Prefix-stem-suffix : un-define-ed.

We see that: the stem have the meaning "define", from which we deduce definition and sentiment of the word. The prefix "un" indicates opposition, while the suffix "ed" indicate the time past-tense of the word. But an approach like this how many possibilities will explore by considering all prefixes, suffixes and root that the English language has? Now for the machine to understand a language it is necessary to construct a map of words, containing also their meanings and interactions with other words; in practice to build a dictionary of words and see where they are located semantically and contextually compared to other words and contexts. Besides to the dictionary it is related the "word embedding": each word is mapped to a set of numbers in a high-dimensional space, in which the similar words are close, while the unlike are far away (Deng and Li 2013; Durrani et al. 2014; Hermann and Blunsom 2014; Mikolov et al. 2013). The embedding can be learned from the machine, and learning varies depending on whether the machine reads a large amount of texts, or just texts about a particular task. Table 1 shows the word nearest to the last one of a sentence.

Then the prediction of a word in a sentence may be achieved with a simple metric: for each word in the dictionary, given a sentence, it assigned the probability that the word appears next in the sentence. For instance let consider a sentence with a part to fill: "I am driving _____". Between the candidate words car has a great probability to appear next, but the word

Table 1 Example of words in dictionary and expressions

Expression	Nearest token
Palace − balconies + windows	Apartment
Ship − commander + company	Entrepreneur
Cellar − dehumidifier + photovoltaic cells	Roof
Reparto − prodotto + ufficio	Documento
Sole − giorno + notte	Luna

tank has very little probability. Then the metric puts the word "car" near the word "driving" as it puts the word "tank" far from this. The next step in language understanding is *language modeling* (Bengio et al. 2003; Larochelle et al. 2007): for instance there are put words together in small sentences (n-grams), grammatically correct, that make sense. The language modeling uses *n-grams*—groups of words—and processes the n-grams further with heuristics; then inserts them into machine learning models. For example, consider the phrase: "I am driving a car"; the 2-grams of the sentence are 4:

"I am", "am driving", "driving a", "a car".

Using a large body of text, we can generate a new phrase by associating 2-grams and 3-grams together and adapting them to other pairs previously seen. But the phrases generated in this way, though grammatically correct, may also be senseless. It is the emergence of deep neural networks that has recently allowed to overcome the n-gram-based models. Word embeddings are usually initialized to random numbers (and learned during the training phase of the neural network), or initialized from previously trained models over large texts like Wikipedia. Embeddings are stored in a simple *lookup table* (hash table), that given a word, returns an array and given a sentence returns a matrix. Figure 1 shows an example.

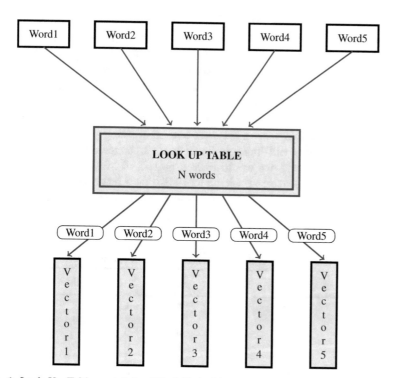

Fig. 1 Look Up Table containing Word Embeddings. It returns a vector given a word, and a matrix given a sentence

3.1 Example of Definition of Word Vectors

We adopt the Word Vectors (Mikolov et al. 2013), representing each word with n-dimensional vector. For instance if $n = 6$ our word is represented as follows:

Undefined $= [\text{- - - - - -}]$

In which the values will be filled in such a way to represent the word context, meaning and semantics. To achieve this is to create a co-occurence matrix. To illustrate how to associate vectors with words we can start from a sentence. For example from the following: I love sea and I like fish.

We can create a word vector for each word by writing a co-occurrence matrix: it contains the number of counts that each word appears to all the other words in the corpus. The matrix for our sentence is shown in Table 2. The rows of the matrix can give us an initialization of word vectors.

$$
\begin{aligned}
\text{I} &= [\ 0\ 1\ 0\ 1\ 1\ 0\ ] \\
\text{Love} &= [\ 1\ 0\ 1\ 0\ 0\ 0\ ] \\
\text{Sea} &= [\ 0\ 1\ 0\ 0\ 1\ 0\ ] \\
\text{And} &= [\ 1\ 0\ 1\ 0\ 0\ 0\ ] \\
\text{Like} &= [\ 1\ 0\ 0\ 0\ 0\ 0\ ] \\
\text{Fish} &= [\ 0\ 0\ 0\ 0\ 1\ 0\ ]
\end{aligned}
$$

The simple matrix of Table 2 allows us to gain useful insights. For example, the words "love" and "like" contain both values for their counts with names Sea and fish. They also have 1 for the count of "I", suggesting that the words are a kind of verb. With a data set larger than a single sentence, you can imagine that this resemblance will become clearer as "like," and "love," and other synonyms will begin to have similar word vectors because of being used in similar contexts. Now, even if this is a good starting point, we notice that the size of each word will increase linearly with the size of the corpus. If we have one million words, we would have a matrix of millions of millions that would be extremely scattered, with a lot of 0. There has been a lot of progress in finding the best ways to represent these word vectors. The most famous is Word2Vec.

Table 2 Co-occurrence matrix of the sentence "I love sea and like fish"

	I	Love	Sea	And	Like	Fish
I	0	1	0	1	1	0
Love	1	0	1	0	0	0
Sea	0	1	0	0	1	0
And	1	0	1	0	0	0
Like	1	0	0	0	0	0
Fish	0	0	0	0	1	0

3.2 Word2Vec

Basically we want to memorize as much as possible in word vector while retaining dimensionality in a manageable scale: from 25 to 1000. Word2Vec works on the idea of predicting the surrounding words of each word. Let's take our previous sentence: "I love sea and I like fish".

We look at the first 3 words of this sentence, adopting 3 will as our window size "m". Now, we will take the center of word, "love", and predict the words that come before and after it. This may be obtained by maximizing/optimizing the function (1), that try to maximize, given the current center word, the log probability of any context word.

$$J(\theta) = \frac{1}{T} \sum_{t=1}^{T} \sum_{-m \le j \le m, j \ne 0} \log p(w_{t+j}|w_t) \tag{1}$$

The above cost function adds the log probabilities of "I" conditioned respect to probability of "Love" and the log probability of "Sea" conditioned respect to probability of "Love", being "Love" the center word in both cases. The parameter T represents the number of training sentences The log probability has the Formula (2).

$$\log p(o|c) = \log \frac{\exp\left(h_o^T v_c\right)}{\sum_{w=1}^{W} \exp\left(u_w^T v_c\right)} \tag{2}$$

Where v_c is the word vector of center word, u_o is the word vector representation when it is the center word, and u_w is the word vector representation when it is used as the outer word (URL). The vectors are trained with *Stochastic Gradient Descent* (SGD). Synthesizing: *Word2Vec* find vector representations of different words by maximizing the log probability of context words, given a center word, and by modifying the vectors with SGD. A great contribution of Word2Vec was the *emergence of linear relationships between different word vectors*. The word vectors, with training, incredibly captures evident grammatical and semantic concepts. Another word vector initialization method is *GloVe* that combines the co-occurrence matrices with Word2Vec.

3.3 Recurrent Neural Networks (RNNs)

Recurrent Neural Networks (RNNs) (Graves 2013; Kalchbrenner and Blunsom 2013; Pascanu et al. 2012). Now we wonder how the word vectors fit into Recurrent Neural Networks (RNNs). RNNs are important for many NLP tasks. They are able to effective use data from the previous time step. In Fig. 2 it is shown a part of RNN.

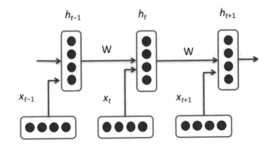

Fig. 2 Recurrent Neural Network

The word vectors are in the bottom and each vector (x_{t-1}, x_t, x_{t+1}) has a hidden state vector at the same time step (h_{t-1}, h_t, h_{t+1}). We call this a module. The hidden state of each module is function of both word vector and hidden state vector at previous time step. The hidden state in each module of the RNN is a *function* of both the word vector and the hidden state vector at the previous time step.

$$h_t = \sigma \left(W^{(hh)} h_{t-1} + W^{(hx)} x_{[t]} \right) \tag{3}$$

In (3) a weight matrix $W^{(hx)}$ are going to multiply with our input, a recurrent weight matrix $W^{(hh)}$ which is multiplied with the hidden state vector at the *previous* time step. These recurrent weight matrices are the same across all time steps. RNN is very different from a traditional 2 layer NN where we normally have a distinct W matrix for each layer (W1 and W2).

3.4 The Deep Neural Networks

Neural networks with multiple layers of neurons, accelerated in the calculation with the use of Graphical Processing Units (GPU), have recently seen enormous successes in many fields. They have passed the previous state of the art in speech recognition, object recognition, images, linguistic modeling and translation.

Figure 3 illustrates a *deep neural network* with three inputs (I_1, I_2, I_3), a first hidden layer ("A") with four neurons, a second hidden layer ("B") with five neurons and two outputs (O_1, O_2), which we indicate with the sequence; 3-4-5-2. This network requires a total of $(3*4)$ weights $+4$ bias $+(4*5)$ weights $+5$ bias $+(5*2)$ weights $+2$ bias $=42$ weights and 11 bias.

The network activation function is the *hyperbolic tangent* for calculating the outputs of the two hidden levels and the *softmax* for calculating the network output. The formulas that calculate the feed-forward are as follows:

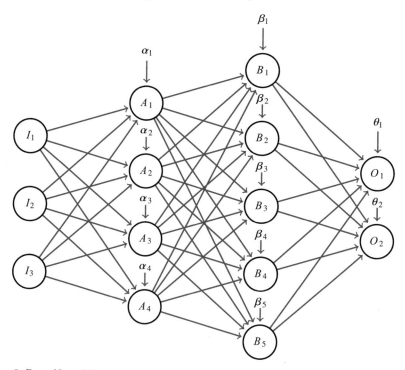

Fig. 3 Deep Neural Network 3452

$$A_i = \tanh(I_1\,p_{1i}+I_2\,p_{2i}+I_3\,p_{3i}+\alpha_i) \qquad \text{hidden layer ``}A\text{''}$$
$$B_i = \tanh(A_1\,p_{1i}+A_2\,p_{2i}+A_3\,p_{3i}+A_4\,p_{4i}+\beta_i) \qquad \text{hidden layer ``}B\text{''}$$
$$O_i = \text{softmax}(B_1\,p_{1i}+B_2\,p_{2i}+B_3\,p_{3i}+B_4\,p_{4i}+B_5\,p_{5i}+\theta_i) \quad \text{outputs.}$$

The training standard of deep NN uses back-propagation. Deep neural network training is more difficult than ordinary neural network training with a single layer of hidden nodes. This factor is the main obstacle to using networks with more than two hidden layers. Back-propagation in this case often fails to deliver excellent results.

4 Deep Learning to Train a Chatbot

A chatbot has to be able to determine the best response for any received message, understand the intentions of the sender's message and the type of response message required, form a grammatically and lexically correct response. Current chatbots are in difficulty facing these tasks. To overcome this, Deep Learning Neural Networks are used. In this case the chatbot uses some variants of the sequence to sequence (*Seq2Seq*) model (Sutskever et al. 2014).

DNNs work well when large labeled training sets are available, but not to map sequences to sequences. In this paper we were inspired Sutskever method by using a multilayered Long Short-Term Memory (LSTM) (Hochreiter and Schmidhuber 1997) to map the input sequence to a vector of fixed dimension; then a deep LSTM to decode target sequence from the vector. The seq2seq model is constituted by an encoder RNN and a decoder RNN. Encoder's task is to encapsulate the input text in a fixed representation, while Decoder's task is to derive from it a variable length text that best responds to it.

RNN contains a number of hidden state vectors, and the final hidden state vector of the encoder RNN contains an accurate representation of the whole input text. In the decoder RNN the first cellÔs task is to take in the vector representation a variable length texts that is the most appropriate for the output response. Mathematically speaking, there are computed probabilities for each words in the vocabulary, and it is chosen the argmax of the values:

$$p(y_1, ..., y_{T'} | x_1, ..., x_T) = \prod_{t=1}^{T'} p(y_t | v, y_1, ..., y_{t-1}) \tag{4}$$

Dataset Selection is fundamental to train the model. For Seq2Seq models, we need a large number of conversation logs. From a high level, this encoder decoder network needs to be able to understand the type of responses (decoder outputs) that are expected for every query (encoder inputs). For this they are available various datasets like *Ubuntu Corpus, Microsoft's Social Media Conversation Corpus*, or *Cornell Movie Dialog Corpus*. But the public data set contains a number of data not always useful. Then it may be better to generate their proper word vectors. To this aim we can use the before seen approach of a Word2Vec model.

Speech Recognition and Speech Synthesis are indispensable steps for completing user interaction with chatbot. These are very developed areas with important results (Auli et al. 2013; Bahdanau et al. 2014; Cho et al. 2014; Devlin et al. 2014; Hermann and Blunsom 2014; Hinton et al. 2012; Mikolov et al. 2013).

5 Conclusion

The paper addresses the problem to build an Artificial Intelligent Chatbot capable of talking to elderly. The system has to understand human language and learn from interactions, increasing its knowledge. Assistive software robots are believed important as interface of elderly to digital technology. It increases quality of life by providing companionship. Chatbot's intelligence is not in the service of e-commerce but of the elderly: creates a smart companion to converse with. The real power of Chatbot is to be self-contained and always present, active 24 h a day, to provide users with help and answers.

We conceived a Chat Bot with integrated intelligent software, and capable to learn from the experience. The means to achieve this is constituted by *machine learning capability*. Recently theoretical results suggest, in order to learn the kind of complicated functions with high-level abstractions, like in natural language, to adopt machine learning algorithms with a *deep architectures*. The tools to design such a Bot are *Machine Learning, speech recognition, NLP*, and *chatbot training* that can be realized with the help of commercial platforms such as Endurance or Pandorabot.

The article, after framing the various tools, analyzed the understanding of natural language aimed to chatbot training. We have analyzed deep learning, Recurrent Neural Networks, Seq2Seq and Sec2Vet language models. It has dealt with Dataset Selection by looking at datasets in net such as Ubuntu corpus, Microsoft's Social Media Conversation Corpus, Cornell Movie Dialog Corpus, et alias. In this way, we created the premises for building our intelligent chatbot with the help a commercial platform such as Endurance, and making it able to learn from experience.

References

Auli, M., Galley, M., Quirk, C., & Zweig, G. (2013). Joint language and translation modeling with recurrent neural networks. In *Proceedings of the 2013 Conference on Empirical Methods in Natural Language Processing (EMNLP 2013)* (pp. 1044–1054), Seattle, Washington, 18–21 October 2013.

Bahdanau, D., Cho, K., & Bengio, Y. (2014). Neural machine translation by jointly learning to align and translate. arXiv:1409.0473.

Bengio, Y., Ducharme, R., Vincent, P., & Jauvin, C. (2003). A neural probabilistic language model. *Journal of Machine Learning Research, 3*, 1137–1155.

Bengio, Y., Lamblin, P., Popovici, D., & Larochelle, H. (2007). Greedy layer-wise training of deep networks. In B. Schölkopf, J. Platt, & T. Hofmann (Eds.), *Advances in neural information processing systems 19* (pp. 153–160). Cambridge: MIT Press.

Bordes, A., Chopra, S., & Weston, J. (2014). Question answering with subgraph embeddings. In *Proceedings of the 2014 Conference on Empirical Methods in Natural Language Processing (EMNLP 2014)*, October 25–29, 2014, Doha (pp. 615–620).

Cho, K., Merrienboer, B., Gulcehre, C., Bougares, F., Schwenk, H., & Bengio, Y. (2014). Learning phrase representations using RNN encoder-decoder for statistical machine translation. arXiv:1406.1078.

Ciresan, D., Meier, U., & Schmidhuber, J. (2012). Multi-column deep neural networks for image classification. In *Proceedings of the IEEE Conference on Computer Vision and Pattern Recognition* (pp. 3642–3649).

Dahl, G. E., Yu, D., Deng, L., & Acero, A. (2012). Context-dependent pre-trained deep neural networks for large vocabulary speech recognition. *IEEE Transactions on Audio, Speech, and Language Processing, 20*(1), 30–42 (Special Issue on Deep Learning for Speech and Language Processing).

Deng, L., & Li, X. (2013). Machine learning paradigms for speech recognition: An overview. *IEEE Transactions on Audio, Speech, and Language Processing, 21*(5), 1060–1089.

Devlin, J., Zbib, R., Huang, Z., Lamar, T., Schwartz, R., & Makhoul, J. (2014). Fast and robust neural network joint models for statistical machine translation. In *Proceedings of the 52nd Annual Meeting of the Association for Computational Linguistics (ACL 2014)* (pp. 1370–1380).

Durrani, N., Haddow, B., Koehn, P., & Heafield, K. (2014). Edinburgh's phrase-based machine translation systems for WMT-14. In *Proceedings of the Ninth Workshop on Statistical Machine Translation* (pp. 97–104).

Graves, A. (2013). Generating sequences with recurrent neural networks. arXiv:1308.0850.

Hermann, K. M., & Blunsom, P. (2014). Multilingual distributed representations without word alignment. In *Proceedings of International Conference on Learning Representations (ICLR 2014)*. arXiv:1312.6173.

Hinton, G. E., Deng, L., Yu, D., Dahl, G. E., Mohamed, A., Jaitly, N., et al. (2012). Deep neural networks for acoustic modeling in speech recognition: The shared views of four research groups. *IEEE Signal Processing Magazine, 29*(6), 82–97.

Hinton, G. E., Osindero, S., & Teh, Y. (2006). A fast learning algorithm for deep belief nets. *Neural Computation, 18*, 1527–1554.

Hochreiter, S., & Schmidhuber J. (1997). Long short-term memory. *Neural Computation, 9*(8), 1735–1780.

Kalchbrenner, N., & Blunsom, P. (2013). Recurrent continuous translation models. In *Proceedings of the 2013 Conference on Empirical Methods in Natural Language Processing (EMNLP 2013)* (pp. 1700–1709), Seattle, Washington, 18–21 October 2013.

Larochelle, H., Bengio, Y., Louradour, J., & Lamblin, P. (2009). Exploring strategies for training deep neural networks. *Journal of Machine Learning Research, 10*(Jan), 1–40.

Larochelle, H., Erhan, D., Courville, A., Bergstra, J., & Bengio, Y. (2007). An empirical evaluation of deep architectures on problems with many factors of variation. In *Proceedings of the 24th International Conference on Machine Learning (ICML 2007)*.

Larochelle, H., Erhan, D., & Vincent, P. (2009). Deep learning using robust interdependent codes. In D. A. Van Dyk & M. Welling (Eds.), In *Proceedings of the 12th International Conference on Artificial Intelligence and Statistics (AISTATS 2009)* (pp. 312–319). JMLR Proceedings 5.

Le, Q. V., Ranzato, M.A., Monga, R., Devin, M., Chen, K., Corrado, G. S., et al. (2012). Building high-level features using large scale unsupervised learning. In *Proceedings of the 29th International Conference on Machine Learning (ICML 2012)* (pp. 507–514).

Li, J., Galley, M., Brockett, C., Spithourakis, G. P., Gao, J., & Dolan, B. (2016). A persona-based neural conversation model. In *Proceedings of the 54th Annual Meeting of the Association for Computational Linguistics* (pp. 994–1003).

Mikolov, T., Chen, K., Corrado, G., & Dean, J. (2013). Efficient estimation of word representations in vector space. arXiv:1301.3781.

Pascanu, R., Mikolov, T., & Bengio, Y. (2012). On the difficulty of training recurrent neural networks. arXiv:1211.5063.

Pouget-Abadie, J., Bahdanau, D., van Merrienboer, B., Cho, K., & Bengio, Y. (2014). Overcoming the curse of sentence length for neural machine translation using automatic segmentation. arXiv:1409.1257.

Salakhutdinov, R., & Larochelle, H. (2010). Efficient learning of deep Boltzmann machines. In *Proceedings of the 13th International Conference on Artificial Intelligence and Statistics (AISTATS 2010)* (pp. 693–700).

Sutskever, I., Vinyals, O., & Le, Q. V. (2014). Sequence to sequence learning with neural networks. In *Proceedings of the 27th International Conference on Neural Information Processing Systems (NIPS 2014)* (pp. 3104–3112).

Vincent, P., Larochelle, H., Bengio, Y., & Manzagol, P.-A. (2008). Extracting and composing robust features with denoising autoencoders. In *Proceedings of the 25th International Conference on Machine Learning (ICML 2008)* (pp. 1096–1103). New York: ACM.

Vincent, P., Larochelle, H., Lajoie, I., Bengio, Y., & Manzagol, P.-A. (2010). Stacked denoising autoencoders: Learning useful representations in a deep network with a local denoising criterion. *Journal of Machine Learning Research, 11*(Dec), 3371–3408.

Yao, K., Zweig, G., & Peng, B. (2015). Attention with intention for a neural network conversation model. arXiv:1510.08565v3.

Yih, W., Chang, M.-W., Meek, C., & Pastusiak, A. (2013). Question answering using enhanced lexical semantic models. In *Proceedings of the 51st Annual Meeting of the Association for Computational Linguistics* (pp. 1744–1753), Sofia, August 4–9, 2013.

Web Resource

www.quora.com/How-does-word2vec-work-Can-someone-walk-through-a-specific-example

The Use of Brain Computer Interface (BCI) Combined with Serious Games for Pathological Dependence Treatment

Natale Salvatore Bonfiglio, Roberta Renati, and Eliano Pessa

1 Introduction

In this contribution we discuss about the possibility of using suitable advanced technological tools to support a clinical treatment of pathological addiction situations, like drug or alcohol assumption and hazardous gambling. Among these tools we focused our attention mainly on the so called "serious computer games" and on Brain-Compute Interfaces (BCI).

The results obtained in a number of pilot experiments performed on a small sample of subjects suggest that these tools could be very useful when coping whit addicted subjects. Then, we can suppose that the tools themselves could rightly cooperate with the medical and psychological experts performing clinical treatments on these subjects.

However, the effectiveness of these tools depends in a crucial way on a deep understanding of the principles underlying the behavior of a human-machine interaction system. The research described in this contribution constitutes a step on the way to be followed in order to achieve this understanding.

N. S. Bonfiglio (✉) · R. Renati
Dipartimento di Scienza del Sistema Nervoso e del Comportamento, Università degli Studi di Pavia, Pavia, Italy
e-mail: salvo.bonfiglio@unipv.it; roberta.renati@unipv.it

E. Pessa
Department of Brain and Behavioral Science, University of Pavia, Pavia, Italy
e-mail: eliano.pessa@unipv.it

G. Minati et al. (eds.), *Systemics of Incompleteness and Quasi-Systems*,
Contemporary Systems Thinking,
https://doi.org/10.1007/978-3-030-15277-2_25

2 Serious Games

The time has come for videogames to become more relevant, more responsible and more meaningful ... in other words, more serious (Rego et al. 2010). For years the industry of games has been developing products that are both educational and pedagogical, focused on learning, rehabilitation, therapy etc. (Connolly et al. 2012; Charsky 2010; Wiemeyer and Kliem 2011; Santamaria et al. 2011), reaching a wider and wider audience. For instance, many games for Nintendo's Wii are now widespread and addressed to audiences ranging from children to whole families.

Despite their circulation, the definition of what a serious game is doesn't find everybody in agreement. Nevertheless, it is generally possible to define it as a computer game whose main aim is not purely entertaining (Michael and Chen 2005). Serious games are entertaining, enjoyable and fun, but their main purpose is other that was conceived by the game designer when designing the game or that the user defined when played the game.

One of the major scopes of use is rehabilitation, a dynamic process of adaptive and programmed change of lifestyle in response to a non-programmed change due to discomfort, disturbance or trauma (Rego et al. 2010). In this respect, the use of games in rehabilitation can stimulate motivation—particularly if subjects/patients are children or less compliant individuals—with an intervention able to work on a multidisciplinary level, i.e. with the support of professionals such as physicians, psychologists, educators, nurses, etc.

In fact, traditional rehabilitation approaches include repetitive and boring exercises, while the use of computer games brings benefits thanks to the physical and cognitive engagement. Furthermore, many games are divided into increasing levels of difficulty giving the "player" a feeling of challenge that make him progress in a way which is adapted to his skills (Burke et al. 2009).

Despite promises and hopes to be found in literature about the positive results of serious games, the main aim should be to discover which basic principles are necessary to design a reliable and effective serious game. For instance, they could be structured in a way to be activating, able to put the user in a position of continuous challenge and also meaningful (Burke et al. 2009).

3 BCI (Brain Computer Interface)

Serious games have enormous potentials and hold unique opportunities that can be emphasized and enhanced by the use of appropriate hardware and device, connected through personal computers and suitable virtual environments. One of the most employed and promising device system is the headset Brain Computer Interface (BCI).

The use of BCI—systems able to catch the EEG brain activity so to interface it with external devices—has been a reality for many years now (Curran and Stokes 2003). The first BCI was described in 1964. Grey Walder, neurophysiologist, linked some electrodes to the motor area of a patient and then asked him to look at a projected presentation of images and to push a button to change slides. The subject's brain activity was registered and later the system was directly connected to the projector. The presentation went on when the brain activity of the patient indicated his intention to push the command. The first BCI with invasive activity of intercortical data collection was created (Graimann et al. 2010).

Future perspective of the use of Brain-computer-interface are those in which one day it may be possible to find a

[...] notice that the room temperature is slightly too high, so you turn your gaze to the temperature display and think it down a few degrees. The room cools to a more acceptable level. Your next task is to send a package to a colleague by 3Dmail. (Adams et al. 2008, p. 6)

These systems make direct communication with the human brain a reality, as described for a long time by science fiction literature and cinema, so to make our thoughts observable.

Even though many problems connected to BCI designing are still far from being solved (Pessa 2009), like, (1) knowing which brain signals are to be associated with specific intentions or internal mental states, (2) selecting the best techniques to identify the above mentioned characteristics even at the presence of artifacts and (3) finding the best way to build "online" the detection-action sequence, practical BCI systems have already been designed and developed, also for non-professional use and entertainment, making possible for a wide audience, with a limited expense, to create a neural interface for leisure purposes.

The rehabilitation and support framework to subjects with severe disabilities is widely benefitting from BCI systems' development. Neuro-prosthesis are artificial devices able to substitute or improve specific functions of the nervous system aiming at creating a human-machine, bi-directional connection. Neural interfaces make the brain-artificial limb connection possible and similar to the natural control of limbs. Regarding psychology, the advent of the BCI technology enabled an important opening towards the possibility of making the user perceptively aware of the changing of his mental state. In this case, neurofeedback is a very useful technique widely used in rehabilitation and in different clinical realities thanks to its non-invasiveness. The association between brain waves and different mental states and the possibility for the subject to "visually see" (on a monitor providing a real-time, continuous feedback) the variations of the mental state has made the development of specific mental training possible and useful for the treatment of a number of psychopathologies. There is more and more clinical evidence regarding the treatment of some neuropsychiatric pathologies (ADHD in particular) and the observable benefits for the attention and memory processes, due to the use of the neurofeedback as a form of training (Heinrich et al. 2007).

Neurofeedback training tends to develop abilities of self-regulation of the brain activity in the subject, based on the idea that the brain is a dynamic and extremely plastic organ where real-time feedback acts like a positive reinforcement.

The use of BCI technologies for neurofeedback and training is widely expanding and diversifying. It is differently applied with clinical aims using specific training, for instance with patients with stroke (Buch et al. 2008) or Parkinson (Turconi et al. 2014).

What we want to highlight in this work is that the use of a BCI system demands a voluntary and deliberated control of the brain activity of which the person is usually unaware and which the individual is unable to recognize and feel. It is like being aware of our hand muscular movement but not of how our brain moves our hand in that same moment (Curran and Stokes 2003).

4 An Example of Training with BCI

"Be Your Brain" is a project aimed at studying training and treatment protocols for multidependent patients and gamblers. The objective is to help patients to manage their levels of self-control and impulsiveness, together with stress and anxiety, by submitting them to training and tests specifically structured with a BCI interface.

If we suppose that self-control, impulsiveness and subsequent stress levels considerably change when we stimulate subjects to act actively on their behavioral response, then the BCI technology can be used to foster learning and a more active and time-enduring behavioral change that uses coping cognitive strategies and problem solving.

Research has shown that the need and search for substance (gambling included) is maintained or exacerbated by environmental stimuli referring to substance which would determine the so called craving status, i.e. an uncontrollable need to use substances linked to a physiological response similar to stress.

This research considers craving as a "conditioning" response where the active behavior (the search for substance) should be considered as a conditioned response (Sharpe 2002; Blaszczynski and Nower 2002). The consequential obsessive-com-pulsive behavior, linked to the search for the substance, would become automatic and unavoidable; the substance itself would be a reinforcement, so that the behavior will be repeated in the future (Rousseau et al. 2002; Vallerand et al. 2003).

The repetition of this mechanism recalls an assimilated learning. It is possible to suppose that the "inverted" mechanism might reduce reactions and answers learnt through a paradigm already experienced in literature, the clue eliciting (Pericot-Valverde et al. 2015), in which the presentation of trigger stimuli reminding of the substance might help the subject to activate strategies to resist and implement self-control in front of the substance. Doing so, the subject is helped, through a training using the BCI technology, to eliminate and de-contextualize stimuli with craving risk. In fact, the subject will learn to manage and manipulate a series of images and visual stimuli associated to craving situations and will be able to improve his self-control and craving management in his daily life.

A specific training has been developed with a "kit" made of a software that manipulates imaging reminding of craving, of a hardware device made of the EMOTIV helmet, a USB device with Bluetooth connection and of a laptop where stimuli are presented (Mazzoleni et al. 2017).

Training protocol is composed by the presentation of a series of images stimulating impulsive behavior and craving in the subject and of the presentation of those same stimuli that the subject can manipulate, i.e.: distance them or move them closer, move them in different directions on the screen (to the right, left, up, down).

First positive results have been obtained on a group of 12 users (8 in experimental group, 4 in control group) with a research design pre-post in which psychometric tests are submitted to measure outcomes at time T0, T1 (1 month after first administration so at the end of the training) and T2 (1 month after the end of the training). Figures show that subjects tend to improve the outcomes typical of the treatment like the ability to resist and the desire for the substance, measured with the SAD (Self-Efficacy and Desire Scale) (Minervini et al. 2011), the stress measured through the PSS (Perceived Stress Scale) (Cohen 1988), impulsiveness and self-control, measured with BIS-11 (Barratt Impulsiveness Scale) (Fossati et al. 2001) or coping strategies, measured through COPE-NVI (Coping Orientation to Problems Experiences-Nuova Versione Italiana) (Sica et al. 2008). These results have provided useful indications for the project and will be useful to make further hypothesis of improvement. Currently, a new experimental condition, i.e. a group of neuro-stimulation with TDCS, has been added to the above mentioned protocol with the aim to reduce the need for the substance and craving by modulating the dorsolateral, pre-frontal area connected with the circuit of gratification (Sauvaget et al. 2015).

5 Human-Machine Interaction in the Use of BCI with Serious Games

The perspective we want to highlight in this work is to consider the user of BCI technology as an element of a wider system where technology itself, but also the conceptual premises at the basis of the training, the software built with specific characteristics, the treatment service in which the user is included, the motivation, the sense of efficiency, the search for performance and the achievement of the result etc., represent the elements of a wide and open system. In such a system, efficiency and good outcome of the training (of the intervention) and of the treatment might be represented by these elements. We need to further understand the importance of each of these elements.

For instance, we should understand how much motivation, sense of self-efficiency, search for performance etc.—subjective variables—are the variables influencing the outcome and on which a serious game developer should focus on. Considering what has been said about serious games, could we consider these the real principles of design and construction of a game and the real variables influencing the system? If so, it is likely that the use of the BCI technology can have great importance in influencing the outcome of the treatment, since the subject would probably give more importance to results achieved only using his own "mind", a factor that might influence the above mentioned psychological variables. Are these emerging properties of a human-machine interaction system? If so, the outcome of the treatment would be an emerging property of emerging properties and we might consider the players of this interaction as systems of a more complex system whose variables are continuously changing.

References

Adams, R. G., Bahr, G. S., & Moreno, B. (2008). Brain computer interfaces: Psychology and pragmatic perspectives for the future. In *AISB 2008 Convention: Communication, Interaction and Social Intelligence* (pp. 1–6). Society for the Study of Artificial Intelligence and the Simulation of Behaviour.

Blaszczynski, A., & Nower, L. (2002). A pathways model of problem and pathological gambling. *Addiction, 97*(5), 487–499.

Buch, E. R., Weber, C., Cohen, L. G., Braun, C., Dimyan, M. A., Ard, T., et al. (2008). Think to move: A neuromagnetic brain-computer interface (BCI) system for chronic stroke. *Stroke, 39*(3), 910–917.

Burke, J. W., McNeill, M. D. J., Charles, D. K., Morrow, P. J., Crosbie, J. H., & McDonough, S. M. (2009). Optimising engagement for stroke rehabilitation using serious games. *The Visual Computer, 25*(12), 1085–1099.

Charsky, D. (2010). From edutainment to serious games: A change in the use of game characteristics. *Games and Culture, 5*(2), 177–198.

Cohen, S. (1988). Perceived stress in a probability sample of the United States. In S. Spacapan & S. Oskamp (Eds.), *The Claremont Symposium on Applied Social Psychology. The Social Psychology of Health* (pp. 31–67). Thousand Oaks, CA: Sage Publications.

Connolly, T. M., Boyle, E. A., MacArthur, E., Hainey, T., & Boyle, J. M. (2012). A systematic literature review of empirical evidence on computer games and serious games. *Computers & Education, 59*(2), 661–686.

Curran, E. A., & Stokes, M. J. (2003). Learning to control brain activity: A review of the production and control of EEG components for driving brain-computer interface (BCI) systems. *Brain and Cognition, 51*(3), 326–336.

Fossati, A., Di Ceglie, A., Acquarini, E., & Barratt, E. S. (2001). Psychometric properties of an Italian version of the Barratt Impulsiveness Scale-11 (BIS-11) in nonclinical subjects. *Journal of Clinical Psychology, 57*(6), 815–828.

Graimann, B., Allison, B., & Pfurtscheller, G. (2010). Brain-computer interfaces: A gentle introduction. In B. Graimann, B. Allison, & G. Pfurtscheller (Eds.), *Brain-computer interfaces* (pp. 1–27). Berlin: Springer.

Heinrich, H., Gevensleben, H., & Strehl, U. (2007). Annotation: Neurofeedback-train your brain to train behaviour. *Journal of Child Psychology and Psychiatry, 48*(1), 3–16.

Mazzoleni, M., Previdi, F., & Bonfiglio, N. S. (2017). Classification algorithms analysis for brain–computer interface in drug craving therapy. *Biomedical Signal Processing and Control.*

Michael, D. R., & Chen, S. L. (2005). *Serious games: Games that educate, train, and inform.* Boston: Course Technology, Incorporated.

Minervini, I., Palandri, S., Bianchi, S., Bastiani, L., & Paffi, D. (2011). Desire and coping self-efficacy as craving measures in addiction: The self-efficacy and desire scale (SAD). *Open Behavioral Science Journal, 5*, 1–7.

Pericot-Valverde, I., García-Rodríguez, O., Gutiérrez-Maldonado, J., & Secades-Villa, R. (2015). Individual variables related to craving reduction in cue exposure treatment. *Addictive Behaviors, 49*, 59–63.

Pessa, E. (2009). Brain-computer interfaces and quantum robots. arXiv: 0909.1508.

Rego, P. A., Moreira, P. M., & Reis, L. P. (2010). Serious games for rehabilitation: A survey and a classification towards a taxonomy. In *2010 5th Iberian Conference on Information Systems and Technologies (CISTI)* (pp. 1–6). Piscataway: IEEE.

Rousseau, F. L., Vallerand, R. J., Ratelle, C. F., Mageau, G. A., & Provencher, P. J. (2002). Passion and gambling: On the validation of the Gambling Passion Scale (GPS). *Journal of Gambling Studies, 18*(1), 45–66.

Santamaria, J. J., Soto, A., Fernandez-Aranda, F., Krug, I., Forcano, L., Gunnard, K., et al. (2011). Serious games as additional psychological support: A review of the literature. *Journal of Cyber Therapy and Rehabilitation, 4*(4), 469–477.

Sauvaget, A., Trojak, B., Bulteau, S., Jiménez-Murcia, S., Fernández-Aranda, F., Wolz, I., et al. (2015). Transcranial direct current stimulation (tDCS) in behavioral and food addiction: A systematic review of efficacy, technical, and methodological issues. *Frontiers in Neuroscience, 9*, 349.

Sharpe, L. (2002). A reformulated cognitive-behavioral model of problem gambling: A biopsychosocial perspective. *Clinical Psychology Review, 22*, 1–25.

Sica, C., Magni, C., Ghisi, M., Altoè, G., Sighinolfi, C., Rocco Chiri, L., et al. (2008). Coping Orientation to Problems Experienced-Nuova Versione Italiana (COPE-NVI): Uno Strumento Per La Misura Degli Stili Di Coping. *Psicoterapia Cognitiva e Comportamentale, 14*, 27–53.

Turconi, M. M., Mezzarobba, S., Franco, G., Busan, P., Fornasa, E., Jarmolowska, J., et al. (2014). BCI-based neuro-rehabilitation treatment for Parkinson's disease: Cases report. In P. Bernardis, C. Fantoni, & W. Gerbino (Eds.), *TSPC2014. Proceedings of the Trieste Symposium on Perception and Cognition, November 27–28* (pp. 63–65). Trieste: EUT Edizioni Università di Trieste.

Vallerand, R. J., Blanchard, C., Mageau, G. A., Koestner, R., Ratelle, C., Léonard, M., et al. (2003). Les Passions de l'âme: On Obsessive and Harmonious Passion. *Journal of Personality and Social Psychology, 85*(4), 756–767.

Wiemeyer, J., & Kliem, A. (2011). Serious games in prevention and rehabilitation–A new panacea for elderly people? *European Review of Aging and Physical Activity, 9*(1), 41–50.

The Management Complexity of Hospital Psychiatric Ward: A "Small World" Approach

Pier Luca Bandinelli and Alvaro Busetti

1 Premise

Tasks and activities can be classified as simple, complicated or complex according to the following criteria (Nason 2017):

- Possibility to define exactly the success criteria: if there is a precise, possibly quantifiable, criteria to say that the task has been completed successfully.
- Same results follow from the same starting conditions by executing the task in exactly the same way (as in physics or in mechanical systems).
- The knowledge needed to successfully execute the task can be codified into a procedure.
- Task's execution needs to be very precise in order to get the desired results.

Let's give a couple of examples: preparing spaghetti is a simple activity, computing the income tax declaration is a complicated activity, teaching a young boy to be polite is a complex activity.

The afore mentioned criteria con be summarized in Table 1.

Complex systems (i.e. systems performing complex activities) exhibit a behavior which cannot be codified into procedures or practices and does not result from the sum of the behaviors (or properties) of their components. The

P. L. Bandinelli (✉)
Dipartimento di Salute Mentale ASL Roma 1, Servizio Psichiatrico Diagnosi e Cura (SPDC), Presidio Ospedaliero San Filippo Neri, Roma, Italy
e-mail: pierluca.bandinelli@aslroma1.it

A. Busetti
Freelance Consultant in "Social Enterprise" Sector, Rome, Italy

© Springer Nature Switzerland AG 2019
G. Minati et al. (eds.), *Systemics of Incompleteness and Quasi-Systems*,
Contemporary Systems Thinking,
https://doi.org/10.1007/978-3-030-15277-2_26

Table 1 Defining simple, complicated and complex activities

	Simple	Complicated	Complex
Possibility to define exactly the success criteria	Yes	Yes	No
Codified knowledge of the success factors	Yes	Yes	No
Predictable and reproducible results from the same conditions	Yes	Yes	No
Precision requested in the task's execution	No	Yes	Not applicable

behavior of a complex system emerges mostly from the interaction of its components and therefore strongly depends on the way its components are related to each other rather than on their characteristics. This property is called emergence and is typical of complex systems.

2 Why the SPDC Is a Complex System

With reference to the premise, in an Hospital Psychiatric Ward, in Italy SPDC (*Servizio Psichiatrico Diagnosi e Cura*), which is represented by a hospitalization department for patients with acute psychiatric disorders within a general hospital:

1. Success criteria cannot be defined in precise, quantifiable way because they depend on the patient's psychopathological conditions, the reasons for hospitalization, the extended context in which the patient lives, the chronicity of the disorder, the good adherence to the treatment, and the ability of the mental health center to have an appropriate therapeutic plan for the patient's needs, the possible coexistence of medical problems or serious socio-economic discomfort.
2. Successful procedures (processes/organization) are hardly codifiable because the system of the turnovers of most of the staff is an obstacle to the continuity of processes that often appear inconsistent, frayed or unclear (to the patient, family members, colleagues in the area, and between the SPDC's operators). Shifts in the 24 h staff cause difficulty in maintaining coherence and continuing the information chain, with the result that particularly intense emotional states or very critical situations experienced by staff tend to lose or increase intensity, to be deformed, or become incomprehensible in their real and original nature, in the upcoming shifts (Cohen et al. 2006; Rodhes 1995).

3. Activities' outcome is not predictable or reproducible even starting from the same psychopathological conditions; due to the premises outlined above, any clinical situation evolves in a difficult, unpredictable way depending on the different intersecting of the variables listed in point (Table 1).

4. It is meaningless to speak of executing precision for the reasons outlined below in point (a). Indeed, while in a department of medicine or surgery, the variations of vital parameters, the response to therapy and the diagnostic procedures are fundamental to decide the outcome of the hospitalization and discharge of the patient (and this can still be considered a linearity that defines the patient is a "simple"), in psychiatry, every patient has different complexities, and above all they are kept on completely different registers, which should be harmonized. The feeling that is often felt is that for this level of complexity and for the disharmonious intersecting of the different levels, there is no coincidence or fluidity in the interventions that take place both internally to the SPDC and the outside, with the sensation of unproductive work overload, and especially for situations that seem to be "suspended" regardless of anyone's will.

3 The SPDC as System of Systems and Its Critical Points

SPDC complexity arises from the fact that it has been attempted to manage complex processes by integrating (in a complicated way!) different systems which in turn are managed as complicated systems.

1. The *"System of Psychiatric Condition Clinical Aspects"*, that should represent the primary center of delivery of care for every single patient, but in reality is often overlooked by the emergency or the prevalence of other systems (listed below), or, in some cases, the patient's baseline clinical condition is intrinsically ambivalent, if not contradictory.

    In these cases, due to the psychopathology of some patients, and the relative unpredictability of their mental states, with continual changes in their goals and decisions, there is the possibility of a continuous remodeling of the set project, with changes in programs that are often not clear in the communications between operators.

2. The *"Patient Medical Criticism System"* and everything related to managing relationships with other colleagues in the hospital of other specialties.

3. The *"Family System"* represented by all interactions between SPDC staff and the patient's family members, and its criticalities.

4. The *"System of other structures of the Mental Health Department"* that intervenes in the patient's therapeutic design and whose main problems

are primarily the perception of different times related to the management of the hospitalization or the assessment of the severity of the patient's psychopathology and of the decision to take shelter.

5. The "*System of internal conflict within the team*" or "between the SPDC team and the various external agencies", so for some patients, in particular, there are completely different points of view, both in terms of diagnostic, therapeutic, but also related to the care project.

This causes further levels of growing complexities of internal and institutional contrasts within the Department of Mental Health. In this sense then, it is of crucial importance, but of difficult quantification, what I call "The hidden factor", which represents the whole set of the human dimensions and relationships within the team, which are intuited or known, but "of which one cannot speak" (a kind of unconscious system).

6. "*Other Systems*", i.e. all the involved agencies which are external with the respect of the Department of Mental Health, but which still concern our patients as judicial system or police, embassies and cultural mediators, health management and emergency assistance of the Hospital which need to be contacted in situations involving many of our patients.

4 The Components of Complexity

As we said in the premise, the "emergent" behavior of a complex system is defined by the interaction of its components, more precisely the complexity of a system (and its behavior) depends on:

1. The *actors* (factors) intervening in the processes: all people inside and outside the organization who are involved and/or intervene in the organization's processes.

2. The *relationships* between the actors (factors): the relationships either formal or informal, codified or not that connect people when carrying out the processes and daily activities.

3. The *ability of actors* (factors) to adapt to different, unforeseen situations: the ability of people to change the way they act or behave in the occurrence of unforeseen circumstances.

Thus, every complex system can be represented as a network whose nodes represent the actors (factors) and whose arcs represents the relationships between actors (factors) (Barabási 2002). The resulting net structure changes as actors face new situations and the relationships between them change to adapt to new situations (Cross and Parker 2004; Cross et al. 2001) (Fig. 1).

System properties can then be studied by analysing the properties of the network representing the system. For example, the speed at which information is transferred between two nodes in a complex system depends on the distance (i.e. the length of the shortest path) between the nodes.

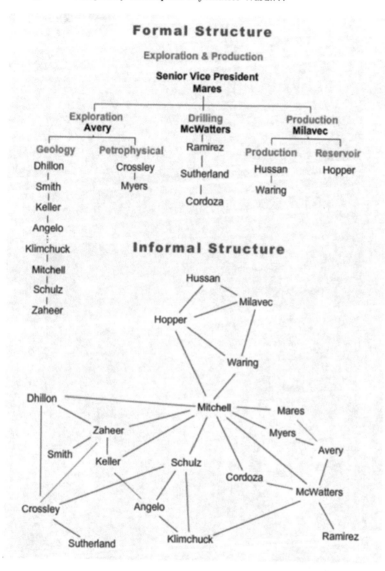

Fig. 1 Formal and informal structures: adapted from Cross et al. (2001) in Hunter (2015): the org chart of a company is compared to network of real working relationships between people (A is connected with B if they have a good working experience together) (Cross et al. 2001; Hunter 2015)

In the same way the critical nodes (i.e. critical actors) are the nodes that connect important parts of the network and are on the shortest path connecting important parts of the network; in fact, when removed, no path between one or more nodes on the network may be found or the path(s)

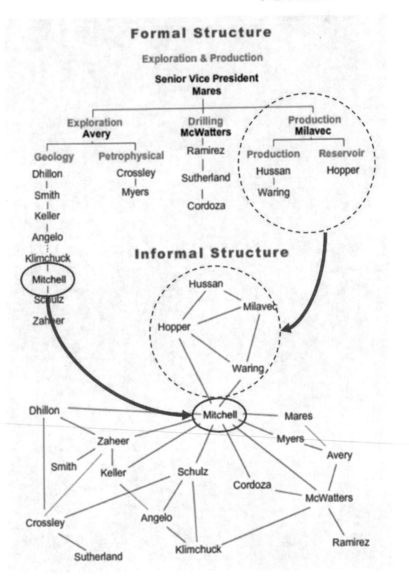

Fig. 2 Formal and informal structures: adapted from Cross et al. (2001): removing Mitchell will disconnect the production department from the rest of the organization ...

remaining may be too long to be effective and may therefore seriously affect the way the entire system behaves (Fig. 2).

In the SPDC:

1. The *actors* of the SPDC are very heterogeneous in training, roles and skills. They are doctors, psychologists, social workers, nurses, and an employee. There is also present external staff represented by specialists in training and trainee psychologists. All this staff is numerically about 35–40 people.

2. The *relationships* are partially codified with respect to individual roles and competencies, but not always; given the complexity of the interventions to be performed on the individual patient, boundaries or relationships between people follow rigidly coded paths. To complicate this picture is the fact that much of the staff is turning, while the professional figures that ensure continuity are few.

3. The *ability of actors* to adapt depends on the individual clinical or procedural situation that may arise when some people may intervene instead of others depending on their ability to handle some critical situations they are facing.

Network connections, i.e. who has interacted with who and how, can be obtained from patient clinical records and from "the communication notebook" (which does not exist in other departments for medical and surgical conception) which is a central work tool comparable to the clinical record.

The progression of the segregation segments in the various subsystems of the SPDC (see above "The SPDC as System of Systems and its Critical Points") can be well represented in it in an extremely synthetic, simple, clear, unambiguous and exhaustive way, considering it as metacognitive notation, and what we can take for granted because we know it, should be clear to everyone. The doctor who goes into the round must read on the patient notebook what is to be done during his working hours.

Moreover, the relationships between people of different subsystems that make up a SPDC originate from the daily activities of patient management: two people will have a positive relationship between them if, during their activities, they have had a positive experience of cooperation in the management of patients when they operated in the same turn.

Note that this applies both between people of the same subsystem and between people of different subsystems. The combination of these three factors gives rise to SPDC's system behavior ("emergence" of behavior).

5 The SPDC System Representation as "Small World Network"

The SPDC's system is therefore representable as a network (which has actors/factors as nodes and their relationships as connections) which grows (because of the different actors' capacity to adapt/relate) as a "small world" network (Buchanan 2002).

Small world networks are a typical structure in social groups: they form spontaneously between people who interact in carrying out tasks that require flexibility and adaptability (complex tasks). Characteristics of a "small world" network:

1. *Low degree of separation*: each node is connected through a few links (steps) to any other node.
2. *Low connectivity*: there are few arcs respect to the possible ones. Most nodes are connected with a few others, only a few (hubs) are connected to many other nodes (for SPDC see paragraph 3: nodes belonging to the same (sub)system interacting in SPDC's daily activities).
3. *High clustering*: two nodes connected to the same node tend to be connected to each other (triangles). If A and B are working well with C usually they also work well with each other regardless of whether they belong or not to the same (sub)system. Note that clusters do not consist only of people belonging to the same subsystem formally defined (department or group) but may, and usually do, consist of people belonging to different subsystems that usually are performing well when working together (informal organization).

6 Managing Complexity

The goal in managing a complex system (i.e. "small world network" of people) is to prepare the conditions to let desired behaviors "emerge" naturally:

- Realize that you are dealing with a complex system and not with a "simple" nor a "complicated" one. You have to recognize the activities that can be encoded into procedures (complicated ones: e.g. administrative ones) and the activities that must be resolved according to the contest and which cannot be foreseen (complex ones: e.g. dealing with the family).
- Approach the problem using a continuous management perspective and not try to find a "once and for all" solution, remembering that people will always try to solve new problems or react to unforeseen situations using their personal connections even if this violates the procedures. It can also happen that complying with the procedures in some cases can turn out to be harmful.
- Try and learn, being adaptive in your approach: learn from new situations, remembering that people always have a "good reason" (at least from their point of view) to behave the way they do.
- Transfer a mindset of complexity to people (codified procedures are useless or even dangerous): people have to be aware of the fact that the activities they are dealing with are complex.
- Enhance and develop the connections between people (see example below) in order to facilitate the emergence of the desired behavior in the network.

Example: People on duty must be chosen making sure that the resulting (sub)network is connected (i.e. no person is isolated and each node can be reached with a minimum number of connections). This can be done by paying attention to the fact that at least the key people on duty are aware that the (sub)network they are part of must have all (or at least most of) the people connected by positive collaborative relationship or, alternatively, know where these connections are missing and are prepared to act accordingly.

References

Barabási, A.-L. (2002). *Linked: The new science of networks*. Cambridge, MA: Perseus.

Buchanan, M. (2002). *Small world: Uncovering nature's hidden networks*. London, New York: Weidenfeld & Nicolson.

Cohen, T., Blatter, B., Almeida, C., Shortliffe, E., & Patel, V. L. (2006). Distributed cognition in the psychiatric emergency department: A cognitive blueprint of a collaboration in context. *Artificial Intelligence in Medicine, 37*(2), 73–83.

Cross, R., & Parker, A. (2004). *The hidden power of social networks: Understanding how work really gets done in organizations*. Boston, MA: Harvard Business School Press.

Cross, R., Parker, A., Prusak, L., & Borgatti, S. P. (2001). Knowing what we know: Supporting knowledge creation and sharing in social networks. *Organizational Dynamics, 30*(2), 100–120.

Hunter, S. D. (2015). Combining theoretical perspectives on the organizational structure-performance relationship. *Journal of Organization Design, 4*(2), 24–37. http://www.jorgdesign.net/jod/article/view/16781

Nason, R. (2017). *It's not complicated: The art and science of complexity in business*. Toronto: University of Toronto Press.

Rodhes, L. A. (1995). *Emptying beds. The work of an emergency psychiatric unit*. Berkeley, Los Angeles: University of California Press.

Natural Rates of Teachers' Approval and Disapproval in Italian Primary and Secondary Schools Classroom

Francesco Sulla, Eusebia Armenia, and Dolores Rollo

1 Introduction

Since the 1960s, researchers have been demonstrating the power of teacher behaviour on the behaviour of both individual students and whole classes (Sulla et al. 2013). Behavioural research and demonstration studies carried out over the past 50 years or so would appear to suggest that by manipulating the nature and quantity of feedback given to pupils, especially the use of praise and reprimand then the behaviour of pupils would change. However, this gives rise to a series of questions concerning teachers' use of feedback to their pupils and the relationship that this may have to their behaviour in those classes: How often do teachers praise their pupils? How often do they tell them off? What effect does the frequency of both these types of verbal feedback have on the pupils' behaviour? A research literature relating to non-experimentally manipulated or "naturalistic" rates answered those questions. Over the years, there have been a number of investigations that have centred on what might be called naturalistic or existing rates: descriptive studies on the ways in which teachers typically deploy praise in the classroom.

White (1975) reported the findings of 16 separate studies in the United States. She found those teachers of the youngest children gave more approval to their pupils than disapproval. However, the opposite appeared to be the case for teachers of older pupils. Teachers gave highest rates of approval for academic behaviour, while for social behaviour the reverse was true. Indeed the rate of teacher approval was so low for social behaviour that White described it as "almost non-existent" (p. 369). The results of other early in-

F. Sulla (✉) · E. Armenia · D. Rollo
Department of Medicine and Surgery, University of Parma, Parma, Italy
e-mail: francesco.sulla@unipr.it

© Springer Nature Switzerland AG 2019
G. Minati et al. (eds.), *Systemics of Incompleteness and Quasi-Systems*,
Contemporary Systems Thinking,
https://doi.org/10.1007/978-3-030-15277-2_27

vestigations (Heller and White 1975; Thomas et al. 1978) tended to support White's findings. However, in the late 1980s, a shift to more teacher approval than disapproval was recorded.

In a study they carried out in Los Angeles, Nafpaktitis et al. (1985) found approval to be more frequent than disapproval in Grades 6 to 9. Wyatt and Hawkins (1987) carried out a study in the United States. Although like White, they found mean rates of both approval and disapproval were highest in classrooms for the youngest pupils, they found that in all age groups approval was more common than disapproval. However, academic behaviour of students was much more likely to attract teacher praise than social behaviour. Positive feedback by teachers was positively related to compliant pupil behaviour as measured by pupil on-task behaviour, while disapproval showed a negative correlation with on-task behaviour. This relationship shows a very consistent pattern across the studies.

Wheldall et al. (1989) collected data in a total of 258 British classrooms, covering the pupil age range from 5 to 16, via an observation system termed OPTIC (Observing Pupils and Teachers in Classrooms) (Merrett and Wheldall 1986). Like Wyatt and Hawkins (1987), they found approval was given primarily for academic behaviours and disapproval for social behaviours and approval to be higher than disapproval at all school levels. They found a negative correlation between disapproval to academic behaviour and on-task behaviour, and a negative correlation between teachers disapproval to social behaviour and on-task behaviour.

Wheldall and Beaman (1994)—using OPTIC schedule—have given an account of work with teachers in Sydney, Australia. They found that their sample of 36 Australian primary school teachers gave very similar proportions of verbal feedback as the British counterparts as reported by Merrett and Wheldall (1986).

Following the earlier work in this area, Harrop and Swinson (2000) sought to examine teacher approval and disapproval in 10 British classes at each level of infants, junior, and secondary schooling. Their results were generally in line with the investigations of the 1980s, where approval rates were higher than disapproval rates at each school level. Similarly, they found that, overall, teachers gave higher rates of approval for academic behaviours than for social behaviours and higher rates of disapproval for social behaviours than for academic behaviours. A significant positive correlation between teacher approval and on-task behaviour was found at the infant school level.

Apter et al. (2010) conducted a nationwide survey further 10 years later after the last study in the context of British primary classrooms. They carried out observations in over 140 classes in England, Wales and Scotland using a revised version of Merrett and Wheldall's OPTIC (1986). They found approval to be more common than disapproval. The one finding that did show a slight difference was in terms of the proportion of positive feedback directed to pupils' behaviour. Earlier studies had found comparatively low percentages of this type of feedback (Merrett and Wheldall 1986, 6%; Harrop and Swinson 2000, 4%), whereas in the study by Apter et al. (2010) it was assessed at almost 15%. They found a positive relationship between teacher's approval for academic behaviour and pupil on-task behaviour. However, no statistically significant link was found between negative comments about academic work and student on-task behaviour.

Whilst a huge amount of research has been carried out over the past 50 years in English-speaking countries on the link between teacher verbal behaviour and pupil behaviour, no previous studies have investigated the existing rate of teacher approval and disapproval in Italian classes. This study aimed to address this gap in current knowledge and add to the international literature on this subject.

2 Method

2.1 Participants

The experimenter (first author) served as the primary data collector. Three undergraduate students and five graduate students in Psychology who were trained by the experimenter served as secondary data collectors. A total of 314 observations (134 in primary schools and 180 in secondary schools) , the equivalent of 9420 min (177 h), were conducted across the country. The schools were allocated in zones that were convenient in logistical terms to the observers, and were representative of the main areas (i.e., north, middle, south, isles) and local types (inner city, suburban, rural/village) of the country.

2.2 Procedure

As in the aforementioned study by Apter et al. (2010), the Partial Time Interval Observation recording sheets for 6 Subjects (PTIObs6s) (Apter 2013) was used for the observations. The PTIObs6s is a paper-pencil tool, which entails a repeated timed observation being made of a small sample of five randomly chosen students in a class. Observations were alternated with recordings of teacher verbal behaviour. The students were observed as being on-task or off-task. Teachers' verbal behaviours were tallied under five headings: INX: Instructions, explanations or expositions; TPP: Task Performance Positive comments; SBP: Social Behavioural Positive comments; TPC: Task Performance Criticism; SBC: Social Behavioural Criticism.

Each observer was trained into the use of the proforma using brief tapes showing teacher-pupils interactions. This procedure was continued until percentage observer agreement reached above 80% on two successive occasions. From that point, the observers entered the classrooms and scored the actual lessons by themselves but were aware that at least the 30% of the lessons, taken at random, would be scored independently by the researcher and observer agreement calculated. Agreement rates were calculated using Cohen's kappa. Kappa was calculated at between 0.77 and 0.89 for joint observations, with a mean value of 0.85.

The observers were often introduced to the class who were then told to ignore their presence. When observing the class each observer sat at the back of the room in a position where they could observe all the pupils. Once students had been observed for 1 min in each cycle using momentary time sampling, the teacher's verbal behaviour was recorded using partial-interval recording. Observers returned pairs of observations of classes being taught by the same teacher, one in the morning and one in the afternoon. Class observations were for 30 min each with a 1 min time-interval for each cycle.

3 Results

The percentages of positive and negative feedback directed by teachers to their pupils' academic work or social behaviour in primary and secondary schools are presented, respectively, in Tables 1 and 2.

Table 1 Percentages of teacher feedback as assessed in 134 lessons in Italian primary schools

Behaviour	Approval	Disapproval	Total
Academic	24	19	43
Social	1	56	57
Total	25	75	100

Table 2 Percentages of teacher feedback as assessed in 180 lessons in Italian secondary schools

Behaviour	Approval	Disapproval	Total
Academic	20	21	41
Social	0	59	59
Total	20	80	100

In primary schools, the majority of feedback was of a negative nature and directed in response to pupils' behaviour (ratio 1:3). Most positive feedback was directed towards pupils' work and very little to pupils' behaviour (Table 1).

As in primary schools, in secondary schools, the majority of feedback was of a negative nature and directed in response to pupils' behaviour (ratio 1:4). Positive feedback was directed toward pupils' work, while positive feedback directed towards pupils' behaviour was very seldom observed (Table 2).

In terms of the rates of approval and disapproval, in primary school, approval occurred about 16 times every hour; disapproval occurred at the rate of about 49 responses per hour. In secondary school, approval occurred about eight times every hour; disapproval occurred at the rate of about 32 responses per hour. Both approval and disapproval rates seem to decline as the age of the pupils increases.

There was a statistically significant difference in the on-task time of pupils in different age cohorts ($F_{[12,261]} = 1.839$, $p < 0.05$). There was a significant linear trend ($F_{[1,11]} = 8.083$, $p < 0.01$) indicating that as the students grow up, their time on-task increased. Central to this study was the nature of the verbal behaviour of teachers in classrooms and the association that there might be with the way children work, operationalised as percentage of time spent on-task (following teacher's instructions): This was correlated with the frequency of different categories of teacher verbal behaviour (see Table 3).

There was a statistically significant link ($\rho = 0.298$, $p < 0.001$) between the teachers' neutral verbal behaviour (INX), and time on-task (Table 3). Teachers' neutral verbal behaviour was associated with high on-task rates. This result was similar to the one found by Apter et al. (2010) in British primary schools, and supportive of the hypothesis that highly verbal teachers are more successful in keeping students on-task.

There was a statistically significant negative correlation between teachers' rates of disapproval and on-task behaviour ($\rho = -0.383$, $p < 0.001$). In classes with higher rates of on-task behaviour, there were lower rates of negative feedback (Table 3).

Table 3 Correlations between Italian students' on-task behaviour as a percentage of the time they were observed, and rates of teacher feedback

| | Neutral verbal behaviour | Praise | | Criticism | |
		Task perform.	Social behaviour	Task perform.	Social behaviour
On-Task	0.298[a]	0.011	0.018	−0.118	−0.383[a]

[a]Correlation is significant at the 0.001 level (two-tailed)

4 Discussions

The proportionality of different types of feedback used by Italian teachers appears to be more similar to the one found in the earlier investigations carried out in the 1970s, than to the pattern found in studies from the 1980s to date. Students received more total teacher disapproval in every grade; for social behaviour, teacher disapproval far outweighed teacher approval, the latter being almost non-existent. As everybody working in education in Italy would probably say, the pattern of feedback given by teachers was as expected.

This pattern of behaviour may have its roots in our cultural heritage. Techniques based on approval and positive reinforcement in general have been, and continue to be basically ignored and misunderstood. Explanations for this misunderstanding may be grounded in a basic cultural ethos: We live in a society in which individuals are free to do as they wish—as long as they do so in a socially appropriate manner—without coercion. In this context, coercion is simply the absence of external pressure—being internally motivated to behave well. This societal value contributes to the widespread acceptance of a punishment mentality that ignores data indicating the effectiveness of techniques based on approval. Techniques based on positive reinforcement are often perceived to threaten individuals' freedom as autonomous human beings. The functional definition of positive reinforcement frequently does not help some teachers get past the stereotypical notion that it is a manipulative tool created to coerce students into behaving appropriately. Consequently, reinforcement continues to be viewed by some educators as synonymous with bribery, undermines students' abilities to become self-directed, and represses internal motivation (Kohn 1993). Ironically, punishment, which is the opposite of positive reinforcement, appears much more acceptable because of the perception that it does not threaten individuals' autonomy—people believe they are free to choose to behave in responsible ways to avoid punishment (Maag 2001).

A punishment paradigm has evolved, and been advocated for, since biblical times and is reflected in the proverb "spare the rod and spoil the child". Besides having history on its side, a punishment mentality has been perpetuated for the simple reason that the behaviour of punishing students has traditionally been highly reinforced. The effects of reprimanding a child who misbehaves are immediate—the negative reinforcement in the form of cessation of the annoying behaviour effectively and naturally teaches us to punish one another. On the other hand, the effects of verbal praise are usually delayed, making it difficult for us to learn to use praise.

As regards teacher verbal behaviour: we found that both approval and disapproval rates per minute declined and neutral verbal behaviour (INX) increased as the age of the pupils increases. There was a clear difference between primary school and secondary school teachers. The reason for this difference may be the fact that a profound conviction remains in Italian schools whereby relational, educational, psychological competences,—hence skills of classroom management—are important for teaching small children, but much less so for teaching secondary school pupils, where a good knowledge of the subjects to be taught would be more than sufficient (Ostinelli 2009). Therefore, teachers in secondary school spend a lot of time talking in order to transfer knowledge to pupils.

Students' time on-task increased as their age increases. This may be explained both from a developmental perspective: as children develop, their self-regulatory skills become more sophisticated (Blair and Diamond 2008), and from an educational perspective: throughout the years, children's classroom experience increase. As we saw, teacher feedback is fundamental in children's classroom experience. While causality between teacher behaviour and student behaviour cannot be established, a strong relation between teacher approval and disapproval and students' time on-task has been demonstrated, even in secondary school.

As in previous studies, teacher disapproval to social behaviour was negatively related to students' time on-task. Again, one must be cautious in assuming a causality. However, the fact that low levels of disapproval were recorded in classes with high on-task rates is hardly surprising. If pupils are getting on with their work, then there is no need to tell them off. Alternatively, if they are off-task then one would expect a higher rate of disapproval. This explanation is of course one that portrays the teacher in a very passive role responding to the pupils' behaviour rather than attempting to change behaviour through use of feedback. Brophy (1981) makes this point at some length. Whatever the explanation, one thing is perfectly evident from the data, if teachers want to improve the behaviour of pupils repeatedly telling them off is not a strategy that according to this data is likely to work.

References

Apter, B. (2013). *PTIObs6S – Partial Time Interval Observation recording sheet for 6 Subjects, (unpublished INSET Course Materials)*. Wolverhampton: Wolverhampton City Council/University of Wolverhampton.

Apter, B., Arnold, C., & Swinson, J. (2010). A mass observation study of student and teacher behaviour in British primary classrooms. *Educational Psychology in Practice, 26*(2), 151–171.

Blair, C., & Diamond, A. (2008). Biological processes in prevention and intervention: The promotion of self-regulation as a means of preventing school failure. *Development and Psychopathology, 20*(03), 899–911.

Brophy, J. (1981). Teacher praise: A functional analysis. *Review of Educational Research, 51*(1), 5–32.

Harrop, A., & Swinson, J. (2000). Natural rates of approval and disapproval in British infant, junior and secondary classrooms. *British Journal of Educational Psychology, 70*(4), 473–483.

Heller, M. S., & White, M. A. (1975). Rates of teacher verbal approval and disapproval to higher and lower ability classes. *Journal of Educational Psychology, 67*(6), 796–800.

Kohn, A. (1993). *Punished by rewards: The trouble with gold stars, incentive plans, A's, praise, and other bribes.* Boston: Houghton-Mifflin.

Maag, J. W. (2001). Rewarded by punishment: Reflections on the disuse of positive reinforcement in schools. *Exceptional Children, 67*(2), 173–186.

Merrett, F., & Wheldall, K. (1986). Observing Pupils and Teachers In Classrooms (OPTIC): A behavioural observation schedule for use in schools. *Educational Psychology, 6*(1), 57–70.

Nafpaktitis, M., Mayer, G. R., & Butterworth, T. (1985). Natural rates of teacher approval and disapproval and their relation to student behavior in intermediate school classrooms. *Journal of Educational Psychology, 77*(3), 362–367.

Ostinelli, G. (2009). Teacher education in Italy, Germany, England, Sweden and Finland. *European Journal of Education, 44*(2), 291–308.

Sulla, F., Perini, S., & Rollo, D. (2013). Analisi funzionale dei comportamenti di approvazione e disapprovazione nell'interazione insegnante–allievo: una rassegna [Functinal Analysis of teachers' approval and disapproval in teacher-pupils interactions: A review]. *Psicologia dell'Educazione, 7*(2), 193–216.

Thomas, J. D., Presland, I. E., Grant, M. D., & Glynn, T. L. (1978). Natural rates of teacher approval and disapproval in grade-7 classrooms. *Journal of Applied Behavior Analysis, 11*(1), 91–94.

Wheldall, K., & Beaman, R. (1994). An evaluation of the WINS (Working Ideas for Need Satisfaction) training package: Report submitted to the New South Wales Department of School Education, 1993. Special Education Centre, Macquarie University. *Collected Original Resources in Education, 18*(1), fiche 4 E01.

Wheldall, K., Houghton, S., & Merrett, F. (1989). Natural rates of teacher approval and disapproval in British secondary school classrooms. *British Journal of Educational Psychology, 59*(1), 38–48.

White, M. A. (1975). Natural rates of teacher approval and disapproval in the classroom. *Journal of Applied Behavior Analysis, 8*(4), 367–372.

Wyatt, W. J., & Hawkins, R. P. (1987). Rates of teachers' verbal approval and disapproval relationship to grade level, classroom activity, student behavior, and teacher characteristics. *Behavior Modification, 11*(1), 27–51.

On Some Open Issues in Systemics

Gianfranco Minati

1 Introduction

As it is well known Ludwig von Bertalanffy is credited as the founder of the so-called General System(s) Theory—GST—(Von Bertalanffy 1968) with other co-founders, including Ashby, Boulding, and Von Foerster (Minati and Pessa 2006, pp. 1–41). The essence of the concept of system is well expressed by von Bertalanffy in a short sentence

> The meaning of the somewhat mystical expression "the whole is more than the sum of its parts" is simply that constitutive characteristics are not explainable from the characteristics of isolated parts. (Von Bertalanffy 1968, p. 55).

Composing elements are supposed to have a set of relationships between them (Hall and Fagen 1956) and

> A system can be defined as a set of elements standing in inter-relations. (Von Bertalanffy 1968, p. 55).

The concept of system is assumed suitable

- to represent something that is not reducible to its component parts, e.g., natural systems such as properties of life, the climate, and ecosystems;
- to constitute a way to build something that is not reducible to its composing parts, e.g., inert electronic circuits becoming systems such as radios, computers, amplifiers, and TV sets when powered on, i.e., making components interact, to function.

G. Minati (✉)
AIRS / Italian Systems Society, Milano, Italy
e-mail: gianfranco.minati@AIRS.it

© Springer Nature Switzerland AG 2019
G. Minati et al. (eds.), *Systemics of Incompleteness and Quasi-Systems*,
Contemporary Systems Thinking,
https://doi.org/10.1007/978-3-030-15277-2_28

The fortune of the concept of system is related to its capability to represent, explain the process to acquire and lose properties by sets of elements in some ways constructivistically (Steffe and Thompson 2010) considered as associated (in the systemic view: in relationship to each other, e.g., having some order, positions in configurations, or interrelated, i.e., one's behaviour affects other behaviours).

The concept of system is ubiquitous, interdisciplinary and used everywhere.

In this contribution we examine whether other conceptual approaches not necessarily based on systems may be considered to effectively design, explain, induce, model, and vary the acquisition of through sets of elements not necessarily related or inter-related. Moreover, it seems quite unlikely that only a mono-dimensional approach is possible, as often the uniqueness of thought and absolute truths are questionable.

This argumentation is considered here conceptually as being coupled with currently unsolved problems of Systemics possibly requiring new unconventional approaches. For both cases, i.e., non-systemic elementary acquisition of properties and current open issues in systemics, one should consider the constructivist role of the observer which is realized in multiple non-equivalent cognitive approaches (Butts and Brown 1989; Steffe and Thompson 2010; Von Glasersfeld 1991a,b, 1995).

2 Non-systemic Elementary Acquisition of Properties

In this section we mention some well-known elementary examples of the process of acquisition of properties by sets of elements, that is, collective properties, not necessarily based on the concept of system, i.e., on the relations or inter-relations among their supposed constituent elements.

2.1 Capillarity

While absorbency is the property of a material to absorb liquids, capillarity is the tendency of a liquid in a sufficiently thin tube to rise as result of surface tension. Capillarity occurs because of intermolecular forces between the liquid and the surrounding surface. When the diameter of the tube is sufficiently small, the combination of surface tension, given by cohesion within the liquid, and the adhesive forces between the liquid and container, will be sufficient to raise the liquid level (de Gennes et al. 2003).

Cohesive and adhesive forces are responsible for the acquisition of a new collective property, i.e., capillarity, by liquid molecules.

2.2 Compositions

We refer here to the process of building, for instance, new molecules in chemistry, e.g., in biotechnology and bio-engineering, with new structural properties (Ho and Gibaldi 2013). An elementary example is given by compositions of patterns, shapes, and pictures. Sequences of dynamic patterns may specify styles of fashion, sequences of shapes may specify behaviours, while sequences of pictures may tell a story such as with films.

2.3 Density Variations

We consider here how solubility and density variations may produce variations in properties. For instance, there are maximum and minimum limits of densities beyond which a swarm ceases to be a swarm, i.e., collapsing into a single entity at very high densities and dissolving at very low densities. The same occurs for the population of cities or molecules within a liquid. We face the two opposite processes of dilution and accumulation which, respectively, cancel or greatly increase, the validity of relationships and interactions.

In both the cases they become irrelevant as new, predominant properties are acquired. High dilutions lead to diluent properties becoming prevalent, whereas high concentrations cancel degrees of freedom and lead to min and max constraints coinciding to generate a resultant immobility with new properties. One example is the consideration of simulated collective behaviours (URL) configured with extreme values, i.e., very high or very low space constraint radii.

Real examples include processes of disaggregation, dissolution, e.g., in the environment, dissolution of links, and of collapsing, e.g., one single property becomes valid with given fixed parameters.

In chemistry the limit for high dilution is represented by Avogadro's number. However, there is an intense debate about possible properties acquired by elements diluted beyond Avogadro's number, studied by the physics of high dilutions (Sukul and Sukul 2010; IJHD 2017; IRG 2017). Moreover, this is also the context of the controversial practice of homeopathy which professes the medicinal validity of high dilutions of suitable chemicals.

In the same way, one can consider the temporal density of events on a suitable scale, either being so rare, or diluted, to become irrelevant, or being so intensely aggregated, concentrated, as to be indistinguishable, collapsing into a single resulting event. One may consider, in this vein, respectively, rare cosmic events with geological sedimentation.

2.4 Optical Properties

The example considered here is the well-known Newton's disc. When a disc with segments in the colours of the rainbow, violet, indigo, blue, green, yellow, orange and red, is rotating the colour perceived is white.

In this case there is the acquisition of a new property due to the contextual superimposition of specific single properties, the absorption/reflection of different wavelengths.

Another example of this nature would be the production of green watercolours by mixing yellow and blue watercolours when painting on white paper.

In this case the process of superimposition is not due to rotation, but to the resulting changes in the reflection/absorption of different wavelengths of light.

2.5 Percolation

In materials science, percolation refers to the movement of fluids through porous materials which can be used for filtering. The study of percolation relates to the finding, building, and keeping of the way(s) to pass through the porous material.

The ability to percolate is a property collectively acquired by molecules of a fluid forced by gravitation or pressure differences to find ways of passing through the porous material while respecting the constraints of that material.

This respect of the multiple constraints of the porous material to pass through and gravitation /pressure differences are responsible for the collective acquired percolation behaviour acquired by the fluid.

This is studied by percolation theory (Grimmett and Kesten 2012) which allows one to model the various dynamic ways by which a fluid may percolate through a porous material having possibly variable and/or non-homogeneous properties. Percolation theory has interdisciplinary applications such as in biology, chemistry, complex networks, epidemiology, materials science, and physics.

2.6 Phase Transitions

Passage between solid, liquid and gaseous states of matter are called first-order phase transitions intended as changes in the arrangement of the constituent atoms, whereas the change from paramagnetic to magnetic is a second-order phase transition. In classical physics phase transitions are considered to derive from the change in some external condition, such as pressure or temperature.

The acquisition of new properties or their loss, e.g., at the Curie point, is also considered to be due to external conditions and not to the systemic interaction among constituent elements (Minati and Pessa 2006, pp. 201–290; Solé 2011).

2.7 Quantum Field Theory

Quantum Field Theory (QFT) may be considered as the new comprehensive corpus, as the theoretical context for a general system theory able to act as a theory for complexity and emergence (Minati and Pessa 2018).

2.8 Sloppiness

> Sloppy is the term used to describe a class of complex models exhibiting large parameter uncertainty when fit to data (Transtrum et al. 2015, p. 2).

The concept relates to models having behaviour controlled by a small number of parameters. As in mechanics, where we can model behaviours by using only a few parameters of the device under study, it is possible to model the behaviour of large aggregates of components without considering them as complex systems. For instance,

> This explains why an effective theory, or an oversimplified "cartoon" microscopic theory, can often make quantitatively correct predictions. Thus, while three dimensional liquids have enormous microscopic diversity, in a certain regime (lengths and times large compared to molecules and their vibration periods), their behavior is determined entirely by their viscosity and density. Although two different liquids can be microscopically completely different, their effective behavior is determined only by the projection of their microscopic details onto these two control parameters. (Transtrum et al. 2015, p. 7).

In this case we do not need to consider the systemic interactions among molecules to model the system behaviour.

This conceptually corresponds to well know cases occurring in thermodynamics when considering pressure and temperature to model the properties of gases.

3 Some Current Systemic Open Issues

This Section includes some examples of current open issues in systemics mainly related to the phenomena of emergence, network representation, the questionable and increasing no-need-for-theories approach, incompleteness and quasiness, and human systems.

3.1 Emergence

There is a lack of suitable general approaches for the phenomena of emergence. For instance, we consider here:

- How to induce a process of emergence from a generic population of incoherently interacting elements, i.e., how to induce coherence(s)?
- How to modify established on-going processes of emergence? How to modify their parameters?
- How to prevent or even deactivate established on-going unwanted processes of emergence?

One possibility is to design suitable environmental changes and insert suitable artificial perturbations.

3.2 Networks

Systems may be suitably represented within the science of networks (Barabási 2002; Cohen and Havlin 2010; Estrada 2016). We argue here the need for effective representations and approaches to modify primary systemic properties such as adaptation and logical openness (Licata 2008; Minati et al. 1998).

3.3 Theory-Less Systems

As introduced in my other contribution in this volume, we are modelling and using models of systems and systemic properties by using concordances and correspondences in a data deluge (Anderson 2008; Calude and Longo 2017), i.e., Big Data, without theories (Von Foerster 2003), that is, without explicit symbolic formal representations as opposed to non-explicit, non-symbolic approaches such as statistical, network, and sub-symbolic approaches.

Big Data (Sagiroglu and Sinanc 2013) refers to the enormous quantity of data now available:

- Real, coming from measurements and observations, and

- Generated, by models of real phenomena.

The problem is how to generate a usage of such very large amounts of data.
This typically occurs when dealing with phenomena and properties of complexity, particularly emergence.

3.4 Incompleteness and Quasi-Systems

As introduced in the call for papers of this conference we need suitable representations, resources, possibly theories (?), and models to deal with theoretical incompleteness (Minati 2016). Incompleteness represents the equivalent varieties of modalities by which the coherence(s) of processes of emergence may be established. On the other hand, incompleteness is a characteristic of quasi-systems related to their continuous structural meta-stable changes.

3.5 Human Systems

The complexity of Human Systems is of a different nature than the complexity of inanimate systems studied in physics, such as double pendulums, lasers, networks of oscillators, signaling traffic, or living systems which are assumed to have no or only simple cognitive systems, e.g., bacterial colonies, cells, macromolecules, anthills, flocks, and swarms.

This is considered to be due to various specific factors, such as sophisticated cognitive processing allowing learning and producing knowledge suitable for designing the environment, varying natural situations, or evolving interactions.

In some ways it seems that Human Systems cannot be reduced to systems (Minati 2017), except for some local specific activities for which organisations and manipulations are sufficient.

4 Conclusions

So-called GST is in reality a corpus of a multitude of approaches all based on the concept of system. Within this theoretical context we have here considered issues such as incompleteness, logical openness, multiple usages of models, and quasiness to be applied to systemic approaches.

We have also considered how this may be reflexively applied to the GST itself and its corpus of a multitude of approaches all based on the concept of system.

Also mentioned is how the processes of the acquisition of properties may be performed without classical systemic mechanisms based on interactions and relations, taking into account how GST may evolve towards a possible generic theory of acquisition of properties, since other mechanisms are possible within the possibly multiple and non-equivalent constructivist roles of the observer.

Correspondingly we have listed some current systemic open issues in which one might find possible aspects of a generic theory of the acquisition of properties without systems.

References

Anderson, C. (2008). The end of theory: The data deluge makes the scientific method obsolete. *Wired Magazine, 26*(6). https://www.wired.com/2008/06/pb-theory

Barabási, A. L. (2002). *Linked: The new science of networks*. Cambridge, MA: Perseus.

Butts, R., & Brown, J. (Eds.). (1989). *Constructivism and science*. Dordrecht: Kluwer.

Calude, C. S., & Longo, G. (2017). The deluge of spurious correlations in big data. *Foundations of Science, 22*, 595–612.

Cohen, R., & Havlin, S. (2010). *Complex networks: Structure, robustness and function*. Cambridge: Cambridge University Press.

de Gennes, P.-G., Brochard-Wyart, F., & Quéré, D. (2003). *Capillarity and wetting phenomena: Drops, bubbles, pearls, waves*. New York, NY: Springer.

Estrada, E. (2016). *The structure of complex networks: Theory and applications*. Oxford: Oxford University Press.

Grimmett, G. R., & Kesten, H. (2012). *Percolation theory at saint-flour*. New York: Springer.

Hall, A. D., & Fagen, R. E. (1956). Definition of a system. *General Systems Yearbook, 1*, 18–28.

Ho, R. J. Y., & Gibaldi, M. (2013). *Biotechnology and biopharmaceuticals: Transforming proteins and genes into drugs*. Hoboken, NJ: Wiley.

Licata, I. (2008). Logical openness in cognitive models. *Epistemologia, 31*, 177–191.

Minati, G. (2016). Knowledge to manage the knowledge society: The concept of theoretical incompleteness. *Systems, 4* (3), 26. http://www.mdpi.com/2079-8954/4/3/26/pdf

Minati, G. (2017). The past, present and possible future for systems. *International Journal of Systems and Society, 4*(1), 1–9.

Minati, G., Penna, M. P., & Pessa, E. (1998). Thermodynamic and logical openness in general systems. *Systems Research and Behavioral Science, 15*, 131–145.

Minati, G., & Pessa, E. (2006). *Collective beings*. New York, NY: Springer.

Minati, G., & Pessa, E. (2018). *From collective beings to quasi-systems.* New York, NY: Springer.

Sagiroglu, S., & Sinanc, D. (2013). Big data: A review. In *2013 International Conference on Collaboration Technologies and Systems (CTS)* (pp. 42–47). IEEE Xplore Digital Library. Available at: https://www.researchgate.net/publication/261456895_Big_data_A_review

Solé, R. V. (2011). *Phase transitions.* Princeton, NJ: Princeton University Press.

Steffe, L. P., & Thompson, P. W. (Eds.). (2010). *Radical constructivism in action: Building on the pioneering work of Ernst von Glasersfeld.* London: Routledge.

Sukul, N. C., & Sukul, A. (2010). *High dilution effects: Physical and biochemical basis.* Dordrecht: Kluwer.

Transtrum, M. K., Machta, B., Brown, K. S., Daniels, B. C., Myers, C. R., & Sethna, J. P. (2015). Perspective: Sloppiness and emergent theories in physics, biology, and beyond. *The Journal of Chemical Physics, 143*(1), 010901-1–13.

Von Bertalanffy, L. (1968). *General system theory. Development, applications.* New York, NY: George Braziller.

Von Foerster, H. (2003). *Understanding understanding: essays on cybernetics and cognition.* New York, NY: Springer.

Von Glasersfeld, E. (Ed.). (1991a). *Radical constructivism in mathematics education.* Dordrecht: Springer.

Von Glasersfeld, E. (1991b). Knowing without metaphysics. Aspects of the radical constructivist position. In F. Steier (Ed.), *Research and reflexivity* (pp. 12–29). London, Newbury Park, CA: Sage.

Von Glasersfeld, E. (1995). *Radical constructivism: A way of knowing and learning.* London: Falmer Press.

Web Resources

https://sourceforge.net/projects/msp3dfbsimulator/?source=navbar.

International Journal of High Dilution Research, at http://www.highdilution.org/index.php/ijhdr. Accessed September 6, 2017.

International Research Group on Very Low Dose and High Dilution Effects, at http://giri-society.org//. Accessed September 6, 2017.

Printed in the United States
By Bookmasters